Lecture Notes in Control and Information Sciences

Edited by A. V. Balakrishnan and M. Thoma

For information about Vols. 1–21 please contact your bookseller or Springer-Verlag.

Lecture Notes in Control and Information Sciences

Edited by M. Thoma and A. Wyner

95

Optimal Control

Proceedings of the Conference on
Optimal Control and Variational Calculus
Oberwolfach, West-Germany, June 15-21, 1986

Edited by
R. Bulirsch, A. Miele, J. Stoer and K. H. Well

Springer-Verlag Berlin Heidelberg GmbH

Editors

Roland Bulirsch
Technische Universität München
Mathematisches Institut
Arcisstraße 21
8000 München 2
Germany

Angelo Miele
Rice University
Aero-Astronautics Group
230 Ryon Building
P. O. Box 1892
Houston, TX 77251
USA

Josef Stoer
Universität Würzburg
Institut für Angewandte Mathematik
und Statistik
Am Hubland
8700 Würzburg
Germany

Klaus H. Well
DFVLR
Deutsche Forschungs- und Versuchsanstalt
für Luft- und Raumfahrt
Institut für Dynamik der Flugsysteme
Oberpfaffenhofen
8031 Wessling
Germany

ISBN 978-3-540-17900-9 ISBN 978-3-540-47907-9 (eBook)
DOI 10.1007/978-3-540-47907-9

Originally published by Springer-Verlag Berlin Heidelberg New York in 1987

2161/3020-543210

PREFACE

The conference on "Optimalsteuerungen und Variationsrechnung - Optimal Control" takes place approximately every five years. The aim of the last year meeting was to review recent developments in optimal control theory and computational methods, in deterministic differential games, oscillatory control, in deterministic control of uncertain systems, nonlinear singularly perturbed optimal control problems, and control of systems with distributed parameters. In addition, practical applications to various technical problems such as flight path control, robot control, control of water resources, and control of flexible structures were presented.

This volume contains selected papers presented at the conference. It is divided into six sections:

1. Theory and Computational Methods

2. Aircraft Trajectory Control

3. Control System Design

4. Robot Control

5. Water Resources Managment

6. Control of Flexible Structures

The organizers of the meeting would like to thank the authors for their contributions. Financial support of the "Gruppe Technomathematik der Universität Kaiserslautern" is appreciated.

R. Bulirsch
A. Miele
J. Stoer
K. Well

February 1987

LIST OF PARTICIPANTS

Prof. Dr. M.D. Ardema
Santa Clara University
School of Engineering
Dept. of Mechanical Eng.
Santa Clara, Ca. 95053
USA

Dr. H.G. Bock
Universität Bonn
Institut f. Angew. Mathematik
Wegelerstr. 6
5300 Bonn 1
FRG

Frau Prof. Dr.
Houria Bourdache-Siguerdidjane
CNRS/ESE
Laboratoire des Signaux & Syst.
Plateau du Moulon
F-91190 Gif-sur-Yvette
France

Prof. Dr. J.V. Breakwell
Stanford University
Dept. of Aeronautics & Astronautics
Stanford, Ca. 94305
USA

Prof. Dr. M. Brokate
Universität Augsburg
Institut f. Mathematik
Memminger Str. 6
8900 Augsburg
FRG

Prof. Dr. R. Bulirsch
Technische Universität München
Mathematisches Institut
Arcisstr. 21, Postfach 20 24 20
8000 München 2
FRG

Prof. Dr. A.J. Calise
Georgia Institute of Technology
School of Aerospace Eng.
Atlanta, Ga. 30332
USA

Dr. G.S. Christensen
University of Alberta
Dept. of Electrical Eng.
Edmonton, Alberta T6G2G7
Canada

Prof. Dr. E.M. Cliff
Virginia Polytechnic Institute
and State University
Aerospace and Ocean Eng.
Blacksburg, Va. 24061
USA

Prof. Dr. K.H. Elster
Technische Hochschule Ilmenau
Am Ehrenberg
Postfach 327
DDR-6300 Ilmenau
GDR

Prof. Dr. G. Feichtinger
Technische Universität Wien
Inst.f.Ökonometrie und
Operations Research
Argentinierstr. 8/119
A-1040 Wien
Austria

Dipl. Math. W. Grimm
DFVLR FF-DF
Oberpfaffenhofen
8031 Wessling
FRG

Dipl. Math. P. Hiltmann
DFVLR FF-DF
Oberpfaffenhofen
8031 Wessling
FRG

Prof. Dr. K.H. Hoffmann
Universität Augsburg
Institut f. Mathematik
Memminger Str. 6
8900 Augsburg
FRG

Frau Dr. M.K. Horn
MBB LKE
Postfach 80 11 60
8000 München 80
FRG

Prof. Dr. J.L. de Jong
University of Technology
Dept. of Math. & Comp. Sc.
P.O. Box 513
NL-5600 MB Eindhoven
The Netherlands

Prof. Dr. H.J. Kelley
Virginia Polytechnic Institute
and State University
Aerospace and Ocean Eng.
Blacksburg, Va. 24061
USA

Prof. Dr. R. Klötzler
Karl-Marx-Universität Leipzig
Sektion Mathematik
Karl-Marx-Platz
DDR-7010 Leipzig
GDR

Frau Dipl. Math.
M. Buchberger-Kolb
Technische Universität München
Arcisstr. 21
8000 München 2
FRG

Prof. Dr. A. Kowalewski
University of Mining &
Metallurgy
Al. Mickiewicza
Pl-30-059 Cracow
Poland

Dr. D. Kraft
DFVLR FF-DF
Oberpfaffenhofen
8031 Wessling
FRG

Dipl. Math. B. Kugelmann
Technische Universität München
Mathematisches Institut
Arcisstr. 21
8000 München 2
FRG

Dipl. Math. U. Leiner
Technische Universität München
Mathematisches Institut
Arcisstr. 21
8000 München 2
FRG

Prof. Dr. G. Leitmann
University of California
Dept. of Mechanical Eng.
Berkeley, Ca. 94720
USA

Prof. Dr. R.W. Longman
Columbia University
Dept. of Mechanical Eng.
New York, N.Y. 10027
USA

Dr. U. Mackenroth
MBB Zentralbereich Technik
Postfach 80 11 09
8000 München 80
FRG

Prof. Dr. Chr. Marchal
ONERA DES/SA
B.P. 72
F-92322 Chatillon Cedex
France

Prof. Dr. L. Meirovitch
Virginia Polytechnic Institute
and State University
Dept. of Eng. Science & Mechanics
Blacksburg, Va. 24061
USA

Prof. Dr. A. Miele
Rice University
Dept. of Mechanical Eng.
and Material Sciences
230 Ryon Building
Houston, Texas 77001
USA

Dr. L. Mikulski
Technische Universität Krakau
z.Zt. Technische Universität München
Mathematisches Institut
Arcisstr. 21
8000 München 2
FRG

Prof. Dr. H.J. Oberle
Universität Hamburg
Inst. f. Angew. Mathematik
Bundesstr. 55
2000 Hamburg 13
FRG

Dr. H.J. Pesch
Technische Universität München
Mathematisches Institut
Arcisstr. 21
8000 München 2
FRG

Prof. Dr. B.L. Pierson
Iowa State University
Dept. of Aerospace Eng.
304 Town Engineering Building
Ames, Iowa 50011
USA

Dr. R. Polis
MBB Zentralbereich Technik
Postfach 80 11 09
8000 München 80
FRG

Prof. Dr. P. Rentrop
Universität Kaiserslautern
Fachbereich Mathematik
Erwin-Schrödinger-Str.
6750 Kaiserslautern
FRG

Prof. Dr. E. Sachs
Universität Trier
Fachbereich IV-Mathematik
Postfach 38 25
5500 Trier
FRG

Dr. K. Schilling
Dornier System GmbH
Abt. RGW
Postfach 1360
7990 Friedrichshafen
FRG

Dipl. Math. J. Schlöder
Universität Bonn, SFB 72
Inst. f. Angew. Mathematik
Wegelerstr. 6
5300 Bonn 1
FRG

Dipl. Math. K. Schnepper
DFVLR FF-DF
Oberpfaffenhofen
8031 Wessling
FRG

Dr. G.C. Shau
DFVLR FF-DF
Oberpfaffenhofen
8031 Wessling
FRG

Prof. Dr. J. Sprekels
Universität Augsburg
Inst. f. Mathematik
Memminger Str. 6
8900 Augsburg
FRG

Prof. Dr. J. Stoer
Universität Würzburg
Inst. f. Angew. Mathematik
Am Hubland
8700 Würzburg
FRG

Prof. Dr. Ing. G. Szefer
Technical University Cracow
Institute for Mechanics
Ul. Warszawska 24
Pl-31-155 Cracow
Poland

Frau Prof. Dr. I. Troch
Technische Universität Wien
Inst. f. Analysis, Technische
Mathematik und Versicherungs-
mathematik
Wiedner Hauptstr. 6-10
A-1040 Wien
Austria

Prof. Dr. R. Walden
Universität GH Paderborn
FB Mathematik-Informatik
Warburger Str. 100
4790 Paderborn
FRG

Dr. K.H. Well
DFVLR FF-DF
Oberpfaffenhofen
8031 Wessling
FRG

Dipl. Math. U. Wever
Universität Kaiserslautern
Fachbereich Mathematik
Erwin-Schrödinger-Str.
6750 Kaiserslautern
FRG

TABLE OF CONTENTS

THEORY AND COMPUTATIONAL METHODS

AIRCRAFT TRAJECTORY CONTROL

CONTROL SYSTEM DESIGN

ROBOT CONTROL

WATER RESOURCES MANAGMENT

CONTROL OF FLEXIBLE STRUCTURES

THEORY AND COMPUTATIONAL METHODS

SINGULAR PERTURBATIONS AND ASYMPTOTIC EXPANSIONS IN NONLINEAR OPTIMAL CONTROL

Mark D. Ardema
Santa Clara University, Santa Clara, CA 95053, U.S.A.

1. Introduction

Application of the necessary conditions for optimal control of systems defined by ordinary differential equations results in a two-point boundary value problem. In many applications, including those involving atmospheric flight mechanics, the boundary-value problem is of great complexity and, consequently, interest persists in finding accurate approximations.

In this paper, we review a rational method of constructing approximate solutions to nonlinear optimal control problems. The method is based on the singular perturbation theory of ordinary differential equations and employs the techniques of matched asymptotic expansions (MAE) to obtain solutions. Much of the material that follows is abstracted from References 1-3.

When confronted with a system of prohibitive computational complexity, one of the most logical and common approaches is to neglect terms in the equations which are thought to have only small effects on the solution. In the usual case, the approximate system has the same behavior as the original system. For example, consider the following initial value problem where x is a scalar function and ε is a "small" scalar parameter:

$$dx/dt = f(x,t) + \varepsilon g(x,t); \qquad x(\varepsilon,0) = x_o$$

Under certain hypotheses, the solution of the system with $\varepsilon = 0$ will give a good approximation to the solution of the original problem uniformly in the interval of interest; in particular, the initial condition can be met. This is termed a regular perturbation problem.

Now consider the system

$$dx/dt = f(x,y,t); \qquad x(\varepsilon,0) = x_o$$

$$\varepsilon(dy/dt) = g(x,y,t); \qquad y(\varepsilon,0) = y_o$$

where x and y are scalar functions and $\varepsilon>0$ is a "small" scalar parameter. We call this the "exact" system and the system with ε set to zero the "reduced" system. It is obvious at once that in general the reduced solution will not be able to satisfy both initial conditions and thus, at least locally, the behavior of the reduced solution will be radically different from that of the exact solution. In fact, the best that can be hoped for is that the reduced solution gives a good approximation for x uniformly in the domain of interest and for y everywhere except near $y(\varepsilon,0) = y_o$. This loss of boundary condition and consequent loss of uniform approximation is characteristic of singular perturbation problems. In spite of this radical change in solution behavior, singular perturbations are attractive because of the considerable simplification resulting from decreased system order.

Singular perturbation theory is concerned with the relation between the exact and reduced solutions of singularly perturbed systems of ordinary differential equations and with constructing asymptotic series representations of the exact solution.

Largely independent of the development of singular perturbation theory for ordinary differential equations has been the development of asymptotic methods to solve certain fluid mechanics problems involving partial differential equations. These methods, most notably the method of matched asymptotic expansions (or method of inner and outer expansions), have their origin in boundary-layer concept. In problems concerning viscous flow past a solid body (Fig. 1), the viscosity is a parameter (usually small) multiplying the highest derivative in the Navier-Stokes equations. If this parameter is set equal to zero, the hydrodynamic system of equations results (reduced system); the solution of this system violates the no-slip boundary condition at the body surface. Thus, in a thin layer of fluid near the surface of the body - the boundary layer - the velocity varies rapdily from zero on the surface of the body to the value given by the hydrodynamic solution.

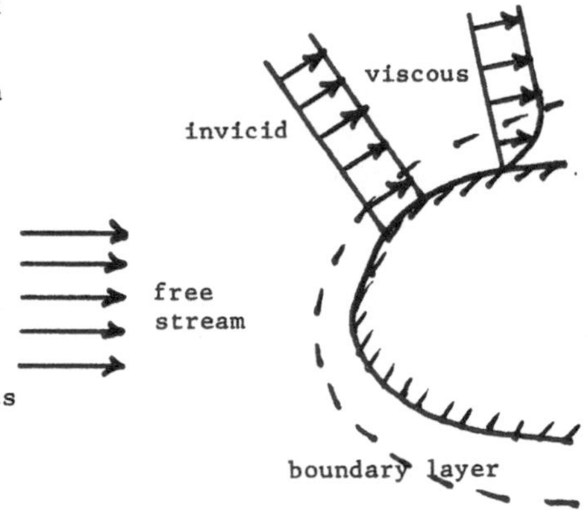

Fig. 1 Boundary layers in fluid mechanics.

The phenomenon of boundary layers occurs in all singular perturbation problems. In such problems, the solution is sought in two (or in some cases, several) separate regions. In the outer region, the variables are relatively slowly varying, resemble the reduced solution, and do not in general satisfy boundary conditions. In the inner region near the boundary (boundary layer) the variables are relatively rapidly varying, asymptotically stable and satisfy appropriate boundary conditions.

A standard technique of obtaining approximate solutions of mathematical problems is to introduce perturbations about a nominal solution. This technique is particularly useful in problems in which a "small parameter" is present, because in this case the nominal solution and the method of introducing the perturbations are suggested in an obvious way. In some problems, no small parameter appears on physical grounds; such a parameter may be artificially inserted to suppress terms in the equation which are expected to have relatively small effects.

In the MAE method, separate solutions are obtained for the inner and outer regions by asymptotic expansion techniques. These asymptotic expansions need not be convergent and in fact often are not convergent in applications. The unknown constants are determined by "matching" the two solutions; the ability to do this depends on the existence of an overlap region of common validity. (Recall that the outer solution is not required to satisfy the boundary conditions). If desired, the inner and outer solutions may then be combined to give a uniformly valid asymptotic representation of the solution.

A few definitions will be needed. If a function $f(\varepsilon)$ has an asymptotic expansion $\sum_{i=o}^{\infty} a_i \varepsilon^i$ we will call $\sum_{i=o}^{\infty} a_i \varepsilon^i$ the "n-th order approximation to $f(\varepsilon)$". By definition of an asymptotic expansion, $|f(\varepsilon) - \sum_{i=o}^{n} a_i \varepsilon^i| = 0(\varepsilon^{n+1})$, and consequently the n-th order approximation is accurate to order $n+1$. We say $f(\varepsilon) = 0(\varepsilon^n)$ if $f(\varepsilon)/\varepsilon^n$ is bounded as $\varepsilon \to 0^+$.

To illustrate the use of asymptotic methods, consider the scalar, linear, constant coefficient, initial value-problem

$$\frac{dx}{dt} = ax \ , \qquad x(o) = x_o$$

If a is small, $o<a<<1$ it is logical to attempt a solution expansion in powers of a,

$$x(t) = \sum_{i=o}^{\infty} x_i(t)\, a^i$$

Substituting into the differential equation and equating coefficients of like powers of a results in the sequence

$$x(t) = x_o(1 + at + \frac{a^2 t^2}{2} + \dots)$$

This is, in fact, the expansion of the exact solution, $x(t) = x_o e^{at}$; as $a \to o$, the sequence converges uniformly to the exact solution, a characteristic of a regular perturbation problem. Note that, however, for large t the sequence may not give a useful approximation.

Now suppose a is large, $a \gg 1$, so that the appropriate expansion is

$$x(t) = \sum_{i=o}^{\infty} x_i(t) \left(\frac{1}{a}\right)^i$$

Then substitution into the differential equation gives

$$x(t) = o + o + o + \ldots$$

Although this may be an accurate approximation for $a \gg 1$ and $t \gg o$, it is not valid for $t = o$, exhibiting the nonuniform convergence associated with a singular perturbation problem.

In both cases, the difficulty can be resolved by formulating the problem on its proper time scale. Make the change of variable $\tau = at$. In the case $o < a \ll 1$, this is a shrinking transformation, and for $a \gg 1$ it is a stretching transformation. Then the problem becomes

$$\frac{dx}{d\tau} = x , \qquad x(o) = x_o$$

and expansion gives the solution as

$$x(\tau) = x_o e^{\tau} + o + o + \ldots$$

Thus the exact solution is obtained with the leading term.

2. Theory of Singularly Perturbed Nonlinear Optimal Control

Consider the system

$$\dot{x} = f(x,y,u,\varepsilon,t) \qquad \varepsilon\dot{y} = g(x,y,u,\varepsilon,t) \tag{2.1}$$

on the interval $0 \le t \le T$ subject to initial conditions

$$x(\varepsilon,0) = x_o(\varepsilon) \qquad y(\varepsilon,0) = y_o(\varepsilon) \tag{2.2}$$

where "·" denotes a derivative with respect to t. It is desired to minimize

$$J = \int_0^T \phi(x,y,u,\varepsilon,t)dt \tag{2.3}$$

where T is prescribed. In these equations, $x(\cdot)$ is a "slow" state vector functions with n_s components, $y(\cdot)$ is a "fast" state vector function with n_f components, $u(\cdot)$ is a control vector function with n_c components, and $\varepsilon > 0$ is a parameter. It is assumed that $f(\cdot)$, $g(\cdot)$, $f_x(\cdot)$, $f_y(\cdot)$, $g_x(\cdot)$, and $g_y(\cdot)$ are continuous and that $u(\cdot)$ is piecewise continuous and unconstrained for $0 \le t \le T$, where subscripts denote partial differentiation.

Rewrite (2.1) as

$$\dot{x} = f(x,y,u,\varepsilon,t) \qquad \dot{y} = \frac{1}{\varepsilon} g(x,y,u,\varepsilon,t) \tag{2.4}$$

Let $\lambda_x(\cdot)$ and $\lambda_y(\cdot)$ be any nonzero vector functions of dimensions n_s and n_f, respectively, such that their components satisfy the following linear system of equations. (The usual prime notation for the transpose of a matrix would prove to be cumbersone in the sequel, and we therefore omitt it; it should be obvious from the context whether a matrix or its transpose is implied).

$$\dot{\lambda}_x = -\phi_x\lambda_o - f_x\lambda_x - \frac{1}{\varepsilon} g_x\lambda_y \qquad \dot{\lambda}_y = -\phi_y\lambda_o - f_y\lambda_x - \frac{1}{\varepsilon} g_y\lambda_y \tag{2.5}$$

Define the scalar function $H'(\cdot)$ by

$$H'(\lambda_o,\lambda_x,\lambda_y,x,y,u,\varepsilon,t) = \phi\lambda_o + f\lambda_x + \frac{1}{\varepsilon} g\lambda_y \tag{2.6}$$

Introducing the transformation

$$\lambda_x = \lambda \;; \quad \lambda_y = \varepsilon\mu \tag{2.7}$$

into (2.5) and (2.6) results in

$$\dot{\lambda} = -\phi_x\lambda_o - f_x\lambda - g_x\mu \qquad \varepsilon\dot{\mu} = -\phi_y\lambda_o - f_y\lambda - g_y\mu \tag{2.8}$$

and

$$H(\lambda_o,\lambda,\mu,x,y,u,\varepsilon,t) = \phi\lambda_o + f\lambda + g\mu \tag{2.9}$$

respectively. Applying the well-known Pontryagin Maximum Principle to this problem then gives the following necessary conditions for optimal control.

Theorem 1. (Maximum Principle for Singularly Perturbed Systems)

If the control $u(\cdot)$ minimizes (2.3) and, along with $x(\cdot)$ and $y(\cdot)$, satisfy (2.1) and (2.2) then there exist nonzero functions $\lambda(\cdot)$ and $\mu(\cdot)$ whose components satisfy (2.8) such that

$$\begin{aligned} &\text{(a)} \quad H_u(\lambda_o,\lambda(\varepsilon,t),\mu(\varepsilon,t),x(\varepsilon,t),y(\varepsilon,t),u,\varepsilon,t) = 0 \\ &\text{(b)} \quad \lambda_o = \text{constant} \le 0 \\ &\text{(c)} \quad \lambda(\varepsilon,T) = 0 \text{ and } \mu(\varepsilon,T) = 0. \end{aligned} \tag{2.10}$$

Thus the problem reduces to one of solving the following two point boundary value problem (from now on functional dependence will be omitted when it does not result in a lack of clarity)

$$\dot{x} = f \qquad \varepsilon\dot{y} = g \qquad \dot{\lambda} = -H_x$$

$$\varepsilon\dot{\mu} = -H_y \qquad 0 = H_u \qquad x(\varepsilon,0) = x_o(\varepsilon) \qquad y(\varepsilon,0) = y_o(\varepsilon) \tag{2.11}$$

$$\lambda(\varepsilon,T) = 0; \qquad \mu(\varepsilon,T) = 0$$

where $H(\cdot)$ is given by (2.9). We note that in this formulation of the problem the adjoint variables associated with the slow state variables are themselves slow and the adjoint variables associated with the fast state variables are themselves fast. If it is assumed that a unique optimal control exists, then (2.11) has a unique solution. We shall in fact assume that H_{uu} is negative definite in the subsequent discussion.

We call the system with $\varepsilon = 0$ and the boundary conditions on the fast variables omitted the reduced problem:

$$\dot{x}_r = f_r \qquad 0 = g_r \qquad \dot{\lambda}_r = -H_{x_r}$$

$$0 = -H_{y_r} \qquad 0 = H_{u_r} \qquad x_r(0) = x_o(0) \qquad \lambda_r(T) = 0 \tag{2.12}$$

where, for example, f_r denotes $f(x_r,y_r,u_r,0,t)$; the reduced state, adjoint control variables are of course functions only of t. Because of the two point boundary value nature of this problem, there will generally be two boundary layer problems associated with (2.11), called herein the initial and the terminal.

To obtain the intial zero-order boundary-layer equation (ZOBLE), we substitute

$$\tau = \frac{t}{\varepsilon} \tag{2.13}$$

in (2.11) and set $\varepsilon = 0$ to get

$$\left.\begin{aligned} \frac{dy_b}{d\tau} &= g_b \\ \frac{d\mu_b}{d\tau} &= -H_{yb} \\ 0 &= H_{u_b} \\ y_b(0) &= y_o(0) \end{aligned}\right\} \tag{2.14}$$

where, for example, H_{y_b} denotes $\partial H(\lambda_o,\lambda_r(0),\mu_b,x_o(0),y_b,u_b,0,0)/\partial y_b$. The perturbation equations obtained by linearizing (2.14) about the outer solution evaluated at $t = 0$, which is an equilibrium point of (2.14), has a constant $2n_f \times 2n_f$ coefficient matrix of the form

$$G_{r_o} = \begin{bmatrix} A_{r_o} & B_{r_o} \\ C_{r_o} & -A'_{r_o} \end{bmatrix} \tag{2.15}$$

where

$$\left.\begin{aligned} A_{r_o} &= g_{y_{r_o}} - g_{u_{r_o}} H^{-1}_{uu_{r_o}} H_{uy_{r_o}} \\ B_{r_o} &= -g_{u_{r_o}} H^{-1}_{uu_{r_o}} g'_{u_{r_o}} = B'_{r_o} \\ C_{r_o} &= -H_{yy_{r_o}} + H_{yu_{r_o}} H^{-1}_{uu_{r_o}} H_{uy_{r_o}} = C'_{r_o} \end{aligned}\right\} \tag{2.16}$$

and where subscript r_o indicates that these matrices are to be evaluated on the reduced solution evaluated at $t = 0$.

Since there are n_f boundary conditions specified for the $2n_f$ equations (2.14), we must have at least n_f stable modes, i.e., the matrix G_{r_o} must have at least n_f eigenvalues with negative real parts. However, a well known property of a matrix with the structure of (2.15) is that if s is an eigenvalue then so is $-s$ and the equilibrium point is a saddle with n_f stable modes. Consequently, we will have the property we want if and only if there are no eigenvalues with zero real parts. Since a similar result holds for the terminal boundary layer, the local eigenvalue criterion for boundary layer stability of (2.11) can be stated as

"There are no eigenvalues of $G_r(t)$ with zero real parts on $0 \le t \le T$" (2.17)

Before stating the basic theorem giving the asymptotic properties of the solution of (2.11), one more matrix must be introduced. Let P be a nonsingular $2n_f \times 2n_f$ matrix such that

$$P^{-1}GP = \begin{bmatrix} D_1 & 0 \\ 0 & D_2 \end{bmatrix} \tag{2.18}$$

where D_1 has only eigenvalues with negative real part and D_2 has only eigenvalues with positive real part. If the eigenvalue criterion on G just stated is satisfied on $0 \leq t \leq T$, then such a matrix P exists on $0 \leq t \leq T$. Partition P into the form

$$P = \begin{bmatrix} P_{11} & P_{12} \\ P_{21} & P_{22} \end{bmatrix} \tag{2.19}$$

where all the P_{ij} are $n_f \times n_f$ matrices. We are now ready to state the following result.

Theorem 2.

Consider the system (2.11) and suppose that the following are satisfied: (A) there exists an $\varepsilon_o > 0$ such that f and g are $K + 2$ times and H is $K + 3$ times continuously differentiable with respect to x,y,u,ε and t and x_o and y_o are $k + 2$ times continuously differentiable with respect to ε, for all $0 \leq t \leq T$ and $0 \leq \varepsilon \leq \varepsilon_o$, in a neighborhood of the reduced solution; (B) the reduced system (2.12) has a continuous solution on $0 \leq t \leq T$; (C) the matrix G defined by (2.15) satisfies the eigenvalue criterion (2.17); (D) $P_{11}(0)$ and $P_{22}(T)$ as defined by (2.19) are nonsingular; and (E) the quantities $|y_o(0) - y_r(0)|$ and $|\mu_r(0)|$ are sufficiently small to insure that the initial and terminal boundary conditions are in the domains of influence of the reduced solution evaluated at $t = 0$ and $t = T$, respectively. Then, for $0 \leq \varepsilon \leq \varepsilon_o$: (i) the full system (2.11) has a unique solution; (ii) the solution of (2.11) for x and λ tends to the solution of (2.12) for x_r and λ_r uniformly on $0 \leq t \leq T$ and the solution of (2.11) for y and μ tends to the solution of (2.12) for y_r and μ_r uniformly on any closed subinterval of $0 < t < T$, as ε tends to zero; (iii) there exist solutions to the initial and terminal ZOBLES which are asymptotically stable with respect to the reduced solution and which satisfy all imposed boundary conditions; and (iv) the outer, initial boundary layer, and terminal boundary layer systems associated with (2.11) all possess asymptotically valid expansions in ε up to order K such that, when suitably combined, they give an asymptotically valid expansion of the solution of (2.11) up to order K in ε.

In practice, as has been remarked earlier, the only condition of the Theorem which is generally useful is the eigenvalue criterion (2.17). This condition is relatively easy to check and gives valuable information regarding the behavior of the ZOBLES.

In the case where f and g are scalar functions, the eigenvalue criterion takes on an especially simple form, namely that

$$H_{yy_r} g_{u_r}^2 - 2H_{yu_r} g_{y_r} g_{u_r} + H_{uu_r} g_{y_r}^2 < 0 \tag{2.20}$$

must be satisfied at $t = 0$ and at $t = T$. This is just the strengthened form of the Legendre Clebzch condition of the calculus of variations for the reduced problem.

In some applications, the reduced solution is a sufficiently good approximation. In this case, the natural question arises as to when it is possible to set $\varepsilon = 0$ before applying the Maximum Principle instead of after. The former procedure is attractive because it involves less algebraic manipulation. The following answers this question.

Theorem 3.

In addition to the assumptions of Theorem 2, suppose that the matrix $[g_y g_u]$ has maximum rank (i.e. rank n_f) evaluated along the reduced solution. Then the reduced problem is the same as the problem obtained by the alternative procedure of setting $\varepsilon = 0$ in the state equations and applying the necessary conditions for optimal control to the result.

3. Solution By Matched Asymptotic Expansions

In this section, we will apply the method of matched asymptotic expansions to the nonlinear optimal control problem. The analysis will follow the following pattern: (i) formulate the outer and boundary layer problems, (ii) obtain asymptotic solutions to these problems, (iii) match these solutions to obtain all constants of integration, and (iv) form the additive composite to obtain a uniformly valid asymptotic representation of the solution to the original problem. We assume that a unique, unbounded optimal control exists and that all hypotheses of Theorem 2 hold.

The outer system is simply (2.11) without the boundary conditions; denoting the outer solution by, for example, $x^o(\varepsilon,t)$, we have

$$\begin{aligned} \dot{x}^o &= f^o & \varepsilon\dot{y}^o &= g^o \\ \dot{\lambda}^o &= -H_x^o & \varepsilon\dot{\mu}^o &= -H_y^0 \\ 0 &= H_u^o \end{aligned} \tag{3.1}$$

where, for example, $H_x^o = \partial H(\lambda_o,\lambda^o,\mu^o,x^o,y^o,u^0,\varepsilon,t)/\partial x^o$.

The initial boundary layer system of equations is obatined as before by introducing the stretching transformation

$$\tau = \frac{t}{\varepsilon} \tag{3.2}$$

into (2.11); the result is, denoting the solution by, for example, $x^{i1}(\varepsilon,\tau)$,

$$\begin{aligned}
&\frac{dx^{i1}}{d\tau} = \varepsilon f^{i1} && x^{i1}(\varepsilon,0) = x_o(\varepsilon) \\
&\frac{dy^{i1}}{d\tau} = g^{i1} && y^{i1}(\varepsilon,0) = y_o(\varepsilon) \\
&\frac{d\lambda^{i1}}{d\tau} = -\varepsilon H_x^{i1} && \frac{d\mu^{i1}}{d\tau} = -H_y^{i1} \\
&0 = H_u^{i1}
\end{aligned} \tag{3.3}$$

where, for example, $H_x^{i1} = \partial H(\lambda_o,\lambda^{i1},\mu^{i1},x^{i1},y^{i1},u^{i1},\varepsilon,\varepsilon\tau)\partial x^{i1}$

Similarly, the terminal boundary layer system is obtained by stretching the time-to-go by ε ,

$$\sigma = \frac{T-t}{\varepsilon} \tag{3.4}$$

The result is, denoting the solution by, for example, $x^{i2}(\varepsilon,\sigma)$,

$$\begin{aligned}
&\frac{dx^{i2}}{d\sigma} = -\varepsilon f^{i2} && \frac{dy^{i2}}{d\sigma^{i2}} = -g^{i2} \\
&\frac{d\lambda^{i2}}{d\sigma} = \varepsilon H_x^{i2} && \lambda^{i2}(\varepsilon,0) = 0 \\
&\frac{d\mu^{i2}}{d\sigma} = H_y^{i2} && \mu^{i2}(\varepsilon,0) = 0 \\
&0 = H_u^{i2}
\end{aligned} \tag{3.5}$$

where, for example, $H_x^{i2} = \partial H(\lambda,\lambda^{i2},\mu^{i2},x^{i2},y^{i2},u^{i2},\varepsilon,T-\varepsilon\sigma)/\partial x^{i2}$

To solve (3.1) we express all dependent variables $(x^o,y^o,\lambda^o,\mu^o,u^o)$ in asymptotic power series, for example,

$$x^o(\varepsilon,t) \sim \sum_{j=0}^{k} x_j^o(t)\varepsilon^j; \quad k \le K \tag{3.6}$$

This leads to the sequence of $2n_s$ dimensional problems

$$\begin{array}{lll}
\dot{x}_o^o = f_o^o & \dot{x}_j^o = f_j^o & 0 = g_o^o \\
\dot{y}_{j-1}^o = g_j^o & \dot{\lambda}_o^o = -H_{x_o}^o & \dot{\lambda}_j^o = -H_{x_j}^o \\
0 = -H_{y_o}^o & \dot{\mu}_{j-1}^o = -H_{y_j}^o & 0 = H_{u_o}^o \\
0 = H_{u_j}^o & j = 1,\ldots,k &
\end{array} \tag{3.7}$$

each of which is linear except for the first.

Next, to solve (3.3) we put, for example,

$$x^{il}(\varepsilon,\tau) \sim \sum_{j=0}^{k} x_j^{il}(\tau)\varepsilon^j; \quad k \le K \tag{3.8}$$

in (3.3) to get a sequence of $2n_f$ dimensional problems

$$\begin{array}{lll}
\frac{dx_o^{il}}{d\tau} = 0 & x_o^{il}(0) = x_{oo} & \frac{dy_o^{il}}{d\tau} = g_o^{il} \\
y_o^{il}(0) = y_{oo} & \frac{d\lambda_o^{il}}{d\tau} = 0 & \frac{d\mu_o^{il}}{d\tau} = -H_{y_o}^{il} \\
0 = H_{u_o}^{il} & \frac{dx_j^{il}}{d\tau} = f_{j-1}^{il} & x_j^{il}(0) = x_{oj} \\
\frac{dy_j^{il}}{d\tau} = g_j^{il} & y_j^{il}(0) = y_{oj} & \frac{d\lambda_j^{il}}{d\tau} = -H_{x_{j-1}}^{il} \\
\frac{d\mu_j^{il}}{d\tau} = -H_{y_j}^{il} & 0 = H_{u_j}^{il} & j = 1,\ldots,k
\end{array} \tag{3.9}$$

where the functions such as f_j^{il} are obtained by expansion about $\varepsilon = 0$ as before, and x_{oj} and y_{oj} are given by

$$x_o(\varepsilon) = \sum_{j=1}^{k} x_{oj}\varepsilon^j \qquad y_o(\varepsilon) = \sum_{j=1}^{k} y_{oj}\varepsilon^j \tag{3.10}$$

As before, the first in the sequence of problems (3.9) is nonlinear and the rest are linear.

Finally, the terminal boundary layer systems (3.5) is solved asymptotically by setting, for example

$$x^{i2}(\varepsilon,\sigma) \sim \sum_{j=0}^{k} x_j^{i2}(\sigma)\varepsilon^j; \quad k \le K \tag{3.11}$$

to get the sequence of $2n_f$ dimensional problems

$$\frac{dx_o^{i2}}{d\sigma} = 0 \qquad \frac{dy_o^{i2}}{d\sigma} = -g_o^{i2} \qquad \frac{d\lambda_o^{i2}}{d\sigma} = 0 \qquad \lambda_o^{i2}(0) = 0$$

$$\frac{d\mu_o^{i2}}{d\sigma} = H_{y_o}^{i2} \qquad \mu_o^{i2}(0) = 0 \qquad 0 = H_{u_o}^{i2} \qquad \frac{dx_j^{i2}}{d\sigma} = -f_{j-1}^{i2} \tag{3.12}$$

$$\frac{dy_j^{i2}}{d\sigma} = -g_j^{i2} \qquad \frac{d\lambda_j^{i2}}{d\sigma} = H_{x_{j-1}}^{i2} \qquad \lambda_j^{i2}(0) = 0 \qquad \frac{d\mu_j^{i2}}{d\sigma} = H_{y_j}^{i2}$$

$$\mu_j^{i2}(0) = 0 \qquad 0 = H_{u_j}^{i2} \qquad j = 1,\dots,k$$

only the first of which is nonlinear.

Usually only the first two terms of these expansions are used in practice. This is due not only to the algebraic complexity of higher order terms but also to the fact that if a satisfactory solution is not obtained after two terms then higher order terms are not likely to improve the situation. Our investigation here will be limited to the first term. Consider, therefore, the leading problem of (3.7), (3.9) and (3.12), i.e., the reduced system with the boundary conditions removed and the initial and terminal ZOBLES. Solution of the first of the problem (3.7) will contain $2n_s$ as yet unknown constants of integration, say $x_o^o(0)$ and $\lambda_o^o(T)$. For the leading problem of (3.9) we have $x_o^{i1}(\tau) = x_{oo}$ and $\lambda_o^{i1}(\tau) = \lambda_o^{i1}(0)$, the later an unknown vector constant. The remaining $2n_s$ equations have only n_s boundary conditions. We use the other n_s boundary conditions, say $\mu_o^{i1}(0)$, to suppress the "unstable modes". We know that we will have precisely the right number of free boundary conditions to do this and that the "stable modes" will be able to satisfy all the prespecified boundary conditions, $y_o^{io}(0) = y_{oo}$. Similarly, for the leading problem of (3.12), $x_o^{i2}(\sigma) = x_o^{i2}(0)$ and $\lambda_o^{i2}(\sigma) = 0$, the former an unknown constant. We use the n_f free constants $y_o^{i2}(0)$ to suppress the "unstable modes", leaving just enough "stable modes" to satisfy the prespecified boundary conditions, $\mu_o^{i2}(0) = 0$.

We now can match the slow variables, x and λ, at $t = 0$ and $t = T$ to zero order to obtain the unknown constants of integration in the outer solution. At $t = 0$, the matching rule applied to x and λ to zero order gives simply

$$x_o^o(0) = x_{oo} \qquad \lambda_o^o(0) = \lambda_o^{i1}(0) \tag{3.13}$$

and at $t = T$ a similar rule implies

$$x_o^o(T) = x_o^{i2}(0) \qquad \lambda_o^o(T) = 0 \tag{3.14}$$

The zero order problem is now fully determined. First, solve

$$\left.\begin{array}{lll} \dot{x}_o^o = f_o^o & \dot{\lambda}_o^o = -H_{x_o}^o & \\ \text{subject to} & & \\ 0 = g_o^o & 0 = H_{y_o}^o & 0 = H_{u_o}^o \\ x_o^o(0) = x_{oo} & \lambda_o^o(T) = 0 & \end{array}\right\} \tag{3.15}$$

Next, solve

$$\left.\begin{array}{lll} \dfrac{dy_o^{i1}}{d\tau} = g_o^{i1} & \dfrac{d\mu_o^{i1}}{d\tau} = -H_{y_o}^{i1} & \\ \text{subject to} & & \\ x_o^{i1} = x_{oo} & \lambda_o^{i1} = \lambda_o^o(0) & 0 = H_{u_o}^{i1} \\ y_o^{i1}(0) = y_{oo} & \mu_o^{i1}(0) \text{ selected to suppress instability} & \end{array}\right\} \tag{3.16}$$

And finally,

$$\left.\begin{array}{lll} \dfrac{dy_o^{i2}}{d\sigma} = -g_o^{i2} & \dfrac{d\mu_o^{i2}}{d\sigma} = H_{y_o}^{i2} & \\ \text{subject to} & & \\ x_o^{i2} = x_o^o(T) & \lambda_o^{i2} = 0 & 0 = H_{u_o}^{i2} \\ y_o^{i2}(0) \text{ selected to suppress instability} & & \mu_o^{i2}(0) = 0 \end{array}\right\} \tag{3.17}$$

The problem (3.15) is a $2n_s$ dimensional two-point boundary value problem (2PBVP) on a finite interval T. Problems (3.16) and (3.17) are essentially $2n_f$ dimensional 2PBVPs on an infinite interval but in practice they would be solved on time intervals τ^* and σ^*, respectively, where τ^* and σ^* are sufficiently large such that the transients have become negligably small; τ^* and σ^* of course depend on ε. Thus, in effect, we have approximated the solution to a $2(n_s+n_f)$ dimensional 2PBVP by the solutions to one $2n_s$ and two $2n_f$ dimensional problems.

For forming additive composite solutions which are valid everywhere on $0 \leq t \leq T$ for all variables the common parts will be needed. Since there are two boundary layers, each variable will have two common parts. For the zero order, these common parts are simply the values of the reduced solution variables evaluated at the boundaries:

$$\begin{aligned}
CP^{i1}_{x_o} &= x_{oo} & CP^{i2}_{x_o} &= x^o_o(T) \\
CP^{i1}_{y_o} &= y^o_o(0) & CP^{i2}_{y_o} &= y^o_o(T) \\
CP^{i1}_{\lambda_o} &= \lambda^o_o(0) & CP^{i2}_{\lambda_o} &= 0 \\
CP^{i1}_{\mu_o} &= \mu^o_o(0) & CP^{i2}_{\mu_o} &= \mu^o_o(T) \\
CP^{i1}_{u_o} &= u^o_o(0) & CP^{i2}_{u_o} &= u^o_o(T)
\end{aligned} \tag{3.18}$$

The additive composite solution for each variable is formed according to, for example,

$$\begin{aligned}
x^a_o(\varepsilon,t) = {} & x^o_o(t) + x^{i1}_o(\tfrac{t}{\varepsilon}) + x^{i2}_o(\tfrac{T-t}{\varepsilon}) \\
& - CP^{i1}_{x_o}(\varepsilon,t) - CP^{i2}_{x_o}(\varepsilon,t)
\end{aligned} \tag{3.19}$$

The result is

$$\begin{aligned}
x^a_o(\varepsilon,t) &= x^o_o(t) \\
y^a_o(\varepsilon,t) &= y^o_o(t) + y^{i1}_o(\tfrac{t}{\varepsilon}) + y^{i2}_o(\tfrac{T-t}{\varepsilon}) - y^o_o(0) - y^o_o(T) \\
\lambda^a_o(\varepsilon,t) &= \lambda^o_o(t) \\
\mu^a_o(\varepsilon,t) &= \mu^o_o(t) + \mu^{i1}_o(\tfrac{t}{\varepsilon}) + \mu^{i2}_o(\tfrac{T-t}{\varepsilon}) - \mu^o_o(0) - \mu^o_o(T) \\
u^a_o(\varepsilon,t) &= u^o_o(t) + u^{i1}_o(\tfrac{t}{\varepsilon}) + u^{i2}_o(\tfrac{T-t}{\varepsilon}) - u^o_o(0) - u^o_o(T)
\end{aligned} \tag{3.20}$$

We note that the additive composition to the zero order for the slow variables x and λ is just the reduced solution, which is independent of ε. For the fast variables and the control, the composite solution consists of, for example for y, the reduced solution $y_o^o(t)$ augmented by boundary layer corrections due to the initial layer $[y_o^{i1}(\frac{t}{\varepsilon}) - y_o^o(0)]$, and to the terminal layer, $[y_o^{i2}(\frac{T-t}{\varepsilon}) - y_o^o(T)]$.

We now use (3.20) to evaluate $y_o^a(\varepsilon,0)$ and $\mu_o^a(\varepsilon,0)$ as a check to see if their boundary conditions are satisfied. The result is

$$y_o^a(\varepsilon,0) = y_{oo} + [y_o^{i2}(\tfrac{T}{\varepsilon}) - y_o^o(T)]$$

$$\mu_o^a(\varepsilon,T) = [\mu_o^{i1}(\tfrac{T}{\varepsilon}) - \mu_o^o(0)] \qquad (3.21)$$

Because of boundary layer stability,

$$\lim_{T/\varepsilon\to\infty} y_o^{i2}(\tfrac{T}{\varepsilon}) = y_o^o(T)$$

$$\lim_{T/\varepsilon\to\infty} \mu_o^{i1}(\tfrac{T}{\varepsilon}) = \mu_o^o(0) \qquad (3.22)$$

Thus the bracketed terms in (3.21) will be asymptotically negligible although not in general numerically zero and the boundary conditions on y and μ will not be met exactly; the larger the value of T/ε the smaller will be the error. This error in boundary conditions is a consequence of "each boundary layer not knowing of the other's existence". In the practical case in which the boundary layer integrations are performed on finite intervals τ^* and σ^*, the boundary conditions will be met exactly provided that $\tau^* < T/\varepsilon$ and $\sigma^* < T/\varepsilon$, i.e. provided that each boundary layer has "died out" before the other boundary has been reached. It is logical to make this condition a requirement, since for the asymptotic solution to give a good numerical approximation requires that the boundary layer motion be relatively insignificant compared to the outer motion, or to put it another way, a "strong separation between the slow and fast variables" and a relatively long time interval. Thus we impose the requirements

$$T - \varepsilon\tau^*(\varepsilon) > 0 \qquad\qquad T - \varepsilon\sigma^*(\varepsilon) > 0 \qquad (3.23)$$

The larger the values of $T - \varepsilon\tau^*(\varepsilon)$ and $T - \varepsilon\sigma^*(\varepsilon)$, the better will the asymptotic solution numerically approximate the exact solution. We do, however, allow the boundary layers to "overlap", that is it is possible that $\varepsilon\tau^* + \varepsilon\sigma^* > T$.

4. References

1. Ardema, M.D., "Singular Perturbations in Flight Mechanics", NASA TM-62,380, August 1974 (revised July 1977).

2. Ardema, M.D., "Solution of the Minimum Time-to-Climb Problem by Matched Asymptotic Expansions", AIAA Journal, Vol. 14, No. 7, July 1976, pp. 843-850.

3. Ardema, M.D., "An Introduction to Singular Perturbations in Nonlinear Optimal Control", in Singular Perturbations in Systems and Control, M.D. Ardema, ed., International Centre for Mechancial Sciences, Courses and Lectures No. 280, 1983, pp. 1-92.

REDUCTION OF DETERMINISTIC DIFFERENTIAL GAMES TO PROBLEMS OF OPTIMIZATION THE METHOD OF SUCCESSIVE APPROXIMATE STRATEGIES

C. Marchal
Office National d'Etudes et de Recherches Aérospatiales
BP 72. F - 92322 Châtillon Cedex, France

Abstract

The deterministic differential games are presented and classified among the other games. They have many singularities, discontinuities and subdivisions. Sufficient conditions of continuity of the Value of the game in terms of the initial conditions are given.

Most deterministic differential games have a very complex and difficult solution, this gives its interest to the approximate strategy method that reduces the game to a succession of ordinary problems of optimization and allows a systematic step by step improvement of the strategies.

Two good opposite strategies give close upper and lower bounds of the Value of the game.

If the Hamiltonian of the game is either convex or concave with respect to the adjoint vector the game is equivalent to a problem of optimization.

The reference 14 is a french translation of this paper with more examples, more demonstrations and longer developments.

Introduction

The problems of optimization can be considered as one-player games and their natural development are the multi-player games met in diplomacy, business, war, economy, etc... However undeterminism arises in most of these multi-player games and only two-player zero-sum games and their equivalent can be deterministic.

The two-player zero-sum games have been the subject of numerous studies of all kinds : theoretical studies [1-7], studies of simple games : evader and pursuer, the two identical cars, the homicidal chauffeur, the isotropic rocket, etc... [1, 2, 8, 9], studies of more complex games more or less related to realistic problems : air or sea fights, interception of a bomber or a missile etc... [10-12].

A characteristic of most of these games, even the simplest, is a great variety of singularities. These singularities are called transition surfaces, universal surfaces, dispersal surfaces, barriers, focal lines, equivocal lines or surfaces... and it is generally very difficult to obtain a full analysis of a game with many state parameters.

The lack of convergence of theoretical methods and/or the oversight of a singularity has led very often to misleading results and it is then very useful to dispose of simple approximate methods allowing to compute as near as desired upper and lower bounds of the Value of the game.

A surprising result is that the solutions leading to an upper bound are generally very different from those leading to a lower bound even if for all these solutions the value of the game is almost the same : most games have generally an infinite number of "optimal" solutions.

1. The classification of games

We will call a game any situation in which people have to take decisions and judge the results of those decisions. It is of course very often the case in day-to-day life.

Games are classified with respect to the number of players, or teams of players, to the respective purposes of these players and to the physical and mathematical properties of the game.

There are one-player, two-player, three-player, many-player games, the first are also called optimization problems ; the last are generally related to problems of economy.

There are cooperative games in which all players have the same purpose (for instance to avoid collisions in air traffic). In these games, if the information is complete and infinitely rapid, the players form a perfect team and the game becomes a one-player game. Thus, ordinarily cooperative games are related to inaccuracies of information and stochastic process.

There are competitive games in which the interests of the players are opposite ; for instance, they share a fixed amount of goods or money. These games can be called "fixed-sum games" ; and, by a simple translation, are generally called "zero-sum games". The most important case of competitive games is the two-player zero-sum game (war, fight, negotiation, etc...).

There are composite games in which the interests of the players are partially converging and partially diverging (as happens very often in problems of economy). These games are usually called "non-zero-sum games". Let us note that a three-player zero-sum game, in which one of the three players is completely passive, is almost identical to a two-player non-zero sum game.

With respect to the physical and mathematical properties of a game there are differential and discrete games, games with complete, incomplete, delayed informations, with deterministic or stochastic rules, etc...

The game theory can be summarized by the question "How to play optimally ?", the word "optimally" being of course related to the purpose of the player of interest but also to the expected strategies of other players - strategies themselves related to their own purposes. We must of course consider the word "strategy" in its largest meaning, which includes the possibility of agreements and coalitions (secret or not). An optimal strategy can be deterministic or stochastic ; but in the latter case it can lead to a stochastic result, and the player has to make a balance between the risks of the different results.

The domain of deterministic games is rather restricted. Indeed three and many-player games are non deterministic as soon as a coalition can bring more to each member of the coalition than an individual play. However some deterministic results can be obtained (Nash equilibrium, Pareto equilibrium...).

Two-player non-zero-sum games can have a "coalition against nature", but if a player thinks that the coalition is insufficiently favourable to him he can threaten to break the coalition... Aside the optimization problems only two-player zero-sum games and equivalent games can be deterministic.

The hidden sources of undeterminism are generally the reasons of lack of convergence of numerical methods and it is necessary to look for all the conditions of determinism of two-player zero-sum games :

A) The rules of the game must be deterministic and completely known to both players.

B) Random choices must be avoided even in the issues of the game : if some issues involved a final lottery and if a player preferred the risk while its opponent the security the ambiguities of non-zero-sum games would reappear.

C) The players should be forced to play and should consider any advantage of their opponent as a personal disadvantage, or at least they should consider the game as a really interesting intellectual challenge. In the seventies Professor Breakwell [2] discovered that his experiments on behavior in a zero-sum game (two studients had to share 20 dollars according to the result of the game) were sometimes biased by a previous agreement of the two players : they played only at random and settled that bargain at 10 dollars each !

D) A more hidden source of undeterminism can be the absence of value of the game if the game has no end. The value of the game should be defined in all cases even if the two players play indefinitely, if not the ambiguities of non-zero-sum games are found again.

E) There remains some complicated conditions of determinism that we will see in sections 6 and 7.

2. Usual presentation of two-player zero-sum differential games

Let us consider the game of Fig. 1. The essentials are the possible issues of the game :

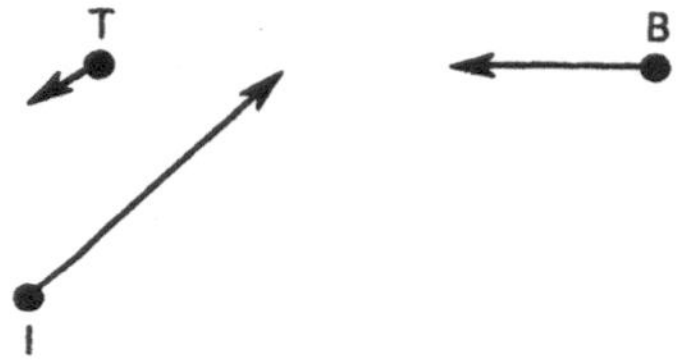

Fig. 1 — The bomber B tries to reach the target T (warship) defended by the interceptor I.

A) Target destroyed, bomber saved.
B) Target and bomber destroyed (Kamikaze)
C) Target and bomber saved.
D) Target saved, bomber destroyed.

The interceptor classifies the issues from A (less desirable) to D (most desirable) and we assume that the bomber do the opposite (zero-sum-game).

More generally issues are classified with a "performance index" I (also called Value or payoff or cost function), the two players are the maximizor M that tries to maximize I and its opponent the minimizor m.

The differential games are characterized by an evolution of the state with respect to a "description parameter" t, generally the time, until some final state where the issue is known.

The present state of the game is defined by n "state parameters" $x_1, x_2, \dots x_n$ (the parameters that are relevant at each instant to a player making decisions as how to play : the positions and velocities of the target, the bomber and the interceptor etc...). We will put :

$$(x_1, x_2, \dots x_n) = \overrightarrow{X} = \text{state vector} \tag{1}$$

The two players M and m control the game through their own controls $\overrightarrow{M(t)}$ and $\overrightarrow{m(t)}$ and through the Borelian equation of motion (also called control function)

$$d\overrightarrow{X}/dt = \overrightarrow{V} = \overrightarrow{V}(\overrightarrow{X},\overrightarrow{M},\overrightarrow{m},t) \tag{2}$$

The controls $\overrightarrow{M}(t)$ and $\overrightarrow{m}(t)$ can be chosen at will (measurable functions of t) in the corresponding closed "control domains" $\mathcal{D}_M(t)$ and $\mathcal{D}_m(t)$, Borelian functions of t.

The analytical singularities are avoided (i.e. one and only one solution $\overrightarrow{X}(t)$ corresponds to a given initial state $\overrightarrow{X_o}$, t_o and given measurable $\overrightarrow{M}(t)$ and $\overrightarrow{m}(t)$) if the control function $\overrightarrow{V}(\overrightarrow{X},\overrightarrow{M},\overrightarrow{m},t)$ is both locally bounded and locally Lipschitz with respect to $\overrightarrow{X}$:

$$\left.\begin{array}{l} \forall\, B, \text{ bounded set of the } \overrightarrow{X}, t \text{ space } \exists\, h, k \text{ real and such} \\ \text{that : } \left\{\begin{array}{l}(\overrightarrow{X},t)\in B\ ;(\overrightarrow{X_2},t)\in B \\ \overrightarrow{M}\in \mathcal{D}_M(t)\ ;\ \overrightarrow{m}\in\mathcal{D}_m(t)\end{array}\right\} \Rightarrow \left\{\begin{array}{l}\|\overrightarrow{V}(\overrightarrow{X},\overrightarrow{M},\overrightarrow{m},t)\| \leq h \\ \|\overrightarrow{V}(\overrightarrow{X},\overrightarrow{M},\overrightarrow{m},t)-\overrightarrow{V}(\overrightarrow{X_2},\overrightarrow{M},\overrightarrow{m},t)\| \leq k\|\overrightarrow{X}-\overrightarrow{X_2}\|\end{array}\right.\end{array}\right\} \tag{3}$$

The solution $\overrightarrow{X}(t)$ is then the locally Lipschitz function starting at $\overrightarrow{X_o}$, t_o and verifying for almost all t :

$$d\overrightarrow{X}(t)/dt = \overrightarrow{V}\left(\overrightarrow{X}(t),\overrightarrow{M}(t),\overrightarrow{m}(t),t\right) \tag{4}$$

However notice that (3) does not forbid the "escapes to infinity in a bounded interval of time" as when $dx_1/dt = x_1^2$ or $dx_1/dt = \exp(x_1)$.

The $\overrightarrow{X}$, t space is a $\mathbb{R}^{n+1}$ space and contains the closed and well defined "playing set" $\mathcal{E}$ (e.g. the altitude of the bomber and the interceptor must be positive, etc...). The boundary $\mathcal{C}$ of $\mathcal{E}$ is the "terminal subset" (e.g. the bomber enters the "capture zone" of the interceptor).

In order to avoid topological singularities we will assume that $\mathcal{C}$ is also the boundary of $\mathring{\mathcal{E}}$ (interior of $\mathcal{E}$) and thus that $\mathcal{E}$ is "the closure of its interior" as is usually the case.

The game starts at an initial point $\overrightarrow{X_o}, t_o$ either given or chosen is some closed "initial domain" by a well determined player.

The game stops at $\overrightarrow{X_f}, t_f$, the first exit out of $\mathcal{E}$. The issue of the game is $I(\overrightarrow{X_f}, t_f)$ and is given by the performance index function $I(\overrightarrow{X}, t)$ defined in the terminal subset $\mathcal{C}$.

We have already notice that the function I should be defined in any occurrence ; for instance it can be I_∞ either finite or infinite, if the game has no end.

Remark 1 . A more general I function is sometimes encountered :

$$I = I_1(\overrightarrow{X_f}, t_f) + \int_{t_o}^{t_f} I_2(\overrightarrow{X}, t)\, dt \tag{5}$$

Fortunately this case equivalent to the case when $I = I(\overrightarrow{X_f}, t_f)$; it is sufficient to define a new component x_{n+1} :

$$x_{n+1} = \int_{t_o}^{t} I_2(\overrightarrow{X}, \theta)\, d\theta \tag{6}$$

and, with $\overrightarrow{X} = (x_1, x_2, \dots, x_{n+1})$ the performance index function I becomes :

$$I = I_1(\overrightarrow{X_f}, t_f) + x_{n+1} = I(\overrightarrow{X_f}, t_f) \tag{7}$$

We will systematically use a such $I(\overrightarrow{X_f}, t_f)$ function and the corresponding vocabulary : "playing set", "terminal subset" etc...

Remark 2. Some authors consider that the end of the game is obtained at the first arrival at $\mathcal{C}$. However this leads to singularities when the interest of a player is to follow the border for a while.

Remark 3. R. Isaacs [1] noticed the difference between the "games of kind" (the performance index I has only a finite number of possible values) and the "games of degree" (I is continuous in terms of $\overrightarrow{X_f}, t_f$). Fortunately this is only a small difference and, for instance, a game of degree is the limit of a succession of game of kind.

3. The points of the terminal subset

The points of the terminal subset $\mathcal{C}$ can be classified into five types :

A) The terminal points

At these points, each player can induce the end of the game even against the will of its opponent. The game stops there because this is the interest of at least one player.

B) The points favourable to the maximizor

The maximizor can, at will, either induce the end of the game or prevent it. The local value of I $(\overrightarrow{X}, t)$ is a lower bound of the Value of the game.

C) The points favourable to the minimizor

These points are symmetrical to the previous ones.

D) The points with impossible exit.

Even if the two players agree with each other they cannot induce the end of the game and the local value of I $(\overrightarrow{X}, t)$ has no interest.

E) The complex points

The exit is possible if the two players agree with each other but each of them can prevent it.

These points are sometimes the end of the game if this is the interest of both players as in the following example :

<u>Game Nr 1</u>

$$\left.\begin{array}{l} \text{State variables : } x \text{ and } y \text{ ; Initial conditions } x_o = y_o = t_o = 0 \\ \text{Playing set : } y \geqslant 0 \text{ ; } t \leq 1 \text{ ; Performance index : } I = x_f \\ \text{Control functions : } dx/dt = m - M \text{ ; } dy/dt = M + m - 1 \\ \text{Control domains : } 0 \leq M \leq 2 \text{ ; } 0 \leq m \leq 2 \end{array}\right\} \quad (8)$$

The initial point is also a point of the terminal subset $\mathcal{C}$, it is a complex point since the exit is possible (with $M + m < 1$) but both players can prevent it even against the will of their opponent (by the choice of M or m = 2).

The Value of the game is obviously 0, indeed with $M \equiv 0$ the maximizor obtains certainly $0 \leq I \leq 2$ and symmetrically with $m \equiv 0$ the minimizor obtains certainly $-2 \leq I \leq 0$, however these two "strategies" $M \equiv 0$ and $m \equiv 0$ lead to an exit at the initial point.

4. <u>Deterministic games with incomplete information</u>

A classical case of deterministic differential games is the following :

A) One of the two players, for instance the maximizor M, knows perfectly the rules of the game but has only poor means for the measure of $\overrightarrow{X}(t)$ at any time (inaccuracies, delays, etc...).

B) This player should choose a "strategy" (or "pure strategy" or "closed-loop strategy") based on his informations :

$$\vec{M} = \vec{M}\,(\vec{X}(\theta),\ \vec{M}(\theta), t)\ ;\ t_o \leq \theta < t \qquad (9)$$

C) He should indicate his choice to his opponent before the departure.

If these three conditions are satisfied the opponent faces an ordinary optimization problem and the first player chooses his strategy accordingly.

However the conditions B and C are generally unrealistic and the choice of a good "super-strategy" (or "mixed-strategy") can improve very much the situation of the first player.

A super-strategy is the definition of several strategies and the random choice, as late as possible, between them. This of course suppresses the determinism of the game as in the following example.

Game Nr. 2

$$\left.\begin{array}{l}\text{State variable : } x \text{ ; Initial state } x_o = t_o = 0 \\ \text{Playing set : } 0 \leq t \leq \pi \text{ ; Performance index : } I = \cos x_f \\ \text{Control function : } dx/dt = M + m \text{ ; Control domains : } |M| \leq 1 \text{ ; } |m| \leq 1\end{array}\right\} \qquad (10)$$

If the maximizor has no information on the intermediate values of $x(t)$, he can only choose an "open-loop strategy" $M(t)$; but this cannot prevent the minimizor, knowing $M(t)$, to reach either $x_f = \pi$ or $x_f = -\pi$ and thus $I = -1$.

However if the maximizor can choose a super-strategy, he will for instance choose :

$$\left.\begin{array}{ll}\text{For } 0 \leq t \leq 2\pi/3 & : M(t) \equiv 0 \\ \text{For } 2\pi/3 < t \leq \pi & : \text{one chance over two for } M(t) \equiv 1 \\ & \ \ \text{one chance over two for } M(t) \equiv -1\end{array}\right\} \qquad (11)$$

With a random choice at $t = 2\pi/3$ the maximizor has at least one chance over two to reach $|x_f| \leq 2\pi/3$ and thus $I > -0.5$.

5. Relaxation, chattering and super-strategy

As seen in the previous section, the super-strategies ruin the determinism of a game but this is not the case for the neighbouring phenomena called relaxation and chattering.

One of the singularities of the ordinary optimization problems is the existence of cases in which a relaxation or a chattering of the control becomes necessary as in the following example :

Game Nr 3 (optimization problem)

$$\left.\begin{array}{l} \text{Maximize } I = y_f \text{ with } x_o = y_o = t_o = 0 \ ; \ t_f = 1 \ ; \\ dx\ dt = M \ ; \ dy\ dt = M^2 - x^2 \ ; \ |M| \leq 1 \end{array}\right\} \quad (12)$$

Since $I = \int_0^1 (M^2 - x^2)\, dt$ and $M^2 - x^2 \leq 1$ we get $I \leq 1$. On the other hand $I > 1 - \varepsilon^2$ can easily be reached : it is sufficient to choose alternately $M = +1$ and $M = -1$ in order that $|x|$ remains less than ε (Fig. 2).

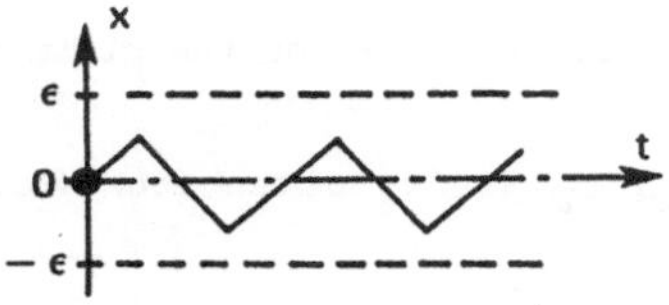

Fig. 2 $dx/dt = M = \pm 1 \ ; |x| < \varepsilon$.

Hence the least upper bound of I is + 1 but that value is not attainable.

In this case and in similar ones, most people choose one of the two following possibilities :

A) The control $\vec{M}$ may "chatter" at a "very high rate" between two or several optimal states $\vec{M_1}, \vec{M_2}, \ldots$ with, for each small interval of time, a well defined proportion in each state.

B) The control can be "relaxed" : a linear composition (with positive proportions) of two or several controls becomes considered as admissible. For instance in (12) it becomes possible to have at the same time $dx/dt = 0$ and $dy/dt = 1 - x^2$ by a linear composition with equal proportions of $M = +1$ and $M = -1$.

This operation is sometimes called "convexisation of the vectogram or of the maneuverability domain" and the limit value $I = 1$ becomes attainable.

It can be demonstrated [15] that, when the conditions (3) are satisfied, these two methods are equivalent and the "relaxed solutions" are limit of suitable sequences of "chattering solutions".

6. Deterministic games with complete and infinitely rapid information

We thus arrive to the only deterministic and realistic case of differential games with more than one player : the two-player, zero-sum games with deterministic rules and issues and with complete and infinitely rapid information.

However an essential point must be clarified : what happens when the two players are led to opposite chatterings ? what is then the meaning of the word relaxation ?

Let us analyse the possibilities of the two players and the inertia of their controls, let us define the following short intervals of time :

A) T_M will be the small duration that is necessary to the maximizor for the measure of the present state of $\vec{X}$ and for a reaction to that measure.

B) τ_M will be the small duration of a chattering between two or several controls $\vec{M_1}, \vec{M_2}, \dots$.

C) T_m and τ_m will be the corresponding small intervals of time for the minimizor.

The comparison of the four durations T_M, τ_M, T_m, τ_m leads to the four following deterministic cases.

I) The maximin case :

$$\tau_m + T_m \ll \tau_M \tag{13}$$

The minimizor has much faster reactions that the maximizor and everything happens as if, at any time, he could choose his control after the maximizor and in terms of the maximizor choice (maximizor first : maximin).

II) The minimax case :

$$\tau_M + T_M \ll \tau_m \tag{14}$$

This case is symmetrical to the previous previous one and then favourable to the maximizor.

III) The neutral case :

$$\tau_M \ll \tau_m + T_M \quad ; \quad \tau_m \ll \tau_M + T_M \tag{15}$$

The chatterings of the two players are independent, they cannot be followed by a chattering of the other player. This case is intermediate.

IV) The separated case :

The control function has the following form :

$$\vec{V}(\vec{X}, \vec{M}, \vec{m}, t) = \vec{V}_M(\vec{X}, \vec{M}, t) + \vec{V}_m(\vec{X}, \vec{m}, t) \tag{16}$$

In this very usual case the three maximin, minimax and neutral cases are equivalent and it is no more necessary to compare T_M, τ_M, T_M, τ_m.

The same happens in the more general case in which :

$$\forall \vec{P} \in \mathbb{R}^n ; \forall (\vec{X}, t) \in \mathcal{E} :$$
$$\sup_{\vec{M} \in \mathcal{D}_M(t)} \left\{ \inf_{\vec{m} \in \mathcal{D}_m(t)} \vec{P} \cdot \vec{V}(\vec{X}, \vec{M}, \vec{m}, t) \right\} = \inf_{\vec{m} \in \mathcal{D}_m(t)} \left\{ \sup_{\vec{M} \in \mathcal{D}_M(t)} \vec{P} \cdot \vec{V}(\vec{X}, \vec{M}, \vec{m}, t) \right\} \tag{17}$$

Several other types of deterministic games can be considered, for instance maximin with respect to x_1 and minimax with respect to $x_2, \ldots x_n$, we will not consider them. On the other hand in the non-separated case if T_M, τ_M, T_m, τ_m are of the same order of magnitude the chatterings of one player can be more or less followed by the other player and the determinism is lost.

These phenomena can be seen in the following simple example :

<u>Game Nr. 4</u>

$$\left.\begin{array}{l} \text{State variable } x \text{ ; Initial conditions } x_o = t_o = 0 \\ \text{Playing set : } t \le 1 \text{ ; Performance index : } I = x_f \\ \text{Control function : } dx/dt = 4 M^2 + M + 2 Mm - 4 m^2 \\ \text{Control domains : } |M| \le 1 \text{ ; } |m| \le 1 \end{array}\right\} \tag{18}$$

It is easy to verify the following analysis of the different cases and their optimal strategies :

I) Maximin case :

$$\left.\begin{array}{l} M \equiv +1 \Rightarrow -1 \le I \le 5.25 \\ m \equiv -\text{sign } M = \pm 1 \Rightarrow -4.6525 \le I \le -1 \end{array}\right\} \Rightarrow m \equiv -1 \text{ ; } I = -1 \tag{19}$$

II) Minimax case :

$$\left.\begin{array}{l} M \equiv \text{sign } (1 + 2m) = \pm 1 \Rightarrow 1 \le I \le 5.25 \\ m \equiv -1 \Rightarrow -4.0625 \le I \le 1 \end{array}\right\} \Rightarrow M \equiv -1 \text{ ; } I + 1 \tag{20}$$

III) Neutral case with two opposite chatterings :

$$\left.\begin{array}{l} M : \{50\% \text{ for } M = +1 \; ; \; 50\% \text{ for } M = -1\} \Longrightarrow 0 \le I \le 4 \\ m : \{25\% \text{ for } m = +1 \; ; \; 75\% \text{ for } m = -1\} \Longrightarrow -4 \le I \le 0 \end{array}\right\} I = 0 \qquad (21)$$

Notice that the Value of the neutral case is above this of the maximin case (favourable to the minimizor) and below this of the minimax case. Also notice that at large scale, contrarily to the super-strategies, the chatterings give deterministic results.

7. Discontinuities of the Value of the game, the superior limit game, the inferior limit game

A final source of undeterminism, and of lack of convergence of numerical methods, is given by the discontinuities of the performance index function $I\,(\overrightarrow{X_f},\, t_f)$ or by these of the Value of the game in terms of the initial conditions $\overrightarrow{X_o}$, t_o.

Let us consider for instance the following very simple example :

Game Nr. 5

$$\left.\begin{array}{l} \text{State variable : } x \; ; \text{ Initial conditions : } x_o,\ t_o \text{ given.} \\ \text{Playing set : } 0 \le t \le 1 \; ; \text{ Performance index } I \begin{cases} x_f = 0 \Rightarrow I = 1 \\ x_f \ne 0 \Rightarrow I = x_f \end{cases} \\ \text{Control function : } dx/dt = M + m \; ; \text{ Control domains : } |M| \le 2 \; ; \; |m| \le 1 \end{array}\right\} \qquad (22)$$

The game is of separated type and theoretically the optimal strategy of the maximizor is obvious, he chooses M = + 2 everywhere except in the triangle OAB of Fig. 3 where the optimal value of M is - 2.

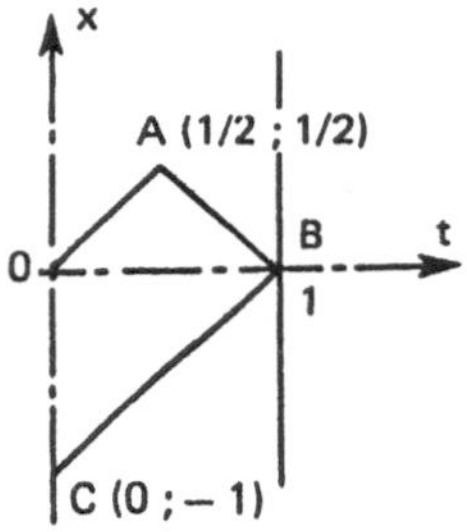

Fig. 3 – The game N° 5.

With this strategy the value of the game will be at least $x_o + 1 - t_o$ outside the trapezium OABC and it will theoretically be 1 in that trapezium. However in the latter case the accuracy of the control of the maximizor must go to infinity when the point (x, t) approaches B and a final small swerve of the minimizor could give $I \simeq 0$ only.

This difficulty is overcome by the following convention :

Let us consider the equation of motion (2) that is defined at least in a vicinity of the playing set $\mathcal{E}$ and let us add two small terms in the right-hand member :

$$d\vec{X}/dt = \vec{V}(\vec{X}, \vec{M}, \vec{m}, t) + \vec{\delta_M} + \vec{\delta_m} \tag{23}$$

In a "superior game" $\vec{\delta_m}$ is identically zero and $\vec{\delta_M}$ is integrable and chosen by the maximizor within the only condition :

$$\int_{t_o}^{t_f} ||\vec{\delta_M}|| \, dt \leq \epsilon \tag{24}$$

The inferior games have a symmetrical definition and when ϵ goes to zero we obtain the superior and inferior limit games whose values are essential in the appreciation of a game. For instance in the trapezium OABC of Fig. 3 these two values are 1 and $x_o + 1 - t_o$ respectively.

Notice that :

A) The notions of superior and inferior limit games are independent of these of maximin, minimax, neutral and separated games. The determinism requires a clear analysis, for instance : "this game is an inferior limit game of minimax type".

B) Discontinuities of the Value in terms of $\vec{X_o}$, t_o occur in three cases (provided that conditions (3) are satisfied, at least in a vicinity of the playing set $\mathcal{E}$) :

B.1. When the performance index function I $(\vec{X_f}, t_f)$ has discontinuities.

B.2. When some optimal solutions are unbounded.

B.3. When some optimal solutions meet the terminal subset $\mathcal{C}$ a first time before their final point.

C) The superior (inferior) limit games do not fit the definition of differential games presented in section 2. Practically their main effect is to substitute to the performance index I $(\vec{X_f}, t_f)$ and to the value of the game V $(\vec{X_o}, t_o)$ their local least upper bound (greatest lower bound). They have few real effect on the classification of the points of the terminal subset $\mathcal{C}$ (section 3).

However there are some singularities for the optimal solutions $\vec{X}(t)$ that meet $\mathcal{C}$ a first time before their terminal point and that are such that all neighbouring solutions have also this property.

D) This leads to the following conditions of continuity of the Value of the game in terms of the initial conditions $\overrightarrow{X_o}$, t_o (in this case the superior and inferior limit games are identical and useless).

The value V (X_o, t_o) is continuous if the conditions (3) are satisfied and if :

$$\left.\begin{array}{l}\text{D.1. The performance index function } I\ (\overrightarrow{X_f}, t_f) \text{ is continuous} \\ \text{D.2. The unbounded solutions are either impossible (bounded playing set } \mathcal{E}) \\ \quad\text{or uninteresting} \\ \text{D.3. The solutions meeting the terminal subset } \mathcal{C} \text{ a first time before their final} \\ \quad\text{point are either impossible (e.g. all points of } \mathcal{C} \text{ are terminal points) or} \\ \quad\text{uninteresting.}\end{array}\right\} \quad (25)$$

8. <u>The approximate strategy method</u>

We have seen that most games have a very complex solution with many singularities (e.g. [1-12]). It is very difficult to solve them directly and a realistic approach is the research of a good "strategy".

If the "first player" chooses a strategy and transmits it to the "second player" the latter faces an ordinary optimization problem (that can be discontinuous...) and is led to an optimal solution with a value V_1. The first player has then the certainty of reaching at least V_1 with his strategy.

The successive improvements of the strategy and a similar research for the second player will lead to excellent upper and lower bounds of the true Value of the game of interest.

The definition of a strategy depends on the type of the game of interest.

A) Strategies for the minimizor

A.1.) Game of minimax type

The maximizor has much less inertia and can, at any time, choose its control in terms of this of the minimizor.

Hence a strategy for the minimizor is a measurable function of $\overrightarrow{X}$ and t only :

$$\overrightarrow{m} = \overrightarrow{m}\,(\overrightarrow{X}, t) \qquad ; \qquad \overrightarrow{m} \in \mathcal{D}_m(t) \tag{26}$$

This function can use here or there some "chatterings" between two or several controls $\overrightarrow{m_j}$.

The maximizor can answer to this control by suitable controls $\vec{M}$ and thus at a point $(\vec{X}, t)$ where, for instance, the minimizor uses a chattering between $\vec{m}_1$ and $\vec{m}_2$ with proportions a_1 and a_2 (with $a_1 + a_2 = 1$), the maximizor will obtain the following possible velocities $\vec{V}$ whose set is its own "vectogram" or "maneuverability domain" :

$$\left.\begin{array}{l} 1°)\ \vec{V}_{\vec{M}_1\vec{M}_2} = a_1.\vec{V}(\vec{X}, \vec{M}_1, \vec{m}_1, t) + a_2.\vec{V}(\vec{X}, \vec{M}_2, \vec{m}_2, t) \\ \quad \text{with } \vec{M}_1 \text{ and } \vec{M}_2 \in \mathcal{D}_M(t) \\ 2°)\ \text{All velocities obtained by a "relaxation" between these velocities } \vec{V}_{\vec{M}_1,\vec{M}_2} \\ \text{(convexisation of the maneuverability domain).} \end{array}\right\} \quad (27)$$

A.2.) Game of neutral type

The strategies are again of the type (26) but now the maximizor cannot follow the chatterings of the minimizor and (27) becomes :

$$\left.\begin{array}{l} 1°)\ \vec{V}_{\vec{M}} = a_1.\vec{V}(\vec{X}, \vec{M}, \vec{m}_1, t) + a_2\,\vec{V}(\vec{X}, \vec{M}, \vec{m}_2, t) \\ \quad \text{with } M \in \mathcal{D}_M(t) \\ 2°)\ \text{All velocities obtained by a "relaxation" between these velocities} \\ \vec{V}_{\vec{M}}\text{(convexisation of the maneuverability domain).} \end{array}\right\} \quad (28)$$

The maneuverability domain of (28) is obviously a subset of this of (27).

A.3.) Game of maximin type

In this type the minimizor has much less inertia and can choose his control in terms of maximizor's control, hence a strategy for the minimizor is a function of $\vec{X}$, $\vec{M}$ and t :

$$\vec{m} = \vec{m}(\vec{X}, \vec{M}, t) \quad ; \quad \vec{m} \in \mathcal{D}_m(t) \tag{29}$$

This strategy can use chatterings even for a single $\vec{M}$.

The maneuverability domain of the maximizor at $(\vec{X}, t)$ is now given by the following velocities :

$$\left.\begin{array}{l} 1°)\ \vec{V}_{\vec{M}} = \vec{V}(\vec{X}, \vec{M}, \vec{m}(\vec{X}, \vec{M}, t), t) \qquad ; \vec{M} \in \mathcal{D}_M(t) \\ 2°)\ \text{All velocities obtained by a "relaxation" between these velocities } \vec{V}_{\vec{M}} \\ \text{(convexisation of the maneuverability domain).} \end{array}\right\} \quad (30)$$

A.4. Game of separated type

It is then possible to use any of the three above cases ; they will lead to the same limit results.

B) Strategies for the maximizor

The situation of the maximizor is perfectly symmetrical to that of the minimizor (just change the sign of the performance index...).

9. Discontinuities in the strategies

The main drawback in the method of the approximate strategy is the existence of mandatory discontinuities in the good strategies.

Let us consider the simple following example :

Game Nr. 6

$$\left.\begin{array}{l}\text{State variables : } x,\ y\ ;\ \text{Initial conditions : } x_o,\ y_o,\ t_o \text{ given}\\ \text{Playing set : } 0 \le t \le 2\ ;\ \text{Performance index : } I = y_f\\ \text{Control functions : } dx/dt = M + m\ ;\ dy/dt = x^2\\ \text{Control domains : } |M| \le t\ ;\ |m| \le 1.\end{array}\right\} \quad (31)$$

Since $I = y_o + \int_{t_o}^{t_f} x^2 . dt$, the solution is almost obvious : the maximizor tries to maximize x^2 at each instant and the minimizor tries to minimize it.

A good strategy for the maximizor is then

$$x \geq 0 \Rightarrow M = + t \quad ; \quad x < 0 \Rightarrow M = - t \quad (32)$$

and for the minimizor :

$$m = - \text{sign } x \quad (33)$$

These opposite strategies are optimal since they always lead to the same value of the game (for instance $I = 0.05$ if $x_o = y_o = t_o = 0$). The corresponding trajectories are drawn in Fig. 4 with a universal line along OA and a dispersal line along AB.

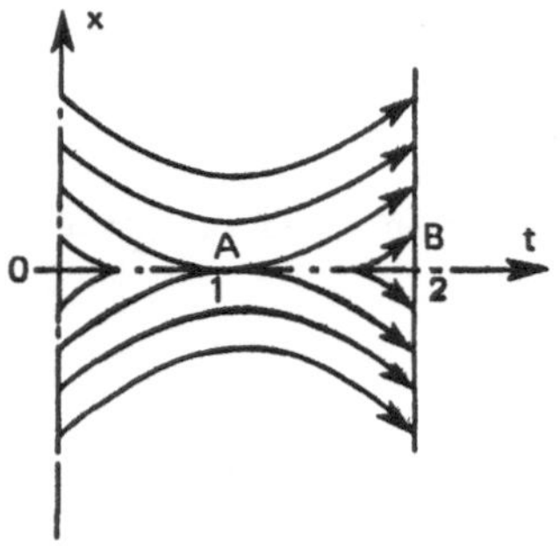

Fig. 4 – Game N° 6. Optimal trajectories in the t, x plane.

Let us analyse the discontinuities :

A) The strategy (32) of the maximizor is necessarily discontinuous at $x = 0$. If for instance it was possible to have $M = 0$ at $x = 0$ it would be possible to follow the segment AB that implies a great loss for the maximizor.

B) The strategy (33) of the minimizor leads to theoretical difficulties (and hence to computational errors). Assume for instance that the answer of the maximizor be $M = t/2$: there is no solution starting from the origin and such that $dx/dt = t/2 - \text{sign } x$.

Of course since the smallest swerve is immediately cancelled the solution $x \equiv 0$ is then the most natural, however in less simple cases the solution will not be so easy especially for computers.

Many papers deal with this question [3-6].
A first solution was proposed in [3] : at the points where the maneuverability domain $D(\vec{X}, t)$ (corresponding to the strategy of interest) is discontinuous with respect to $\vec{X}$, we add into D the velocities $\vec{W}$ that are necessary for the convexity and the upper semi-continuity of D with respect to $\vec{X}$; i.e. the velocities $\vec{W}$ such that :

$$\text{A) } \vec{W} = \lim_{n \to \infty} \vec{V_n} \text{ with, when } n \to \infty, \begin{cases} \vec{X_n} \to \vec{X} \\ \vec{V_n} \in D(\vec{X_n}, t) \end{cases}$$

B) The new maneuverability domain D is convexised.

This first method is excellent for the strategy (33) and the segment OA but it is a catastrophy for the strategy (32) and the segment AB : this segment would become a possible solution.

Finally the suitable method was studied and developed in [4-6].

Let us consider a strategy $\vec{m}(\vec{X}, t)$ given in (26) or $\vec{m}(\vec{X}, \vec{M}, t)$ given in (29) and the corresponding maneuverability domains $D(\vec{X}, t)$ of the maximizor, with possibly chatterings.

To these closed and convex maneuverability domains already correspond many admissible trajectories $\vec{X}(t)$, i.e. locally Lipschitz functions with almost always :

$$d\,\vec{X}(t)/dt \in D\,(\vec{X}\,(t),\, t) \qquad (34)$$

To these domains D $(\vec{X}, t)$ we must now add <u>the final velocity vectors of admissible trajectories ending at $\vec{X}$, t)</u>. As usual these new maneuverability domains must be closed and convexised and this can introduce new admissible trajectories, the same procedure must then be used again and again up to exhaustion.

Notice that at the $(\vec{X}, t)$ points where D is continuous in terms of $\vec{X}$ and t it will remain unchanged. Elsewhere it will never become greater than what is sufficient for the upper semi-continuity.

Also notice that if we used the initial velocity vectors of admissible trajectories starting at $(\vec{X}, t)$ we would find again the objection of segment AB of Fig. 4.

Notice finally that a player has no interest in a pathologically complex strategy since he must himself solve the optimization problem of its opponent in order to know the value of his strategy. It seems that, when the continuity conditions (25) are satisfied, good "piecewise smooth strategies" can have values as near as desired to the true Value of the game.

10. <u>Improvements of an approximate strategy</u>

They are many possible ways for improving a strategy, for instance an initial strategy $\vec{m}\,(\vec{X}, t)$ can be considered as an element of a family of strategies $\vec{m}\,(\vec{X}, C, t)$ and the improvement is the determination of the best value of the parameter C (see the section 12 and also the game Nr. 7 in this section).

If the strategy $\vec{m}\,(\vec{X}, t)$ or $\vec{m}\,(\vec{X}, \vec{M}, t)$ of interest is analysed by the Pontryagin method [16] a systematic improvement can be obtained in the following way.

Let us assume that the maximizor find an optimal solution $\vec{X}\,(t)$ leading from $\vec{X}_o$, t_o to $\vec{X}_f$, t_f.

There are two main cases :

A) If $\vec{X}_f$, t_f is inside the attainable domain from $\vec{X}_o$, t_o and corresponds to a local maximum of the performance index I $(\vec{X}_f, t_f)$ only a large modification of the strategy of the minimizor allows to avoid $\vec{X}_f$, t_f.

B) On the contrary if the optimal $\vec{X}_f$, t_f is at the limit of the attainable domain (as is usually the case), the solution $\vec{X}\,(t)$ is an extremal solution and corresponds to a non-zero adjoint vector $\vec{P}(t)$ with the usual equations :

$$\vec{V} = d\vec{X}/dt = \vec{V}(\vec{X},\vec{M},\vec{m},t) = \text{velocity vector} \tag{35}$$

At $\vec{X}$, t the velocity $\vec{V}$ belongs to the "maneuverability domain" or "vectogram" D $(\vec{X}, t)$ defined in the previous section.

$$\vec{P} = \text{adjoint vector of Pontryagin} \tag{36}$$

$$H = \vec{P}.\vec{V} = \text{"control Hamiltonian"} \tag{37}$$

$$H^* = H^*(\vec{P}, \vec{X}, t) = \sup_{\vec{V} \in D(\vec{X}, t)} \vec{P}.\vec{V}. = \text{"optimal Hamiltonian"} \tag{38}$$

With the discontinuous strategies of differential games the Pontryagin theory can be used if H^* is a piecewise locally Lipschitz function of $\vec{P}$, $\vec{X}$, t with regular pieces, and the Pontryagin equations are :

A) At points where H^* is differentiable :

$$\left.\begin{array}{l} \dfrac{d\vec{X}(t)}{dt} = \dfrac{\partial H^*}{\partial \vec{P}} \; ; \; - \dfrac{d\vec{P}(t)}{dt} = \dfrac{\partial H^*}{\partial \vec{X}} \; ; \; \dfrac{dH^*(\vec{P}(t), \vec{X}(t), t)}{dt} = \dfrac{\partial H^*}{\partial t} \\ \text{(almost always)} \end{array}\right\} \tag{39}$$

B) At points where H^* is continuous but not differentiable :

$$\left.\frac{d}{dt}\left\{\vec{X}(t) \; ; \; -\vec{P}(t) \; ; \; H^*(t)\right\} \in \text{convex hull of local gradients } \frac{\partial H^*}{\partial(\vec{P},\vec{X},t)} \; \text{(almost always)}\right\} \tag{40}$$

C) Finally at points where $H^*(\vec{P}, \vec{X}, t)$ is discontinuous, the function $\vec{X}(t)$ remains continuous and even Lipschitzian but a drift $(-\frac{d\vec{P}}{dt} ; \frac{dH^*}{dt})$ or even a jump $(-\Delta\vec{P}, \Delta H^*)$ directed toward large H^* can be add to the equations (39) or (40) (this case can be considered as the limit of the previous one).

The adjoint vector $\vec{P}$ gives the essential direction of motion and the first-order idea of the neighbouring attainable domain
(i.e. $\vec{P}_o . \delta\vec{X}_o - H^*_o \, \delta t_o \geqslant \vec{P}_f \, \delta\vec{X}_f - H^*_f \, \delta t_f + o(\delta\vec{X}_o, \; \delta t_o, \; \delta\vec{X}_f, \; \delta t_f)$
in the non-singular cases), hence if the minimizor wants to improve its strategy he must :

1) Determine the maximizor's optimal solution or solutions $\vec{X}_j(t)$ that leads to the largest I $(\vec{X}_f, t_f)$.

2) Determine the corresponding adjoint function or functions $\vec{P}_{k.j}(t)$.

$H^* = \sup(\vec{P}.\vec{V})$ for $\vec{V} \in D(\vec{X}, t)$ shows then the possibilities of the maximizor into the essential direction or directions that lead to the largest I $(\vec{X_f}, t_f)$, hence the minimizor improves his strategy if he chooses $\vec{m_2}$ $(\vec{X}, t)$ or $\vec{m_2}$ $(\vec{X}, \vec{M}, t)$ that decreases H^* $(\vec{P}, \vec{X}, t)$ at the points $(\vec{P_{k.j}}(t), \vec{X_j}(t), t)$ and in their neighbourhood.

Notice :

1) The sudden change from the stragegy $\vec{m}$ to $\vec{m_2}$ can be so large that it becomes unfavourable, fortunately a "relaxed" strategy $(1 - q)\vec{m} + q\vec{m_2}$ will be favourable for sufficiently small q.

2) This method allows to choose a better strategy at the points $\vec{X_j}$ (t), t and in their vicinity but it gives no information elsewhere. Indeed elsewhere exists some freedom and we can loose at non optimal points if we gain at essential points. However this method requires the knowledge of all optimal trajectories of the strategy of interest and unfortunately the nature of differential games gives usually a large number of such optimal solutions.

Let us consider for instance the following example :

Game Nr. 7

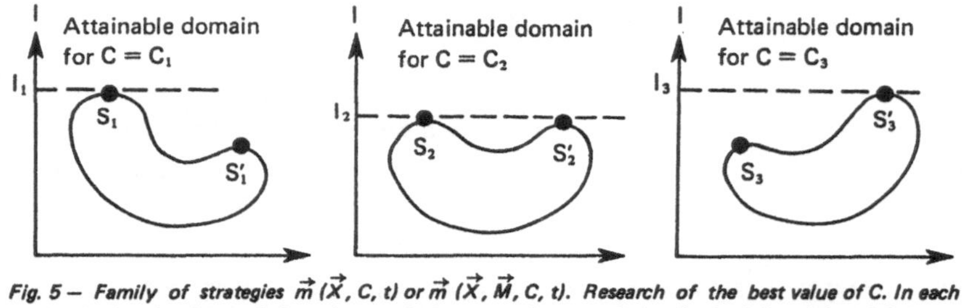

Fig. 5 — Family of strategies $\vec{m}(\vec{X}, C, t)$ or $\vec{m}(\vec{X}, \vec{M}, C, t)$. Research of the best value of C. In each attainable domain, the maximizor chooses the largest performance index (I_1, I_2, I_3) and the minimizor minimizes this index and is led to the choice of $C = C_2$ with two equivalent solutions.

The game number 7 is presented in the figure 5 : the research of the best value of the parameter C in the minimizor's family of strategies $\vec{m}$ $(\vec{X}, C, t)$ or $\vec{m}$ $(\vec{X}, \vec{M}, C, t)$ leads very often to two equivalent solutions : the best C is C_2 if, in the attainable domain drawn by the maximizor with the given strategies, one of the two local maxima of the performance index increases with C while the other decreases.

This phenomenon is very general especially if the family of interest is a multi-parameter family and most differential games have many equivalent optimal solutions.

The game of the equivocal line ([1], page 286) is very simple and has an infinite number of equivalent solutions in spite of its simplicity.

11. Essential questions on the method of the successive approximate strategies

First question.

Let us consider a given deterministic differential game.

Does there exist a couple of opposite and piecewise Lipschitz strategies (one for the maximizor and one for the minimizor) leading both to the same value of the performance index I, value that is then the Value of the game ?

The answer to that first question is generally no, even in simple games as those associated to the "Fuller problem" [13, 14].

Second question.

The strategies of the maximizor give lower bounds I_M of the Value of the game and the strategies of the minimizor give upper bounds I_m.

Is it possible to write :

$$\sup I_M = \inf I_m \ ? \qquad (41)$$

This common value would of course be the Value of the game.

The answer to that second question depends of course of the game of interest. It is generally "no" if the Value of the game $V(X_o, t_o)$ in terms of the initial conditions is discontinuous at the initial conditions X_o, t_o of interest (see conditions (25) and (3)).

This answer can again be "no" in the cases called "pathological" by the engineers (for instance I = 1 if x_{1f} is rational and I = 0 if x_{1f} is irrational). But it seems that the answer be yes in the usual "piecewise Lipschitz" differential games when the discontinuities of the Value of the game are avoided (either because the conditions (25) and (3) are satisfied or because the initial conditions $\vec{X}_o$, t_o are in a region of continuity of the Value $V(X_o, t_o)$).

The mathematical demonstrations remain to be done but at least the method of approximate strategies gives upper and lower bounds of the Value of the game and generally the problem is considered as solved when these upper and lower bounds have a very small difference.

12. Example : The interception of a bomber

An interceptor tries to shot down an escaping bomber [10] and in a first analysis only horizontal motions are considered.

The bomber and the interceptor have their own performances (given maximum velocities and maximum turning rates, limits on accelerations, etc...).

A "capture zone" is defined in front of the interceptor with a corresponding performance index (fig. 6) at some realistic "final time t_f of the game". The sign of the performance index is such that the bomber is the maximizor and the interceptor the minimizor.

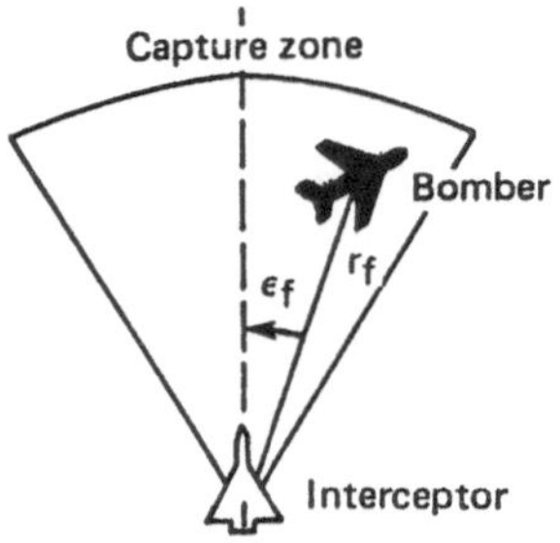

Fig. 6 – The capture zone is defined by $r_f \leq r_{max}$; $|\epsilon_f| \leq \epsilon_{max}$. If at t_f the bomber is in the capture zone, the value of the game is $r_f + \lambda \sin^2(\epsilon_f/2)$, if not this value is infinite.

Four initial positions, with full speed have been considered (fig. 7) and the first tested strategies are the following :

I) Strategies for the interceptor : the proportional navigation.

$$\text{Turning rate of the interceptor} = \text{K.rate de rotation of } \overrightarrow{IB} \tag{42}$$

II) Strategies for the bomber :

$$\text{Turning rate of the bomber} = \text{K}'.\text{Angle}(\text{IB}.\text{V}_B) = \text{K}'\varepsilon' \tag{43}$$

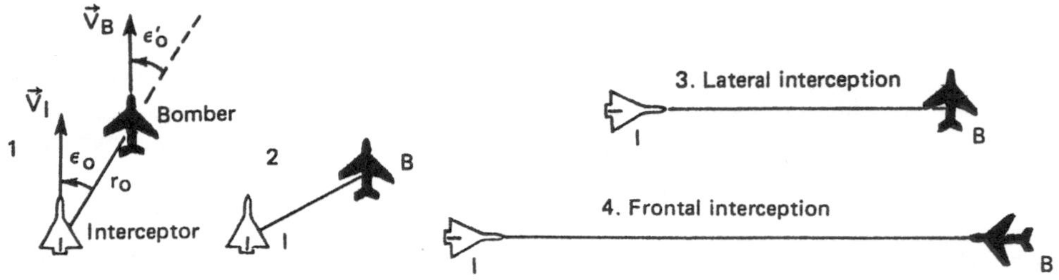

Fig. 7 – The four initial presentations : 1 - $r_o = 1$ km ; $\epsilon_o = \epsilon'_o = 30^o$. 2 - $r_o = 1$ km ; $\epsilon_o = \epsilon'_o = 60^o$. 3 - $r_o = 3$ km ; $\epsilon_o = 0$; $\epsilon'_o = 90^o$. 4 - $r_o = 5$ km ; $\epsilon_o = 0$; $\epsilon'_o = 180^o$.

Of course if these turning rates are above the maximum turning rate they cannot be used and the strategy uses then the maximum turning rate.

The results presented in the reference 10 are the following with $V_{I.max} = 290$ m/s, $V_{B.max} = 240$ m/s, $\lambda = 1$ km, $t_f = 10$ s, large r_{max} and ε_{max}.

	Case 1	Case 2	Case 3	Case 4
Best value of K Performance index	4 542.9 m	4.5 648.3 m	1 1667.5 m	1.4 707.7 m
Best value of K' Performance index	- 0.15 525.4 m	- 0.20 588.9 m	- 0.10 1655.1 m	- 0.10 583.2 m

These upper and lower bounds of the Value of the game are excellent in the case 3 but weak in the case 4 and a slight improvement of the strategy (K decreases linearly from 4.5 to 0 between t_o and t_f) leads to a much better upper bound of this case 4 : 633.2 m instead of 701.7 m.

The figure 8 presents the optimal trajectories of the case 1 for the opposite strategies K + 4 and K' = - 0.15. These trajectories are very different in spite of their almost equal value of the performance index.

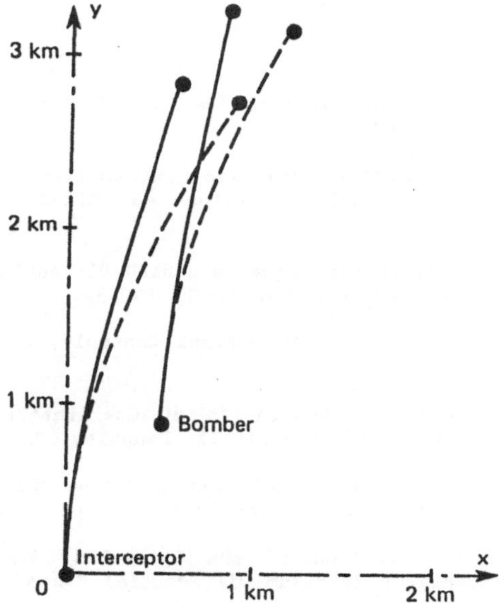

Fig. 8 — The optimal trajectories of the case 1.
Full lines : K = 4. Dotted lines : K' = — 0.15.

This analysis has been improved to much better bounds of the Value of the game (with more complicated strategies) and to three-dimensional motions in the references 11 and 12.

Conclusion

The deterministic differential games are more than a generalization of the optimization problems, they present many new types of singularities and have several specific properties. In most cases the only really deterministic element in these games is the Value of the game and very often the number of "optimal solutions" is infinite.

Most differential games are very difficult because of the great number, the variety and the impredictability of singularities.
For these reasons the method of the approximate strategies can be very useful especially if two good opposite strategies give close upper and lower bounds of the Value of the game.

The methods of improvement of approximate strategies can be very helpful and we must notice that in usual games the two players are indeed looking for a more or less complex strategy giving at least a satisfying result close to the optimum.

References

[1] Isaacs, R., "Differential Games", The Siam Series in Applied Mathematics, John Wiley and Sons (1965).

[2] Breakwell, J.V., "Séminaire sur les jeux Différentiels", Centre d'Automatique de l'Ecole Nationale Supérieure des Mines de Paris, Fontainebleau (1971).

[3] Filippov, A.F., "Differential Equations with Discontinuous Right Hand Side", Doklady Akademy Nauk CCCP 151, pp. 91-126 (1963).

[4] Brunowsky, P., "The Closed Loop Optimal Control", SIAM Journal of Control, (1974).

[5] Masle, J.F., "Problèmes Quantitatifs et Qualitatifs Liés en Jeux Différentiels", Thèse,Université de Paris IX. Dauphine. 22 Juin 1976.

[6] Sentis, R., "Equations Différentielles à Second Membre Mesurable", Compte rendu de l'Académie des Sciences, Série A. 284, pp. 113-116 (1977).

[7] Marchal, C., "Generalization of the Optimality Theory of Pontryagin to Deterministic Two-player Zero-sum Differential Games", ONERA T.P. Nr. 1233-1973.

[8] Bernhard, P., "Linear Pursuit Evasion Games and the Isotropic Rocket Game", Ph. D. Stanford University (1970).

[9] Merz, A.W., "The Homicidal Chauffeur", Standford University (1971).

[10] Nguyen Van Nhan, Aumasson, C., "Interception et Combat Aérien, Recherche Numérique des Solutions Optimales", Rapport Technique ONERA 2/3437, SN, Juillet 1981.

[11] Nguyen Van Nhan, "Application de la Méthode d'Encadrement du Point-selle d'un Jeu Différentiel à un Problème Tridimensionnel de Poursuite-évasion", Rapport Technique ONERA 4/3534 SN, 12 Août 1985.

[12] Aumasson, C., "Evaluation Dynamique des Performances d'un Avion en Combat Aérien par Simulation et Optimisation Numérique des Trajectoires", Rapport Technique ONERA 14/5148 SY, 29 Octobre 1984.

[13] Marchal, C., "Survey Paper ; Chattering arcs and chattering controls", Journal of Optimization Theory and Applications, Vol. 11, Nr. 5, 1973.

[14] Marchal, C., "Résolution des Jeux Différentiels Déterministes ; la Méthode de la Stratégie Approchée", Rapport Technique ONERA 3/3437 SN, 15 Avril 1981 ; also ONERA Publication 1987 to appear.

[15] Marchal, C, "Theoretical Research in Deterministic Optimization", ONERA Publication 139, pp. 38-40, 1971.

[16] Pontryagin, L.S., Boltyanskii, V.G., Gamkrelidze R.V., Mischenko, E.F., "The Mathematical Theory of Optimal Processes". Interscience Publishers, New York, (1962).

Appendix

The reference 14 is a french translation of this paper with more examples, more demonstrations and longer developments.

Let us consider two minor points analysed in this reference 14.

I) The K-strategies of R. Isaacs ([1], p. 38).

The interval (t_o, t_f) is subdivided into many small sub-intervals of the type $(t_o + n\,\varepsilon\ ;\ t_o + (n + 1)\,\varepsilon))$ and at each $t_o + n\,\varepsilon$ one of the two players decides his control $\vec{M}(t)$ or $\vec{m}(t)$ for the next sub-interval in terms of the present state $\vec{X}(t_o + n\,\varepsilon)$ and the time $t_o + n\,\varepsilon$.

This information is given to the second player that now faces a problem of deterministic optimization for the interval of interest (provided that the similar problems of all ulterior intervals be already solved).

The optimal successive choice of the first player are of course function of these conditions and this general problem may seem simpler than the initial differential game.

However this model doesn't allow to approach and surround differential games of the minimax or maximin types (try to solve the game number 4 of the section 6). It also leads to many more singularities than the method of approximate strategies.

II) A case of simplification of differential games.

In the equations (37), (38) were presented the usual definitions of the control

Hamiltonian H and the optimal Hamiltonian H* of an ordinary problem of optimization :

$$d\vec{X}/dt = \vec{V} \in D(\vec{X}, t) \tag{44}$$

D $(\vec{X}, t)$ is the vectogram or maneuverability domain

$$H = \vec{P}.\vec{V} \text{ ; with } \vec{P} = \text{adjoint vector of Pontryagin} \tag{45}$$

$$H^* = H^*(\vec{P}, \vec{X}, t) = \sup_{\vec{V} \in D(\vec{X},t)} \vec{P}.\vec{V} \tag{46}$$

The natural extension of these notions to differential games is obvious :

$$\vec{V} = \vec{V}(\vec{X}, \vec{M}, \vec{m}, t) \text{ ; } \quad \vec{M} \in \mathcal{D}_M(t) \text{ ; } \vec{m} \in \mathcal{D}_m(t) \tag{47}$$

and thus :

A) For a game of minimax type the Hamiltonian of the game will be :

$$H^*_{mM} = H^*(\vec{P}, \vec{X}, t) = \inf_{\vec{m} \in \mathcal{D}_m(t)} \left\{ \sup_{\vec{M} \in \mathcal{D}_M(t)} \vec{P}.\vec{V}(\vec{X}, \vec{M}, \vec{m}, t) \right\} \tag{48}$$

B) For a game of maximin type :

$$H^*_{Mm} = H^*(\vec{P}, \vec{X}, t) = \sup_{\vec{M} \in \mathcal{D}_M(t)} \left\{ \inf_{\vec{m} \in \mathcal{D}_m(t)} \vec{P}.\vec{V}(\vec{X}, \vec{M}, \vec{m}, t) \right\} \tag{49}$$

C) For a game of neutral type, with positive λ_i such that $\Sigma\, \lambda_i = 1$

Either :

$$H^*_N = H^*_{N1} = H^*(\vec{P}, \vec{X}, t) = \sup_{\lambda_i} \left\{ \sup_{\vec{M}_i \in \mathcal{D}_M(t)} \left[\inf_{\vec{m} \in \mathcal{D}_m(t)} \vec{P}.\sum_i \lambda_i \vec{V}(\vec{X}, \vec{M}_i, \vec{m}, t) \right] \right\} \tag{50}$$

Or :

$$H^*_N = H^*_{N2} = H^*(\vec{P}, \vec{X}, t) = \inf_{\lambda_i} \left\{ \inf_{\vec{m}_i \in \mathcal{D}_m(t)} \left[\sup_{\vec{M} \in \mathcal{D}_M(t)} \vec{P}.\sum_i \lambda_i \vec{V}(\vec{X}, \vec{M}, \vec{m}_i, t) \right] \right\} \tag{51}$$

It easy to verify that $H^*_{N1} \leq H^*_{N2}$ and the conditions (3) imply $H^*_{N1} = H^*_{N2}$. On the other hand in a given game the minimax type is more favourable to the maximizor and the maximin type is more favourable to the minimizor. This is in agreement with the following general property :

$$\forall\, (\vec{P}, \vec{X}, t) : H^*_{Mm} \leq H^*_N \leq H^*_{mM} \tag{52}$$

These three quantities are equal in the separated case as written in (17).

The simplification considered in this appendix and developed in the reference 14 is the following.

The optimal Hamiltonian $H^* (\vec{P}, \vec{X}, t)$ considered in (46) is <u>convex with respect to $\vec{P}$</u>, that is :

$\{a \geq 0 \; ; \; b \geq 0 \; ; \; a+b = 1\}$ imply always :

$$a\, H^* (\vec{P}_1, \vec{X}, t) + b\, H^* (\vec{P}_2, \vec{X}\, t) \geq H^* (a\, \vec{P}_1 + b\, \vec{P}_2, \vec{X}, t) \tag{53}$$

and the simplification is :

If a game (of any type) has a Hamiltonian $H^* (\vec{P}, \vec{X}, t)$ <u>convex with respect to $\vec{P}$</u> this game is equivalent to an ordinary one-player problem of optimization.

The corresponding maneuverability domain $D (\vec{X}, t)$ is closed, convex and given by (46) with the given Hamiltonian $H^* (\vec{P}, \vec{X}, t)$. The maximizor can choose any solution $\vec{X} (t)$ that agrees with this maneuverability domain and the minimizor can only forbid the other solutions.

A symmetrical simplification, favourable to the minimizor, occurs if the Hamiltonian of the game is concave with respect to the adjoint vector $\vec{P}$.

LIMIT CYCLES IN ECONOMIC CONTROL MODELS

Gustav Feichtinger
Institut für Ökonometrie und Operations Research
Technische Universität Wien
A-1040 Wien/Austria

1. Introduction

Recent years have seen a growing interest in the question of whether cyclical policies can be optimal in intertemporal economic decision problems. In the present paper we try to study endogenous optimal cyclical policies, which are neither the result of non-autonomous fluctuations, nor of prescribed boundary conditions. To be more specific, we question under what conditions the solutions of an autonomous continuous-time optimal control model with infinite time horizon converges to a limit cycle.

Among the methods used to establish the optimality of periodic solutions in optimal control theory, is the application of the Hopf bifurcation theory (see, e.g., [1], [2]), analytical solution of the canonical system of differential equations (see [3]), and numerical investigations. The main purpose of the following paper is to provide an example of the latter approach. It is shown how a limit cycle can be established in an interesting problem which arises in the dynamics of a representative firm. Moreover, some other examples of optimal periodic solutions of economic control models are reviewed.

2. How an Asymmetric Incentive Scheme Generates Cycles

In [4] the concept of a firm's normal employment level as a weighted average of past employment levels is introduced. Denoting by $L(t)$ the number of workers *actually* employed by a firm at time t, the firm's *normal* employment level defined by the government is computed as an exponentially weighted average of past levels of employment

$$A(t) = m\int_{-\infty}^{t} L(s)e^{-m(t-s)}ds, \tag{1}$$

where the parameter $m > 0$ represents the relative weight given to more recent employment levels. Hence

$$\dot{A}(t) = m[L(t) - A(t)]. \tag{1a}$$

The aim is to study the impact of an incentive scheme in which the firm receives a reward (or pays a penalty) when it deviates above (below) its normal employment level. More specifically, it is assumed that the government prescribes the following linear subsidy/penalty function f:

$$f(L-A) = \begin{cases} \alpha \\ 0 \\ \beta(L-A) \end{cases} \quad \text{for} \quad L \begin{Bmatrix} > \\ = \\ < \end{Bmatrix} A, \tag{2}$$

where α and β are nonnegative constants. Apparently, f has a kink at $L = A$ for $\alpha \neq \beta$.

The firm's objective is to maximize the present value of the profit stream (r is the discount rate):

$$\max_{L(t)} \int_0^\infty e^{-rt}[pF(L(t)) - wL(t) + f(L(t)-A(t))]dt, \tag{3}$$

where F(L) is a strictly concave production function, and $p > 0$, $w > 0$ represents the price of the product,and the wage rate, respectively.

Thus, the firm is faced with the following optimal control problem: (3) subject to the system dynamics (1a). L(t) is the control variable which is restricted to $L(t) \geq 0$, and A(t) is the state variable, whose initial value is assumed to be given: $A(0) = A_o$.

Considering initially the case of high penalty and low subsidy: $\alpha \leq \beta$. In [4], it is shown that the firm's optimal long run employment level, $\hat{A}$, depends on its initial normal employment level:

$$\hat{L} = \hat{A} = \begin{cases} \hat{A}_1 \\ A_o \\ \hat{A}_2 \end{cases} \quad \text{for} \quad A_o \begin{cases} \leq \hat{A}_1 \\ \in (\hat{A}_1, \hat{A}_2) \\ \geq \hat{A}_2, \end{cases}$$

where $\hat{A}_1 < \hat{A}_2$ are certain 'one-sided' stationary levels.

In the case of high subsidy and low penalty ($\alpha > \beta$) the Hamiltonian

$$H = pF(L) - wL + f(L-A) + \psi m(L-A),$$

(ψ is the adjoint variable belonging to A) consists of two concave parts, with a 'convex kink' at the junction point $L = A$ (see [5]). For *non-concave* optimal control models with *one state variable*, a *chattering control* is optimal. Thus, the presumption of the authors [4] that *cyclical behaviour* in this case might be optimal is *not* true.

In [6, p. 164] it is stated that in continuous-time nonlinear autonomous control models with *one* state variable,the *state trajectory* is always *monotonic*. For a rigorous proof of this result under more general assumptions see [7].

The *cyclical* optimal policy in a simple discrete example provided by [4] degenerates to a *chattering* control if the time increments tend to zero. Note that a similar situation occurs in the marketing literature, where periodic advertising expenditures are optimal only in the discrete version [8], while in a continuous setting,at least two state variables are required to generate these cycles (see [9]).

A non-realistic feature of the Long-Siebert model [4], is the fact that the firm can adjust its work force arbitrarily fast and without costs. In the case of high subsidy and low penalty, this assumption leads to chattering control, which means that workers are hired and fired 'at the same time'. It would therefore be more realistic to consider a model with labour being a state variable. The control variable is the hiring and firing rate which is either affected by concave adjustment costs, or is subject to upper and lower bounds. Both assumptions lead to cyclical employment policies.

Hence we consider the following employment model (see also [5]):

$$\max_u \int_0^\infty e^{-rt}[pF(L) - wL + f(L-A) - k(u)]\,dt \tag{4}$$

$$\dot{A} = m(L-A), \quad A(0) = A_0 \tag{5}$$

$$\dot{L} = u - qL, \quad L(0) = L_0 \tag{6}$$

$$L \geq 0. \tag{7}$$

The (new) control variable u denotes the hiring and firing rate, where $u > 0$ refers to recruitment, and $u < 0$ means discharging. $k(u)$ is the labour adjustment cost function (training and integration costs, lay off costs). In [10] and [11] reasons are given for the convexity of k. More specifically, we assume

$$k(0) = 0, \quad k'(u) \begin{Bmatrix} > \\ = \\ < \end{Bmatrix} 0 \text{ for } u \begin{Bmatrix} > \\ = \\ < \end{Bmatrix} 0, \quad k''(u) > 0. \tag{8}$$

The voluntary quit rate is denoted by q: For $F'(0) = \infty$ the state constraint (7) will never become active.

In order to generate a cyclical optimal solution we assume that $\alpha > \beta$.

The firm is faced with an optimal control problem with two state variables (A,L) and one control (u). An additional difficulty, is the non-differentiability of the objective functional, with respect to both states A and L. Since L is a state variable, the nonsmooth function f is now a function of the state variables. Therefore (4) - (7) is not a standard optimal control problem, and a generalized maximum principle has to be used (see, e.g., [12]). Following (5) the Hamiltonian

$$H = pF(L) - wL + f(L-A) - k(u) + \psi m(L-A) + \phi(v-qL)$$

yields the necessary optimality conditions:

$$v = \arg\max_v H, \quad \text{i.e.,} \quad \phi = k'(u) \tag{9}$$

$$(\dot{\psi}-r\psi, \dot{\phi}-r\phi) \in \partial_{(A,L)} H = (m\psi, -pF'(L)+w-\psi m+\phi q) - \partial_{(A,L)} f(L-A), \tag{10}$$

where

$$\partial_{(A,L)} f(L-A) = \begin{Bmatrix} \{(-\beta,+\beta)\} \\ \{(-\gamma,\gamma)\,|\,\beta \le \gamma \le \alpha\} \\ \{(-\alpha,+\alpha)\} \end{Bmatrix} \quad \text{for} \quad L \begin{Bmatrix} < \\ = \\ > \end{Bmatrix} A. \qquad \begin{matrix} (11a) \\ (11b) \\ (11c) \end{matrix}$$

The four-dimensional state-costate space is divided by the hyperplane $L = A$, along which the right hand side of the adjoint "equation" (10) is discontinuous.

Since an explicit solution of the canonical system does not seem to be possible, a *numerical analysis* has to be carried out. For this we specify the functions F and k as

$$F(L) = \sqrt{L}, \quad k(u) = u^2/2 \qquad (12)$$

and choose the following values for the parameters:

$$r = 0.1,\ m = 0.5,\ q = 0.1,\ p = 1,\ \alpha = 1,\ \beta = 0.5,\ w = 1. \qquad (13)$$

Thus, $u = \phi$ because of (9), and the canonical system becomes

$$\left.\begin{aligned} \dot{A} &= 0.5(L-A) \\ \dot{L} &= \phi - 0.1L \\ \dot{\psi} &= (r+m)\psi + \gamma \\ \dot{\phi} &= (r+q)\phi - \frac{1}{2\sqrt{L}} + 1 - 0.5\psi - \gamma \end{aligned}\right\} \qquad (14)$$

where $\gamma = \begin{Bmatrix} \alpha \\ \beta \end{Bmatrix}$ for $L \begin{Bmatrix} > \\ < \end{Bmatrix} A$.

In order to find a cyclical solution of (14), the boundary value problem solver COLSYS was applied (see, e.g. [13]). It turned out that a closed orbit exists. In figures 1 and 2 the projections of this orbit into the (A,L), and (L,u) spaces are depicted as the solid lines.

The interpretation of Fig. 1 is as follows: Let us start in P_1 wich $A = L$ in a situation of a small stock of labour. In this case, the shadow price ϕ of labour L is high, so that it is optimal to increase L in order to obtain the subsidy $\alpha(L-A)$ and to increase the output F(L). Following clockwise the cycle in Fig. 1, we reach point P_2, where L starts to decrease, because its shadow price ϕ has fallen below a certain level. This is a consequence of the fact that L is already larger than required for a profitable production. L was only increased up to that level because of the subsidies. While L decreases now, its weighted average, A, is still increasing until point P_3 is reached. Since ϕ is now at its minimum, labour is discharged at the maximum rate and L falls below A. Hence the firm has to pay taxes $\beta(A-L)$. The stock of labour decreases until we arrive at point P_4 where ϕ is large enough to make L increase again. The reason is that a higher L is needed for a reasonable production F(L), and A is small enough to start the next cycle in which the high subsidies $\alpha(L-A)$ can be obtained.

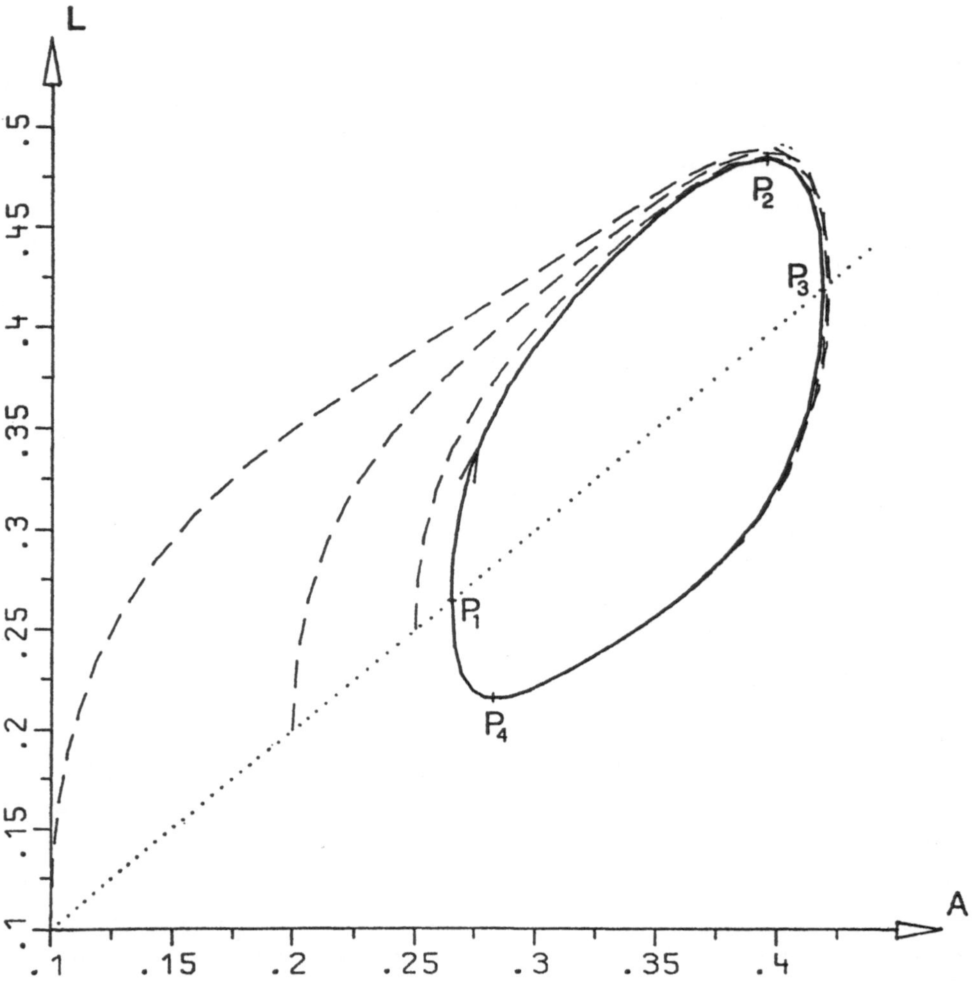

Figure 1: Projection of the limit cycle in the state space (see [5]).

The economic explanation for the periodic behaviour is as follows (cf. [4]). At first, the firm increases its level of employment to obtain the subsidy. But as the subsidy vanishes if there is no change in unemployment level, the firm will reduce its employment level (thus incurring a penalty) in order to increase it again. The cycle described above represents a *tradeoff* between profits by production (for which a stationary medium level of L would be optimal) and the gains from the subsidy/tax scheme (which are higher, the more L and A fluctuate).

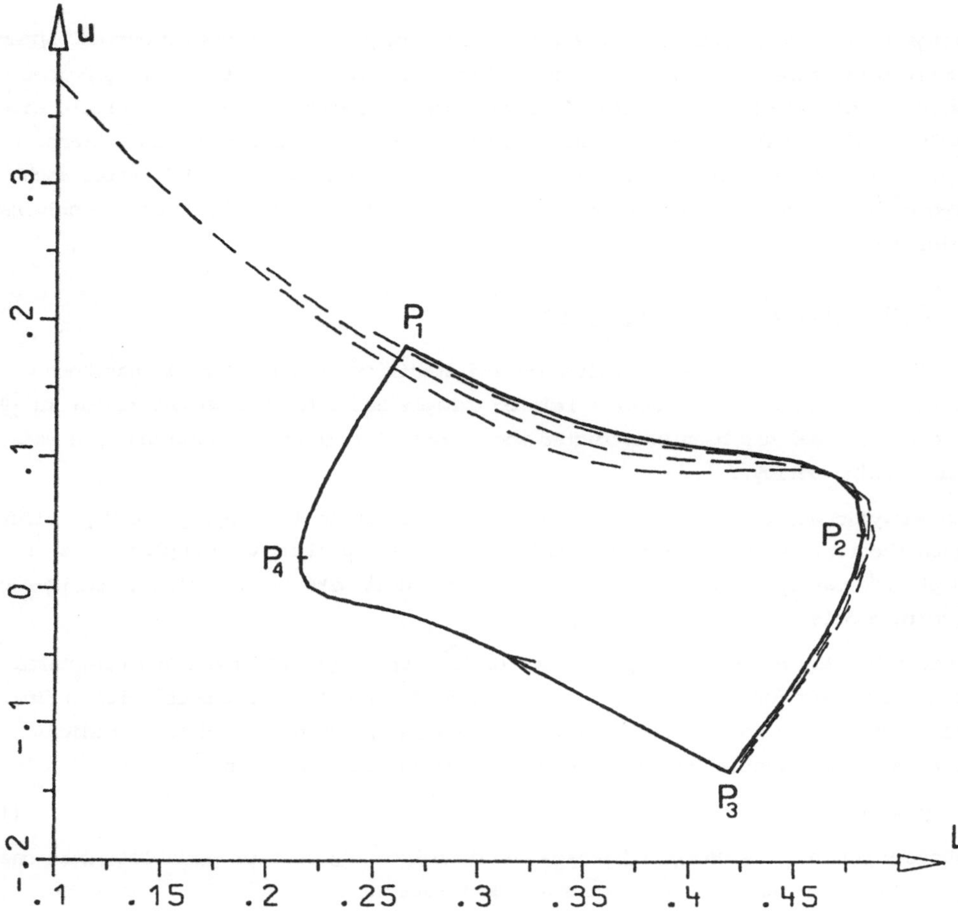

Figure 2: Projection of the limit cycle in the phase plane of actual labour force and hiring/firing rate (see [5]).

In summarizing the result, we show that an incentive scheme, which is asymmetric in the sense that the marginal subsidy for an amployment surplus exceeds the marginal penalty for a deficit of employees may cause cycles. Thus, the government's good intentions could result in periodic hiring and firing.

The cyclical solution takes on the role of a stationary point in a standard concave problem. There exists a two-dimensional manifold in the (A, L, ψ, ϕ) space, which contains those trajectories that converge towards the cycle. In figures 1-2 three sample trajectories that converge to the *limit cycle* are sketched as dashed curves. Apparently this convergence is very fast, so that after one period it almost coincides with the closed orbit. The rate of convergence is characterized by the stable eigenvalue (which is the one within the unit circle) of the linearized Poincaré-map. With our choice of parameters (13) this eigenvalue is approximately 0.001. After each cycle, the distance to the closed orbit is reduced to roughly 1/1000 of the original distance.

3. Application of the Hopf Bifurcation

In the preceding example, the existence and stability of a limit cycle has been established numerically. Another related example has already been mentioned: In [9] a control model has been constructed whose optimal (numerical) solution is a *pulsing* advertising policy.

An analytic way to prove the existence of a limit cycle is to apply the Hopf bifurcation theorem. There are, however, only a few papers dealing with applications of the Hopf bifurcation to economics ([14] and [15] deal with limit cycles in multi-sector growth models).

Let us briefly discuss the application of the *Hopf bifurcation theorem* to optimal control models (cf. [16, Chap. 3]). Consider an optimal control model with infinite time horizon and no path constraints. The maximum principle yield the canonical system of differential equations with continuous right hand side:

$$\dot{y} = g(y;\mu), \tag{15}$$

where y denotes the vector of states and corresponding costate variables. μ is one of the model parameters, e.g., the discount rate r.

To prove the existence and stability of limit cycles, the following steps have to be taken:

(i) Determine the stationary points $\hat{y}(\mu)$ of (15). Calculate the eigenvalues of the linearization of (15) around $\hat{y}(\mu)$.

(ii) Determine the critical parameter value μ_o, where one of the eigenvalues crosses the imaginary axis. The crossing velocity has to be positive.

(iii) Transform the canonical system (15) by means of the central manifold to a two-dimensional normal form, suitable for Hopf bifurcations (ρ and θ are polar coordinates):

$$\left.\begin{aligned} \dot{\rho} &= [D(\mu-\mu_o) + A\rho^2]\rho \\ \dot{\theta} &= \omega + C(\mu-\mu_o) + B\rho^2 \end{aligned}\right\} \tag{16}$$

A, B, C, D have to be calculated from the model parameters, and ω is the imaginary part of the critical eigenvalue.

A sufficient condition for the existence of stable limit cycles is

$$D \neq 0, \; A < 0. \tag{17}$$

The Hopf bifurcation theorem is a local result, in the sense that the existence of limit cycles is guaranteed only in a one-sided neighbourhood of μ_o.

To illustrate the Hopf bifurcation method, we consider the following simple *inventory problem* with constant demand d (see also [2]):

$$\min_{v} \int_{o}^{\infty} e^{-rt}[hx^2 + c(u) + kv^2]\,dt \tag{18}$$

$$\dot{x} = u - d, \; x(0) = x_o \tag{19}$$

$$\dot{u} = v, \qquad u(0) = u_o. \tag{20}$$

Here x denotes the stock of inventory, u the production rate, v the rate of change in production, r the discount rate. The inventory or shortage costs and the production adjustment costs are assumed to be quadratic, i.e. hx^2 and kv^2, respectively.

The production costs are a smooth, *concave-convex* function. More precisely we assume that

$$c''(d) < 0, \; c^{(iv)}(d) > 0. \tag{21}$$

For k = 0 the production rate u acts as control and chatters between 0 and $\tilde{u}$, where $\tilde{u}$ is the production rate for which average costs are equal to marginal costs. It is assumed that the maximal production rate $\bar{u}$ is greater than $\tilde{u}$ and d.

By using the procedure sketched above, the canonical system of problem (18) - (20) can be reduced to (16), where the discount rate r acts as bifurcation parameter μ. It turns out that $D \neq 0$ and that

$$A = -ac^{(iv)}(d) + bc^{(iii)}(d)^2, \tag{22}$$

where a and b are positive constants. Thus, the sufficiency condition (17) is satisfied if and only of

$$c^{(iv)}(d) > \frac{b}{a}c^{(iii)}(d)^2. \tag{23}$$

Hence, the existence of a stable limit cycle is guaranteed, provided that the third derivative of the production cost function evaluated at d is sufficiently small.

References

[1] Benhabib, J. and Nishimura, K., "The Hopf Bifurcation and the Existence and Stability of Closed Orbits in Multisector Models of Optimal Economic Growth", J. Econ. Theory Vol. 21, 421-444, 1979.

[2] Feichtinger, G., "Periodic Optimal Control: Can Oscillations be Optimal in Autonomous Economic Control Models?" Working Paper, Technical University Vienna, 1986.

[3] Näslund, B., "Consumer Behaviour and Optimal Advertising", J. Oper. Res. Soc. Vol. 30, 237-243, 1979.

[4] Long, N.V. and Siebert, H., "Lay-Off Restraints, Employment Subsidies, and the Demand for Labour", in: G. Feichtinger (Ed.) Optimal Control Theory and Economic Analysis 2, North-Holland, Amsterdam, 293-312, 1985.

[5] Steindl, A., Feichtinger, G., Hartl, R. and Sorger, G., "On the Optimality of Cyclical Employment Policies: A Numerical Investigation", Forschungsbericht Nr. 84, Inst. f. Ökonometrie & Operations Research, Techn. Univ. Wien, Januar 1986.

[6] Kamien, M.I. and Schwartz, N.L., "Dynamic Optimization: The Calculus of Variations and Optimal Control in Economics and Management", North-Holland, New York, 1981.

[7] Hartl, R.F., "A Simple Proof of the Monotonicity of the State Trajectories in Autonomous Control Problems", J. Econ. Theory 40, 1987.

[8] Simon, H., "ADPULS: An Advertising Model with Wearout and Pulsation", J. Marketing Res. 19, 352-363, 1982.

[9] Luhmer, A., Steindl, A., Feichtinger, G., Hartl, R. and Sorger, G., "ADPULS in Continuous Time", Forschungsbericht Nr. 91, Inst. f. Ökonometrie & Operations Research, Techn. Univ. Wien, April 1986.

[10] Holt, C.C., Modigliani, F., Muth, J.F. and Simon, H.A., "Planning Production, Inventories and Work Force", Prentice-Hall, Englewood Cliffs, 1960.

[11] Salop, S.C., "Wage Differentials in a Dynamic Theory of the Firm", J. Econ. Theory 6, 321-344, 1973.

[12] Clarke, F.H., "Optimization and Non-Smooth Analysis", Wiley, New York, 1983.

[13] Ascher, U., Christiansen, J. and Russell, R.D., "A Collocation Solver for Mixed Order Systems of Boundary Value Problems", Mathematics of Computation 33, 659-679, 1978.

[14] Benhabib, J. and Nishimura, K., "The Hopf Bifurcation and the Existence and Stability of Closed Orbits in Multi-Sector Models of Optimal Economic Growth", J. Econ. Theory 21, 421-444, 1979.

[15] Medio, A., "Oscillations in Optimal Growth Models", Working Paper, University of Venice, 1986.

[16] Guckenheimer, J. and Holmes, P., "Nonlinear Oscillations, Dynamical Systems, and Bifurcation of Vector Fields, Springer, New York, 1983.

AN APPROACH TO CONTROL THEORY BY FIXED POINT ALGORITHMS

Klaus Schilling
Dornier System GmbH, Abt. ERY
Postfach 13 60
D-7990 Friedrichshafen

1. Introduction

Fixed point principles are a fundamental tool to prove the existence of solutions for nonlinear equations. With respect to control theory there have been discussed applications in the areas controllability [10], [9], [6], [8], observability, parameter estimation [4] and existence of optimal controls [5]. These results were built on fixed point theorems for set valued operators, only based upon compactness and continuity requirements.

For real, set valued mappings this type of problem is a recognized domain for simplicial fixed point algorithms [1], [15]. The aim of this paper is to point out how recent extensions of convergence results to Banach-space problems [13], [12], provide a constructive basis with regard to controllability and optimal trajectories.

In section 2 some relevant fixed point results are summarized. The application of these to prove controllability is discussed in section 3, permitting derivation of an algorithm to compute controllable trajectories. Within the framework of optimal control theory section 4 reviews the computation of optimal trajectories by simplicial fixed point algorithms.

2. Fixed Point Results

A particular advantage of simplicial fixed point algorithms is the capability to even solve problems for set valued, real mappings [1], [15], [12]. The required continuity and compactness conditions allow the transformation of the fixed point problem into a sequence of combinatorial problems through piecewise linear approximations. The special features of the algorithm for piecewise linear structures in combination with projection methods have been exploited in [13], [12] to derive constructive proofs of existence results for set valued operators.

As a typical example, a result based on collectively compact operators in a Banach-space X is referred to below.

2.1 Definition

A ball with radius r and center x is denoted by $B(x,r) := \{y \in X : \|x-y\| \le r\}$. Particular subsets of 2^X, the set of all subsets of X, are described by

$$K(X) := \{\mathcal{Y} \subset X : \mathcal{Y} \text{ is nonempty and compact}\},$$
$$CK(X) := \{\mathcal{Y} \subset K(X) : \mathcal{Y} \text{ is convex}\}.$$

The set valued mapping $F: X \to 2^X$ is said to be upper semicontinuous at $x \in X$, if for any neighbourhood V of Fx, there exists a neighbourhood U of x, such that $FU = \{Fy : y \in U\} \subset V$.

A sequence of mappings $(G_i)_{i \in \mathbb{N}}$ with $G_i : D \to 2^X$ on the bounded, closed, nonempty set $D \subset X$ is called collectively compact, if $\overline{\bigcup_{i \in \mathbb{N}} G_i D}$ is compact.

Let $P_i : X \to X_i$ be a projection onto a finite dimensional subspace $X_i \subset X$ with $\|P_i\| \le M$ for all $i \in \mathbb{N}$. The sequence $(X_i, P_i)_{i \in \mathbb{N}}$ is called projection scheme for X, if for all $x \in X$ follows $\lim_{i \to \infty} \|P_i x - x\| = 0$.

2.2 <u>Theorem</u> (cf. [13], theorem 3.6, 3.7)

Let $D \subset X$ be a bounded, closed set of the Banach-space X and $(x_i, P_i)_{i \in \mathbb{N}}$ be a projection scheme for X. If the conditions

a) $F: D \to CK(X)$ is upper semicontinuous,

b) there exists a $\hat{x} \in \text{int}(D)$, such that for all $y \in \partial D$ follows
$$\nu(\hat{x}-y) \in Fy - y \implies \nu \geq 0,$$

c) $(P_i F)_{i \in \mathbb{N}}$ is collectively compact with respect to D,

are fulfilled, then there exists at least one fixed point $x^* \in Fx^*$, which can be approximated by simplicial algorithms.

The compactness requirement c) can still be relaxed by an approach via approximation-proper mappings, but at the cost of tightening a) to be a requirement for uniform continuity [13].

3. <u>Controllability</u>

An elegant technique to prove global controllability uses a transformation via boundary value problems for differential inclusions to apply fixed point arguments for establishing the existence of solutions [10], [5], [6], [8]. Using this approach it is also possible to derive, on the basis of theorem 2.2, an algorithm for the computation of controllable trajectories. The numerical methods for differential inclusions, required to assist in the solution, are also of interest per se, as use of the theory of differential inclusions is further motivated by examples from economics and biology [2], [3].

For functions $f:[0,T] \to \mathbb{R}^n$ we denote the Banach-spaces

- of continuous functions (with the topology of uniform convergence on compacta) by $C[0,T]^n$,
- of measurable functions by $L_1[0,T]^n$,
- of almost bounded functions by $L_\infty[0,T]^n$.

$AC[0,T]^n$ describes the absolutely continuous functions.
In the following consider the dynamic system

$$(1)\qquad \begin{aligned} &\dot{x}(t) = f(t,x(t),u(t)) \\ &u(t) \in U(t) \subset \mathbb{R}^m \end{aligned} \qquad \text{for almost every } t \in [0,T]$$

characterized by the domain of control $U \subset \mathbb{R}^m$ and by the function $f:[0,T] \times \mathbb{R}^n \times \mathbb{R}^m \to \mathbb{R}^n$.

3.1 Definition

The system (1) is said to be controllable from a set $D_1 \subset \mathbb{R}^n$ to a set $D_2 \subset \mathbb{R}^n$, if there exists an $u \in L_1[0,T]^n$ with $u(t) \in U(t)$ a.e., such that the related trajectory $x \in C[0,T]^n$ satisfies $x(0) \in D_1$, $x(T) \in D_2$.

In his famous paper [7] Filippov proved the lemma, that for continuous f and upper semicontinuous $U:[0,T] \to K(\mathbb{R}^m)$ this controllability is equivalent to the existence of a solution for

$$(2)\qquad \begin{aligned} &\dot{x}(t) \in F(t,x(t)) := f(t,x(t),U(t)) := \{f(t,x(t),v) : v \in U(t)\} \\ &x(0) \in D_1,\ x(T) \in D_2. \end{aligned} \qquad \text{a.e. in } [0,T]$$

Before stating results for this problem, the concept for a solution of boundary value problems with differential inclusions is given.

3.2 Definition

Consider the boundary value problem

$$\begin{aligned} &\dot{x}(t) \in F(t,x(t)) \qquad \text{a.e. in } [0,T] \\ &Mx(0) - Nx(T) = c \end{aligned}$$

determined by the nxn-matrices M, N, the vector $c \in \mathbb{R}^n$, $T \in [0,\infty[$ and the set valued mapping $F:[0,T] \times \mathbb{R}^n \to 2^{\mathbb{R}^n}$.

A function $x:[0,T] \to \mathbb{R}^n$ is called solution of this problem, if there exists a $\xi \in L_1[0,T]^n$ with $\xi(t) \in F(t,x(t))$ a.e. and a $x_o \in \mathbb{R}^n$, such that

$$x(t) = x_o + \int_o^t \xi(s)\, ds \quad \text{for } t\in[0,T]$$

$$Mx_o - N(x_o + \int_o^T \xi(s)\, ds) = c \; .$$

The first result deals with the situation that D_1, D_2 each consists of only a single point.

3.3 Theorem

Let the dynamic system (1) satisfy the conditions

i) $U:[0,T] \to K(\mathbb{R}^m)$ is upper semicontinuous.

ii) There exists a $r\in\mathbb{R}$, such that

$\|f(t,x,u)\| \le r$ for all $(t,x)\in[0,T]\times B(x_o,Tr)\subset\mathbb{R}^{n+1}$, $u\in U(t)$.

iii) f is continuous in $[0,T]\times B(x_o,Tr)\times U([0,T])\subset\mathbb{R}^{n+m+1}$ and $f(t,x,U(t))$ is convex for $(t,x)\in[0,T]\times B(x_o,Tr)\subset\mathbb{R}^{n+1}$.

iv) For all $z\in B(x_o,T_r)\subset C[0,T]^n$ exists a $\xi\in L_1[0,T]^n$, $\xi\in f(\cdot,z(\cdot),U(\cdot))$, such that $x_T - x_o \in \int_o^T \xi(s)\, ds$.

Then the system (1) is controllable from x_o to x_T and a related trajectory can be approximated by simplicial algorithms.

Proof:

Conditions i) and iii) imply the applicability of Filippov's Lemma [7]. As stated above the controllability problem for system (1) can thus be transformed to the equivalent boundary value problem (2). Consider the operator

$$G\colon C[0,T]^n \supset B(x_o,Tr) \to 2^{C[0,T]^n}$$

$$x \to \{y : y(t)=x_o+\int_o^t \xi(s)ds, \xi\in L_1[0,T]^n, \xi(t)\in F(t,x(t)) \text{ a.e.}\}.$$

It follows from this operator, together with iv), that every fixed point of G provides a solution of (2).

The existence of at least one fixed point of G will be established by theorem 2.2, thus also providing the algorithm.

In analogy to [2], p. 129, theorem 1 follows that Gx is compact, convex and nonempty for every $x \in B(x_o, Tr)$ as well as the upper semicontinuity of G.

Further the boundary condition 2.2 b) has to be investigated for $y \in \partial B(x_o, Tr) \subset C[0,T]^n$:

$\nu(x_o - y) \in Gy - y$ implies the existence of a $\xi \in L_1[0,T]^n$ with $\xi(t) \in F(t, x(t))$ a.e., such that

$$(1-\nu)\ (y(\cdot) - x_o) = \int_o^{\cdot} \xi(s)\, ds \qquad \text{and therefore}$$

$$|1-\nu|\ Tr = \left\| \int_o^{\cdot} \xi(s)\, ds \right\| \le Tr \quad . \qquad \text{This leads through}$$

$$|1-\nu| \le 1 \qquad \text{to the required result}$$

$$\nu \ge 0 \ .$$

It remains to study the collective compactness, which will be proved with respect to the following projection scheme:

$X_i := \{y \in C[0,T]^n :\ y$ is linear on every interval $[t_j, t_{j+1}] \subset [0,T]$, with $t_k := k \frac{T}{2^i},\ k = 0, \ldots, 2^i\}$

$P_i : C[0,T]^n \to X_i,\ P_i x(t) := \frac{2^i}{T} \left((t_{j+1} - t)\ x(t_j) + (t - t_j) x(t_{j+1}) \right)$ for $t \in [t_j, t_{j+1}] \subset [0,T],\ j = 0, \ldots, 2^i - 1$

$(P_i G)_{i \in \mathbb{N}}$ is collectively compact, if

$$\mathcal{A} := \{y \in C[0,T]^n :\ y \in P_i Gx,\ i \in \mathbb{N},\ x \in B(x_o, Tr)\}$$

is precompact. This follows from the Arzelà-Ascoli-theorem as

α) $\mathcal{A}(t)$ is bounded for all $t\in[0,T]$:
For $y\in P_i Gx$ it follows that

$$\|y(t)\| \leq \|y\| \leq \|P_i\| \; \|Gx\| = \|Gx\| \leq Tr \; .$$

β) $\mathcal{A}$ is equicontinuous:

For a $y\in P_i Gx$ with $y = P_i \; (x_o + \int_o^{\cdot} \xi(s)\, ds)$, $\xi(t)\in F(t,x(t))$

it follows that

$$\|y(t) - y(\tilde{t})\| \leq \|P_i\| \; \|\int_{\tilde{t}}^{t} \xi(s)\, ds\| \leq r \; |t-\tilde{t}| \; .$$

Therefore all requirements of theorem 2.2 are met, ensuring that a fixed point of G can be approximated by simplicial algorithms.

In [14], using similar techniques based on theorem 2.2, boundary value problems for differential inclusions are studied. These results can be interpreted in the framework of controllability between linear subspaces.

It is assumed that the dynamics can be split into a linear and a bounded part, leading to problems of the form:

(3) $\dot{x}(t) \in f(t,x(t),U(t)) =: A(t)x(t) + F(t,x(t))$ a.e.,

(4) $Mx(0) - Nx(T) = c$,

where M, N, A(t) are real nxn-matrices for all $t\in[0,T]$, $c\in\mathbb{R}^n$ and $F:[0,T]\times\mathbb{R}^n \to 2^{\mathbb{R}^n}$.

If range $(M) \cap$ range $(N) = \{0\}$ then (4) assigns to a solution $x(\cdot)$ a start- and end-condition, described by M (respectively N) and the related components of c.

3.4 Theorem

For every $c \in \mathbb{R}^n$ there exists a solution of the boundary value problem (3), (4), which can be approximated by using simplicial algorithms, if the following conditions hold:

i) $A(\cdot)$ is continuous in $[0,T]$ and the associated homogeneous, linear problem

$$\dot{y}(t) = A(t)y(t) , \qquad t \in [0,T]$$
$$My(0) - Ny(T) = 0$$

has only the trivial solution $y \equiv 0$.

ii) $U: [0,T] \to K(\mathbb{R}^m)$ is upper semicontinuous.

iii) There exists a $r \in \mathbb{R}$ with

$$\|F(t,x)\| := \max_{g \in F(t,x)} \|g\| \le r \text{ for all } (t,x) \in [0,T] \times \mathbb{R}^n.$$

iv) f is continuous in $[0,T] \times \mathbb{R}^n \times U([0,T])$ and $f(t,x(t),U(t))$ is convex for all $(t,x) \in [0,T] \times \mathbb{R}^n$.

The proof is a direct consequence of [14], theorem 3.2. It is already sufficient to restrict iii), iv) with respect to the x-component to a suitable ball in $\mathbb{R}^n$ (cp. [14], 3.3).

Numerical examples of boundary value problems, solved by the proposed method, have been presented in [12], [14].

4. Optimal Control

The necessary conditions for optimal trajectories of the Pontryagin Maximum Principle will be interpreted as a boundary value problem for differential inclusions. This enables the application of theorem 2.2 to derive a new, indirect algorithm for the computation of optimal trajectories. This method is particularily suitable for compact control domains and has advantages if singular subarcs are included and if only weak differentiability assumptions hold.

Following [14], this approach is outlined for the special case of known starting conditions, i.e. consider for $T < \infty$, $x \in AC[0,T]^n$, $u \in L_\infty[0,T]^m$ the problem:

$$\begin{aligned}
&\text{Minimize} \int_0^T \phi(t,x(t),u(t))\,dt \\
&\text{while} \quad \dot{x}(t) = \psi(t,x(t),u(t)) \quad \text{a.e. in } [0,T] \\
&\qquad\qquad x(0) = x_o \\
&\qquad\qquad u(t) \in U \subset \mathbb{R}^m \quad \text{a.e. in } [0,T] \quad .
\end{aligned}$$

Here $\phi : \mathbb{R}^{1+n+m} \to \mathbb{R}$, $\psi : \mathbb{R}^{1+n+m} \to \mathbb{R}^n$ are continuous functions, continuously differentiable with respect to the 2. to (n+1). argument. In [14] it is also shown that mixed, linear start- and end-conditions can be treated in this framework.

In the nondegenerate situation the Hamiltonian is given by $H(t,x,u,p). = \psi(t,x,u)^* p - \phi(t,x,u)$ and an optimal solution (x,u) satisfies, according to the Pontryagin Maximum Principle, together with the adjoint variable $p(\cdot) \in AC[0,T]^n$ the extended boundary value problem:

$$\begin{aligned}
&\dot{x}(t) = \psi(t,x(t),u(t)) && \text{a.e. in } [0,T], \\
(5) \quad &\dot{p}(t) = -\frac{\partial H}{\partial x}(t,x(t),u(t),p(t)) && \text{a.e. in } [0,T], \\
&x(0) = x_o \ , \ p(T) = 0 \ , \\
&H(t,x(t),v,p(t)) \le H(t,x(t),u(t),p(t)) && \text{for all } v \in U \text{ and almost all } t \in [0,T].
\end{aligned}$$

To present u as a function of (x, p) here, generalizing the usual approach, a set valued mapping is defined by

$$R : [0,T] \times \mathbb{R}^{2n} \to 2^{\mathbb{R}^m}$$
$$R(t,x,p) := \{u \in U : H(t,x,u,p) = \sup_{v \in U} H(t,x,v,p)\}.$$

From the assumptions on ϕ, ψ follows that, for compact U, the mapping R is upper semicontinuous and nonempty. This suggests the replacement of the inequality, leading to the equivalent problem

$$(6)\quad \begin{pmatrix} \dot{x} \\ \dot{p} \end{pmatrix} \in \begin{pmatrix} \psi\ (t,x(t),R(t,x(t),p(t))) \\ -\frac{\partial H}{\partial x}\ (t,x(t),R(t,x(t),p(t))) \end{pmatrix} = $$

$$=: A(t) \begin{pmatrix} x(t) \\ p(t) \end{pmatrix} + F(t,x(t),p(t),R(t,x(t),p(t))) \quad \text{for } t \in [0,T],$$

$$x(0) = x_o \ , \ p(T) = 0.$$

Here the dynamics is split similar to 3.4 into a linear and a set valued part.

4.1 Theorem

For a compact control domain U let a continuous 2n x 2n-matrix $A(\cdot)$ and a $F:\mathbb{R}^{1+n+m} \to \mathbb{R}^{2n}$ be selectable in (6), such that

i) The homogeneous linear boundary value problem related to $A(\cdot)$ has only the trivial solution.

ii) F is continuous and $F(y,R(y))$ is convex and bounded for all $y \in [0,T] \times \mathbb{R}^{2n}$.

There then exists a solution of (6), which can be approximated by using simplicial algorithms. This solution meets the necessary conditions for an optimal trajectory of the Pontryagin Maximum Principle.

This is a special case of [14], theorem 4.3, which is proved through transformation to a fixed point problem (by a similar technique as theorem 3.3) and the subsequent application of theorem 2.2.

As the algorithm, proposed here for the computation of optimal trajectories, is based upon compact control domains U, the result obtained thus supplements the classical indirect methods (see [11], chapter 3), tailored for unbounded U. The advantages of this approach are that

- It requires no differentiability of the Hamiltonian with respect to the component u.

- For treating problems with singular arcs no additional information (i.e. about the switching structure) is necessary.

- It includes the potential to compute several distinct solutions.

- Convergence of the method is independent from starting points.

while the disadvantages include:

- The need of an explicitly derived mapping R.
- A cubic increase of computation time with the dimension of the fixed point problem (due to the combinatorial basis of the simplicial algorithms).

Thus for the classical situation, an attractive application could be the exploitation of the global convergence properties for simplicial algorithms to generate starting trajectories for fast convergent, local methods like multiple shooting.

References

[1] Allgower, E.L./Georg, K.
Simplicial and Continuation Methods for Approximating Fixed Points and Solutions to Systems of Equations, SIAM Review 22 (1980), 28 - 85.

[2] Aubin, J.P./Cellina, A.
Differential Inclusions, Springer Verlag 1984.

[3] Aubin, J.P./Sigmund, K. (eds.)
Dynamics of Macrosystems, Springer Verlag 1985.

[4] Carmichael, N./Pritchard, A.J./Quinn, M.D.
State and Parameter Estimation for Nonlinear Systems, Appl. Math. Optim. 9 (1982), 133 - 161.

[5] Cesari, L.
Existence of Solutions and Existence of Optimal Solutions, in: Cecconi, J.P./Zolezzi, T. (eds.), Mathematical Theories of Optimization, Springer Verlag 1983.

[6] Dauer, J.P.
A Controllability Technique for Nonlinear Systems,
J. Math. Anal. Appl. 37 (1972), 442 - 451.

[7] Filippov, A.F.
On Certain Questions in the Theory of Optimal Control, SIAM J. Control 1 (1962), 76 - 84.

[8] Hermes, H.
On the Structure of Attainable Sets for Generalized Differential Equations and Control Systems,
Journal of Differential Equations 9 (1971), 141 - 154.

[9] Lukes, D.L.
Global Controllability of Nonlinear Systems,
SIAM J. Control 10 (1972), 112 - 126 and
SIAM J. Control 11 (1973), 186.

[10] Magnusson, K./Pritchard, A.J/Quinn, M.D.
The Application of Fixed Point Theorems to Global Nonlinear Controllability Problems,
Mathematical Control Theory, Banach Cent. Publ. 14 (1985), 319 - 344.

[11] Polak, E.
A Historial Survey of Computational Methods in Optimal Control,
SIAM Review 15 (1973), 553 - 584.

[12] Schilling, K.
Simpliziale Algorithmen zur Berechnung von Fixpunkten mengenwertiger Operatoren,
Wissenschaftlicher Verlag Trier, 1986.

[13] Schilling, K.
Constructive Proofs of Fixed Point Theorems for Set Valued Operators by Simplicial Algorithms, submitted.

[14] Schilling, K.
Boundary Value Problems for Differential Inclusions and the Computation of Optimal Trajectories, submitted.

[15] Todd, M.J.
Computation of Fixed Points and Applications,
Springer Verlag, 1976.

NUMERICAL SOLUTION OF AN OPTIMAL CONTROL PROBLEM WITH HYSTERESIS

Martin Brokate
Institut für Mathematik
Universität Augsburg
8900 Augsburg, West Germany

1. Introduction

One of the most basic situations in control theory is the forced scalar harmonic oscillator

$$\ddot{x} + x = u \quad , \tag{1}$$

u being the control function. In this paper, we study some problems of optimal control for the dynamical system

$$\ddot{x} + y = u \quad , \quad y = Wx \quad . \tag{2}$$

Here W is the operator defined in [1], which describes an input-output behaviour of hysteresis type. It is illustrated in figure 1.

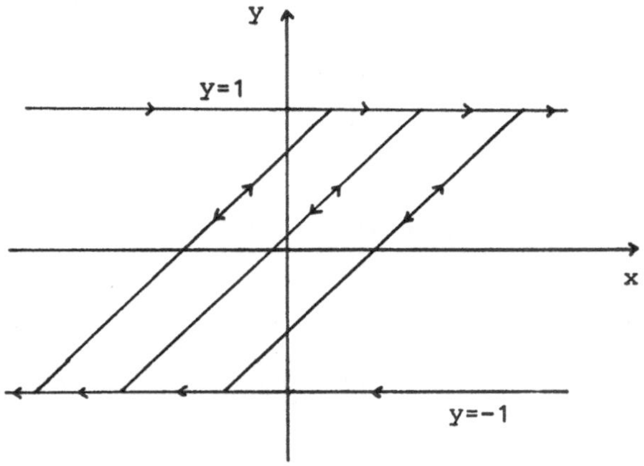

Figure 1.

The output $y = y(t)$ is obtained from the input $x = x(t)$, if one starts at a given (x_0, y_0) and follows the arrows, according to the sign of $\dot{x}(t)$.

Within the context of solid deformations (i.e., if $x(t)$ denotes the strain of a spring at time t), equation (1) results from Hooke's law, whereas system (2) describes an idealized elastic - perfectly plastic situation, where the yield surface (here: $|y| = 1$) does not depend upon the plastic deformation.
We define the operator W formally.

<u>1.1 Definition</u>

Let $(x_o,y_o) \in \mathbb{R} \times [-1,1]$, $x \in C[0,T]$ piecewise monotone with $x(0) = x_o$. If $\{t_i\}$ is a partition of $[0,T]$ such that $x|[t_i,t_{i+1}]$ is monotone, we define inductively

$$(Wx)(0) = y_o$$

$$(Wx)(t) = g(x(t) - x(t_i) + (Wx)(t_i)) \quad , \quad t \in (t_i,t_{i+1}],$$

where

$$g(x) = \begin{cases} \min\{x,1\} & \text{if } x \geq 0 \\ \max\{x,-1\} & \text{if } x \leq 0 \end{cases}$$

□

<u>1.2 Theorem</u> [1]

Let $(x_o,y_o) \in \mathbb{R} \times [-1,1]$. Set

$$X_o = \{ x \in C[0,T] : x(0) = x_o \} .$$

Then W as defined in (1.1) can be extended uniquely to a Lipschitz continuous operator $W: X_o \to C[0,T]$.
<u>Proof</u>: See [2], chapter 1.

□

We replace the second order ordinary differential equation in (2) by a first order system for a vector function $z:[0,T] \to \mathbb{R}^n$. The scalar input x is obtained from z by

$$x = S^T z$$

where $S \in \mathbb{R}^n$ is a fixed vector ($S^T = (1,0)$ in (2)). We consider the following problem of optimal control.

<u>1.3 Problem (K)</u>

Minimize

$$L_T(y(T),z(T),T) + \int_0^T L(y(t),z(t),t,u(t))\, dt$$

subject to

$$\begin{aligned} \dot{z}(t) &= f(y(t),z(t),t,u(t)) \quad , \quad z(0) = z_o \\ y(t) &= (Wx)(t) \quad , \quad x(t) = S^T z(t) \end{aligned} \tag{4}$$

$$u(t) \in U \tag{5}$$

$$f_T(y(T),z(T),T) = 0 \tag{6}$$

The final time T can be fixed or free.

□

This paper is now organized as follows: We present a (Pontryagin-type) maximum principle for problem (K), which has been obtained in [3], and apply it to some control problems with dynamics (2). From the resulting multi-point boundary value problems we conclude some properties of the optimal control and compute solutions numerically with the multiple shooting method (adapted to optimal control problems in [5], [6]), using the program from [7], [8].

2. The maximum principle

If one wants to formulate and derive a maximum principle, i.e. first order necessary optimality conditions, for problem (K), one immediately faces the problem that the operator W is not differentiable and that, for a not piecewise monotone input, W is defined by a limit process only. We do not want to describe here how these difficulties are overcome; in the resulting maximum principle, they are reflected by the fact that the adjoint function jumps at points where the derivative $\dot{x}_*$ of the optimal input x_* changes its sign, and that a part of the adjoint equation loses its meaning when there are infinitely many such sign changes on the boundary, i.e. in

$$\{ t : |y_*(t)| = 1 \} \quad .$$

We list some assumptions concerning problem (K), which are sufficiently general for the control problems discussed in this paper. More general situations are treated in [3], [4].

2.1 Assumptions for problem (K)

(i) The functions f, f_T, L, L_T are twice continuously differentiable;
$f_T\colon \mathbb{R} \times \mathbb{R}^n \times \mathbb{R} \times \mathbb{R}^m \rightarrow \mathbb{R}^k$ etc.

(ii) U is a compact convex subset of $\mathbb{R}^m$.

(iii) $S \in \mathbb{R}^n$, W is defined as in (1.1) with $x_o = S^T z_o$ and $y_o \in [-1,1]$ is given.

(iv) There exists a continuous function c_o such that

$$|f(y,z,t,u)| \le c_o(y,u)(1 + |z|)$$

for all arguments.

(v) f is affine linear w.r.t. u, L is convex w.r.t. u.

□

One looks for solutions (x_*,y_*,z_*,u_*) of problem (K) with

$$u_* \in L^\infty(0,T;\mathbb{R}^m);\ x_*,y_* \in W^{1,\infty}(0,T;\mathbb{R});\ z_* \in W^{1,\infty}(0,T;\mathbb{R}^n) \qquad (7)$$

Since W is nonanticipating and Lipschitz continuous, the system equations (4) have for any given $u \in L^\infty$ a unique solution (x,y,z); if there exists an admissible point (i.e. if the terminal condition (6) can be satisfied by a solution of (4),(5)), then furthermore problem (K) has a solution with regularity (7). This has been proved in [3].

2.2 Definition

Let (x_*,y_*,z_*,u_*) be a solution of (K). We say that x_* is regular, if $x_* \in C^1[0,T]$ and if there is an open neighbourhood $N \subset [0,T]$ of $\{ t \in [0,T] : |y_*(t)| = 1 \}$, such that $\dot{x}_*$ has only finitely many zeros $\{\tau_i\}$ in N.

□

We partition the components of the vector z into those which contribute to the input x and those which do not, So, if

$$S = (s_1,\dots,s_M,0,\dots,0) \quad ,$$

we set

$$z^I = (z_1,\dots,z_M) \quad , \quad z^{II} = (z_{M+1},\dots,z_n) \quad .$$

In the system corresponding to (2) we have $z^I = z_1$, $z^{II} = z_2$.
We furthermore denote the Hamiltonian by

$$H(y,z,t,u,p) = \ell_o L(y,z,t,u) + f(y,z,t,u)^T p$$

and abbreviate derivatives along the optimal trajectory as

$$D_z H(t) = D_z H(y_*(t),z_*(t),t,u_*(t),p(t))$$

$$g'(t) = g'(\ y_*(t) - x_*(\tau_i) + y_*(\tau_i)\ ) \qquad , \quad t \in (\tau_i,\tau_{i+1}] \quad .$$

We now state the maximum principle.

2.3 Theorem (Maximum principle)

Let (x_*,y_*,z_*,u_*) be a solution of (K), let assumption (2.1) be satisfied. Then there exists a $\ell_o \ge 0$, $\ell_1 \in \mathbb{R}^k$, $\ell_2 \in \mathbb{R}$ with $(\ell_o,\ell_1) \neq (0,0)$ and a $p \in BV(0,T;\mathbb{R}^n)$ such that the following assertions hold:

(i) If x_* is regular, then p is absolutely continuous on (τ_i, τ_{i+1}) and satisfies

$$\dot{p}(t) = -(D_z H(t) + D_y H(t) g'(t) S) \quad \text{in } (\tau_i, \tau_{i+1}) \tag{8}$$

$$\begin{aligned} p(T) = {} & \ell_o D_z L_T(T) + D_z f_T(T)^T \ell_1 + \\ & + (\ell_o D_y L_T(T) + D_y f_T(T)^T \ell_1) g'(T) S \end{aligned} \tag{9}$$

$$p(\tau_i^+) - p(\tau_i^-) = \alpha_i S \quad , \quad \alpha_i \in \mathbb{R} \tag{10}$$

(ii) If x_* is not regular, then p^{II} is absolutely continuous on $[0,T]$, and the II-components of equations (8) and (9) hold.

(iii)

$$H(y_*(t), z_*(t), t, u_*(t), p(t)) = \min_{u \in U} H(y_*(t), z_*(t), t, u, p(t)) \tag{11}$$

$$H(t) = -\ell_o D_t L_T(T) - D_t f_T(T)^T \ell_1 - \int_t^T D_t H(s)\, ds + \ell_2 \tag{12}$$

(iv) If T is free, then $\ell_2 = 0$.

Proof: A detailed proof is given in [3], see also [4]. □

2.4 Definition

We say that (x_*, y_*, z_*, u_*) is a Kuhn-Tucker point, if it is admissible for (K) (i.e., (4) - (6) are satisfied), and if there exist multipliers $(\ell_o, \ell_1, \ell_2, p)$ with $\ell_o = 1$ such that the assertions of (2.3) are satisfied.

□

In [3], chapter 7, a controllability condition is formulated which is sufficient for a solution to be a Kuhn-Tucker point. As usual, a numerical method based on the maximum principle computes Kuhn-Tucker points, and for degenerate solutions (i.e. solutions of (K) which satisfy the maximum principle only with $\ell_o = 0$) the maximum principle is not helpful.

3. A time optimal problem

We formulate problem (K1) as a special case of problem (K), with dynamics given by (2).

3.1 Problem (K1)

Minimize T

subject to

$$\dot z_1 = z_2 \ , \quad \dot z_2 = u - y \ , \qquad |u| \le u_{max}$$
$$x = z_1 \ , \quad y = Wx \ ,$$
$$z(0) = z_o \ , \quad y(0) = y_o \ , \quad z(T) = z_T \ , \quad y(T) = y_T \ . \qquad \square$$

We denote by (x,y,z,u) a solution of (K1). We write down the conclusions from the maximum principle (2.3). We have

$$L = 1 \ , \ L_T = 0 \ , \ f(y,z,t,u) = (z_2 \ , \ u - y) \ ,$$
$$f_T(y,z,T) = (z_1 - z_{1T} \ , \ z_2 - z_{2T} \ , \ y - y_T) \ .$$

Conditions (2.3iv) and (2.3iii) imply (the argument t is omitted)

$$0 = H = \ell_o + p_1 z_2 + p_2(u - y) \tag{13}$$

$$u = \begin{cases} u_{max} \ , & p_2 < 0 \\ -u_{max} \ , & p_2 > 0 \end{cases} \tag{14}$$

The adjoint p_2 is absolutely continuous and satisfies

$$\dot p_2 = -p_1 \ , \quad p_2(T) = \ell_{12} \tag{15}$$

If x is regular, the adjoint p_1 satisfies

$$\dot p_1 = g' \cdot p_2 = \begin{cases} p_2 \ , & |y| < 1 \\ 0 \ , & |y| = 1 \end{cases} \tag{16}$$

$$p_1(T) = \ell_{11} + \ell_{13} \cdot g'(T) \tag{17}$$

$$p_1(\tau_i^+) - p_1(\tau_i^-) = \alpha_i \ , \tag{18}$$

where $\{\tau_i\}$ are the points on $\{|y| = 1\}$, where $z_2(\tau_i) = 0$. Now - regardless whether x is regular or not - if $\ell_o = 0$, we may choose $p_1 = p_2 = \ell_{12} = 0$, $(\ell_{11},\ell_{13}) \neq (0,0)$, such that the maximum principle is satisfied trivially, leaving us with no information at all. On the other hand, if $\ell_o = 1$, we see that the finite bang-bang principle holds:

3.2 Lemma

If (x,y,z,u) is a Kuhn-Tucker point for problem (K1), then x is regular, and the optimal control is bang-bang with a finite number of switchings, which are distinct from the $\{\tau_i\}$.

Proof: If $z_2(\tau) = 0$ for some τ, then because of (13) and since p_1 is bounded,

$$|p_1(s)\dot z_2(s)| \ \ge \ \frac{1}{2}$$

in some neighbourhood of τ. Therefore τ is an isolated zero of $\dot x = z$, $p_2(\tau) \neq 0$, and x is regular. If now $p_2(t) = 0$, then again (13) implies that

$$|p_1(s)z_2(s)| \quad \geq \quad \frac{1}{2}$$

in some neighbourhood D of t. Therefore, t is an isolated zero of p_2.

□

Because of this lemma, a Kuhn-Tucker point for (K1) is characterized by the zeros of p_2 and z_2 and by the sign structure of p_2. We fix the initial and terminal conditions as

$$(z_o, y_o) = (0,0,0) \quad , \quad (z_T, y_T) = (2,0,0) \quad . \tag{19}$$

For u_{max} large, the solution (x(t),y(t)) should look like figure 2. Transformed to the time interval [0,1], it should be a solution of the following multi-point boundary value problem:

$$\begin{aligned}
&\dot{z}_1 = z_2 \cdot T \quad , \quad \dot{z}_2 = (u - y) \cdot T \quad , \\
&\dot{y} = \begin{cases} 0 & \text{in } [\xi_1, \xi_3] \\ z_2 \cdot T & \text{elsewhere} \end{cases} \quad , \quad \dot{p}_1 = \begin{cases} 0 & \text{in } [\xi_1, \xi_3] \\ p_2 \cdot T & \text{elsewhere} \end{cases} \\
&\dot{p}_2 = -p_1 \cdot T \quad , \quad \dot{T} = 0 \quad , \quad \dot{\alpha}_1 = 0 \\
&u = \begin{cases} -u_{max} & \text{in } [\xi_2, \xi_4] \\ u_{max} & \text{elsewhere} \end{cases}
\end{aligned} \tag{20}$$

together with the boundary conditions (19) and the switching conditions for the unknown switching points ξ_1, ξ_2, ξ_3, ξ_4

$$\begin{aligned}
&y(\xi_1) = 1 \; , \; p_2(\xi_2) = 0 \; , \; z_2(\xi_3) = 0 \; , \; p_2(\xi_4) = 0 \; , \\
&1 + p_2(\xi_3) \cdot (-u_{max} - 1) = 0 \quad (= H(\xi_3))
\end{aligned} \tag{21}$$

as well as the jump condition

$$p_1(\xi_3^+) = p_1(\xi_3^-) + \alpha_1 \tag{22}$$

If one applies the multiple shooting algorithm in the version of [8], then it turns out that the boundary value problem (19) - (22) indeed leads to a Newton iteration with nonsingular iteration matrix, and that for $u_{max} = 2$ the algorithm computes a Kuhn-Tucker point as in figure 2. If one decreases u_{max}, then for $u_{max} \approx 0.83$ this Kuhn-Tucker point vanishes (since ξ_2 and ξ_3 merge), but already for $u_{max} = 0.86$ another Kuhn-Tucker point (with an additional interval $u = -u_{max}$ at the beginning) yields a lower time T (see figure 3 for $u_{max} = 0.8$). Decreasing u_{max} further, eventually the trajectory touches $y = -1$, and for $u_{max} = 0.52$, a Kuhn-Tucker point with two jumps of the adjoint p_1 emerges (see figure 4). For $u_{max} < 0.5$, the boundary arc with $y = 1$ from (1,1) to (3,1) cannot be traversed within one interval of $\dot{x} > 0$. Therefore, there will be at least one new interior arc with $y < 1$, splitting the boundary arc into two parts.

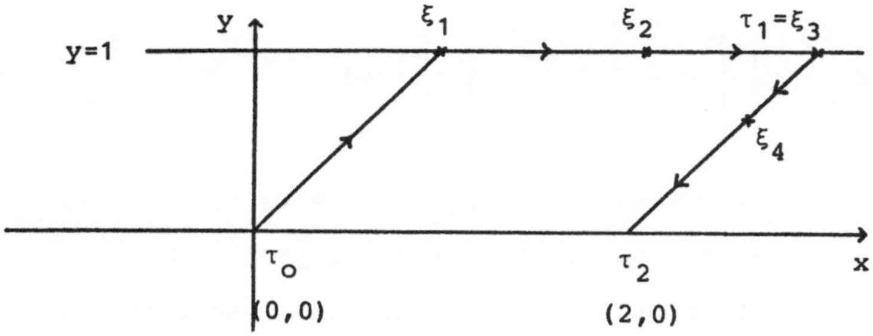

$u_{max} = 2.0$
$T = 3.69$
$\xi_1 = 0.28$
$\xi_2 = 0.44$
$\xi_3 = 0.64$
$\xi_4 = 0.78$

Figure 2.

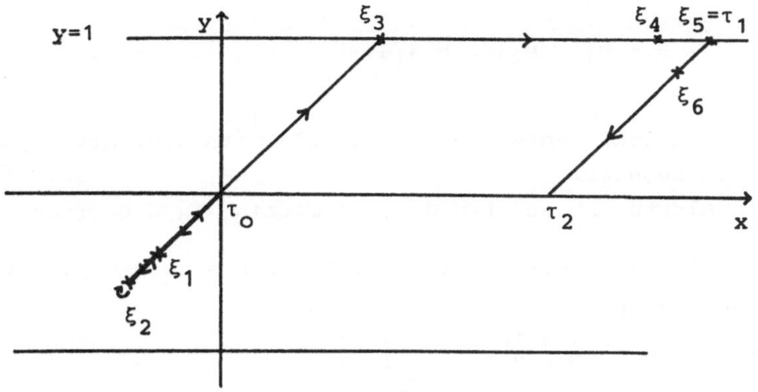

$u_{max} = 0.8$
$T = 7.29$
$\xi_1 = 0.15$
$\xi_2 = 0.22$
$\xi_3 = 0.46$
$\xi_4 = 0.63$
$\xi_5 = 0.72$
$\xi_6 = 0.78$

Figure 3.

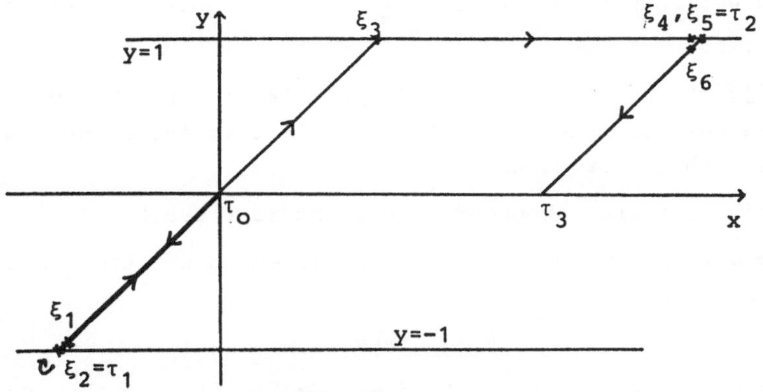

$u_{max} = 0.52$
$T = 9.85$
$\xi_1 = 0.27$
$\xi_2 = 0.29$
$\xi_3 = 0.48$
$\xi_4 = 0.68$
$\xi_5 = 0.71$
$\xi_6 = 0.73$

Figure 4.

4. A minimum norm problem

4.1 Problem (K2)

Minimize $\quad J = \frac{1}{2}\int_0^T u(t)^2\,dt \quad , \quad T$ fixed ,

subject to

$$\dot z_1 = z_2 \ , \quad \dot z_2 = u - y \ , \quad x = z_1 \ , \quad y = Wx \ ,$$

$$z(0) = z_o \ , \quad z(T) = z_T \ , \quad y(0) = y_o \ , \quad y(T) = y_T \ . \qquad \square$$

A weak convergence argument shows that (K2) has a solution $u \in L^2(0,T)$. If such an u is bounded, then (K2) can be subsumed formally under (K), taking U large enough. The maximum principle (2.3) then applies and yields for a Kuhn-Tucker point (x,y,z,u)

$$\ell_2 = H = \frac{1}{2}u^2 + p_1 z_2 + p_2(u - y) \tag{23}$$

$$u = - p_2 \ , \ \dot p_2 = - p_1 \ , \ p_2(T) = \ell_{12} \ .$$

4.2 Lemma

If (x,y,z,u) is a Kuhn-Tucker point for (K2) with boundary conditions (19), then x is regular.

<u>Proof:</u> If we evaluate H at $t = 0$, we obtain $\ell_2 \leq 0$ from (23) and (19).
We set $N = \{t \in [0,T] : |y(t)| > 1/2\}$. If $t \in N$ with $\dot x(t) = z_2(t) = 0$, we have at t

$$p_2(u-y) = \ell_2 - \frac{1}{2}u^2 \leq - \frac{1}{2}u^2 = - \frac{1}{2}p_2^2 \ .$$

If $p_2 = 0$ then $u = 0$ und $|\dot z_2| \geq 1/2$. If $p_2 \neq 0$, then

$$|\dot z_2| = |y - u| \geq \max\{|u|/2 \ , \ |y| - |u|\} \geq \frac{1}{6} \ .$$

Since $\dot z_2$ is continuous in $[0,T]$, z_2 can have only finitely many zeros in N . $\square$

Again, the adjoint p_1 therefore satisfies (16) - (18). For various times T , we have computed with the multiple shooting algorithm Kuhn-Tucker points with the structure

(interior arc , boundary arc , interior arc) .

The resulting optimal controls, again normalized to the time interval [0,1] , are given in figure 5.

5. References

[1] Krasnoselskii, M.A., Darinskii, B.M., Emelin, I.V., Zabreiko, P.P., Lifsic, E.A., Pokrovskii, A.V.: Hysterant operator,

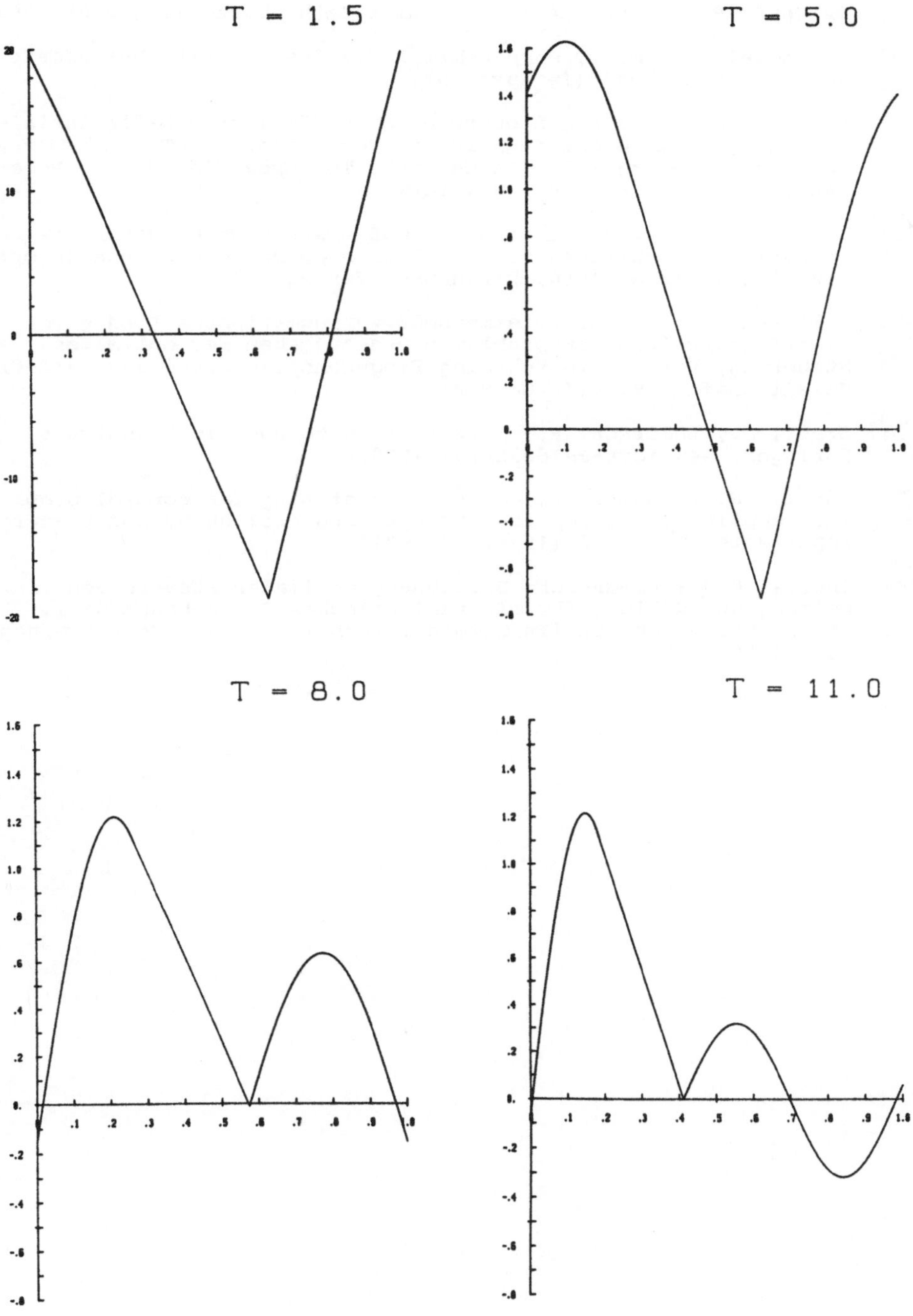

Figure 5.

DAN SSSR 190 (1970), 34 - 37; Soviet Math. Dakt. 11 (1970), 29-33.

[2] Krasnoselskii, M.A., Pokrovskii,A.V.: Systems with hysteresis. Nauka, Moscow 1983 (In Russian).

[3] Brokate, M.: Optimale Steuerung von gewöhnlichen Differentialgleichungen mit Nichtlinearitäten vom Hysteresis-Typ. Habilitationsschrift, Augsburg 1985 (In German). To appear in: Verlag Peter Lang, Frankfurt - Bern - New York.

[4] Brokate, M.: Optimal control of ODE systems with hysteresis nonlinearities. Submitted to: 4th French-German Conference on Optimization at Irsee 1986, Birkhäuser Verlag.

[5] Bulirsch, R.: Die Mehrzielmethode zur numerischen Lösung von nichtlinearen Randwertproblemen und Aufgaben der optimalen Steuerung, Vortrag im Lehrgang Flugbahnoptimierung der Carl-Cranz-Gesellschaft, 1971 (In German).

[6] Stoer, J., Bulirsch, R.: Introduction to numerical analysis. Springer, New York-Heidelberg, 1980.

[7] Oberle, H.J.: Numerical computation of singular control problems with application to optimal heating and cooling by solar energy, Appl. Math. Optim. 5 (1979), 297-314.

[8] Oberle, H.J.: Numerische Berechnung optimaler Steuerungen von Heizung und Kühlung für ein realistisches Sonnenhausmodell, Habilitationsschrift (In German). Technische Universität München, TUM-M8310, 1983.

COMPARISON BETWEEN SEVERAL CONJUGATION CONCEPTS

K.-H. ELSTER and A. WOLF, Ilmenau (GDR)

In the present paper we introduce the concept of P-conjugation using the notion of polarity, which was considered p.e. by EWERS/VAN MAAREN [10] . By the P-conjugation we will compare several wellknown conjugation concepts such as the generalized FENCHEL-conjugation, the Φ-conjugation (DEUMLICH/ELSTER) and others. The following results are closely connected with our paper [9] .

0. Introduction

The concepts of conjugation, which were developed in the last decennium, differ strongly from case to case. Thus, caused by the high level of generalizations, difficulties arise when those concepts should be compared mutually,
In the paper [9] the authors have introduced the concept of P-conjugation (P means polarity) which give a possibility for the comparison of conjugation concepts, which can be described as special cases of the P-conjugation. A assertion was proved concerning the generalized FENCHEL-conjugation, characterized by a coupling functional φ ([8] , [11] , [12] , [15]), and the Φ-conjugation, introduced by DEUMLICH and ELSTER ([2] , [3] , [4] , [5] , [6]).
In the present paper we will prove other comparison theorems concerning the generalized FENCHEL-conjugation, the Φ-conjugation, the F-conjugation ([1]), and the level-set-conjugation. To do this, we have to establish some propositons about the P-conjugation. Finally, a special case of P-conjugation will be discussed, which is of some interest in connection with fractional programming problems.

1. Polarities

Definition 1.1 (10). A polarity between the nonempty sets X,Y

is a mapping $\Delta: 2^X \rightarrow 2^Y$ which satisfies the condition

$$\Delta(\bigcup_{i\in I} A_i) = \bigcap_{i\in I} \Delta(A_i) \tag{1.1}$$

for each family $(A_i)_{i\in I}$, $A_i \subseteq X$, where I is an arbitrary index set.

If for instance Δ is the complement operator of sets, then the corresponding polarity (1.1) is the de MORGAN's rule.

In the following it is often convenient to characterize a polarity by a functional according

Theorem 1.1. $\Delta: 2^X \rightarrow 2^Y$ is a polarity if and only if there is a functional $p: X \times Y \rightarrow \bar{R}$ such that

$$\Delta(A) = \bigcap_{x\in A} \{y \in Y \mid p(x,y) \geq 0\}, \quad A \subseteq X . \tag{1.2}$$

Proof ([10]). Obviously, (1.2) is a polarity for a given functional p. Conversely, suppose $\Delta: 2^X \rightarrow 2^Y$ to be a polarity. Then

$$p(x,y) := \begin{cases} 0 \text{ if } y \in \Delta\{x\} , \\ -1 \text{ if } y \notin \Delta\{x\} \end{cases}$$

generates $\Delta(A)$ in (1.2).//

Definition 1.2. The functional p occuring in Theorem 1.1 is called generating functional of the polarity Δ.

Now let us introduce the dual polarity Δ^* of a polarity Δ.

Definition 1.3. If $\Delta: 2^X \rightarrow 2^Y$ is a polarity and p a generating functional of Δ, then the mapping $\Delta^*: 2^Y \rightarrow 2^X$, where

$$\Delta^*(B) = \bigcap_{y\in B} \{x \in X \mid p(x,y) \geq 0\}, \quad B \subseteq Y,$$

is called the dual polarity of Δ.

Some interesting properties of polarities are given in [10] . There is in general no one-to-one correspondence between a generating functional p and a given polarity Δ.

But an equivalence relation holds.

Definition 1.4. Two functionals $p_1: X \times Y \to \bar{R}$, $p_2: X \times Y \to \bar{R}$ are said to be P-equivalent if

$$\forall (x,y) \in X \times Y : \quad p_1(x,y) \geq 0 \Leftrightarrow p_2(x,y) \geq 0.$$

Obviously two polarities Δ_1, Δ_2 are identical if and only if each generating functional of Δ_1 is P-equivalent to any generating functional of Δ_2.

In the following we treat polarities on sets $X \times \bar{R}$ resp. $Y \times \bar{R}$. The generating functional of such a polarity $\Delta: 2^{X \times \bar{R}} \to 2^{Y \times \bar{R}}$ is denoted by

$$p(x,k,y,l): X \times \bar{R} \times Y \times \bar{R} \to \bar{R}.$$

Definition 1.5. A functional $p(x,k,y,l): X \times \bar{R} \times Y \times \bar{R} \to \bar{R}$ is called epigraphical if

(i) $\forall (x_o,k_o,y_o) \in X \times \bar{R} \times Y \ \exists l_* \in R$:

$$l_* := \min \{ l \in R \mid p(x_o,k_o,y_o,l) \geq 0 \},$$

(ii) $\forall (x_o,k_o,y_o,l_o) \in X \times \bar{R} \times Y \times \bar{R} \quad \forall l_1 \geq l_o$:

$$p(x_o,k_o,y_o,l_o) \geq 0 \Rightarrow p(x_o,k_o,y_o,l_1) \geq 0.$$

A functional $p(x,k,y,l): X \times \bar{R} \times Y \times \bar{R} \to \bar{R}$ is called hypographical if

(iii) $\forall (x_o,k_o,y_o) \in X \times \bar{R} \times Y \quad \exists l^* \in R$:

$$l^* = \max \{ l \times R \mid p(x_o,k_o,y_o,l) \geq 0 \}.$$

(iv) $\forall (x_o,k_o,y_o,l_o) \in X \times \bar{R} \times Y \times \bar{R} \ \forall l_2 \leq l_o$:

$$p(x_o,k_o,y_o,l_o) \geq 0 \Rightarrow p(x_o,k_o,y_o,l_2) \geq 0.$$

Accordingly to this definition we will say that a polarity is epigraphical resp. hypographical if there are epigraphical resp. hypographical generating functionals.

The following characterization of such polarity is of importance for comparison theorems given below.

Theorem 1.2. Let $\Delta: 2^{X\times\bar{R}} \longrightarrow 2^{Y\times\bar{R}}$ be a polarity.

(i) Δ is epigraphical resp. hypographical if and only if there is a generating functional p such that
$\forall(x,k,y,l)\in X\times\bar{R}\times Y\times\bar{R}$: $p(x,k,y,l) = -e(x,k,y) + l$
resp.
$\forall(x,k,y,l)\in X\times\bar{R}\times Y\times\bar{R}$: $p(x,k,y,l) = h(x,k,y) - l$,
where $e: X\times\bar{R}\times Y \rightarrow R$ resp. $h: X\times\bar{R}\times Y \rightarrow R$ are appropriate functionals.

(ii) The functionals e,h in (i) are determined uniquely.

Proof. (i): The functionals $-e(x,k,y) + l$ and $h(x,k,y) - l$ are obviously epigraphical resp. hypographical according to Definition 1.5.
Conversely, let $\Delta : 2^{X\times\bar{R}} \longrightarrow 2^{Y\times\bar{R}}$ be an epigraphical polarity with the generating functional $p(x,k,y,l)$ and assume

$$e(x,k,y) = \min\{l\in R \mid p(x,k,y,l) \leq 0\} \quad .$$

The minimum exists on $X\times\bar{R}\times Y$ because of Definition 1.5 (i).
We have to show
$\forall(x,k,y,l)\in X\times\bar{R}\times Y\times\bar{R}$: $p(x,k,y,l)\geq 0 \Longleftrightarrow -e(x,k,y) + l \geq 0$,
i.e the P-equivalence on $X\times\bar{R}\times Y\times\bar{R}$.
It holds for each $(x,k,y,l)\in X\times\bar{R}\times Y\times R$:

$$\begin{aligned} -e(x,k,y) + l \geq 0 &\Longleftrightarrow -\min\{\bar{l}\in R \mid p(x,k,y,\bar{e})\geq 0\} + l\geq 0 \\ &\Longleftrightarrow l\geq \min\{\bar{l}\in R \mid p(x,k,y,\bar{l}) \geq 0\} \\ &\Longleftrightarrow p(x,k,y,l) \geq 0 \quad . \end{aligned}$$

The last equivalence is true because of Definition 1.5 (i).
(ii): Now let us assume that there are two functionals e_1, e_2 on $X\times\bar{R}\times Y$ such that
$e_1(x_o,k_o,y_o) = e_2(x_o,k_o,y_o) + d$, where $d\in R$, $d\neq 0$.
By the P-equivalence we obtain for
$x=x_o$, $y=y_o$, $k=k_o$, $l_o=e_1(x_o,k_o,y_o) - \frac{d}{2}$

$$\begin{aligned} -e_1(x_o,k_o,y_o)+e_1(x_o,k_o,y_o)-\frac{d}{2}\geq 0 \Longleftrightarrow e_1(x_o,k_o,y_o) &- \\ - e_2(x_o,k_o,y_o) - \frac{d}{2} &\geq 0 \end{aligned}$$

and hence $-\frac{d}{2} \geq 0 \Longleftrightarrow \frac{d}{2} \geq 0$

which contradicts to $d\neq 0$.

There is an analogous proof for hypographical polarities. With respect to the definition of P-conjugate functions we introduce the notion of a symmetrical polarity.

Definition 1.6. Let Δ_1, Δ_2 be two polarities on $X \times \bar{R}$. Δ_1 is called symmetrical to Δ_2 if there exist generating functionals p_1 of Δ_1 and p_2 of Δ_2 such that

$\forall (x,k,y,l) \in X \times \bar{R} \times Y \times \bar{R} : p_1(x,k,y,l) = -p_2(x,k,y,l)$.

For a given polarity Δ there exists one and only one polarity $\tilde{\Delta}$ symmetrical to Δ (cf. [9]).

2. P-Conjugation

Definition 2.1. Let $\Delta: 2^{X \times \bar{R}} \to 2^{Y \times \bar{R}}$ be a polarity, $\tilde{\Delta}: 2^{X \times \bar{R}} \to 2^{Y \times \bar{R}}$ the polarity symmetrical to Δ and $f \in \bar{R}^X$.

The functionals f^{Δ}, $f_{\tilde{\Delta}} \in \bar{R}^Y$ with

$$f^{\Delta}(y) := \inf \{ l \in R \mid (y,l) \in \Delta(\text{graph } f) , \tag{2.1}$$

$$f_{\tilde{\Delta}}(y) := \sup \{ l \in R \mid (y,l) \in \tilde{\Delta}(\text{graph } f) \tag{2.2}$$

are called upper P-conjugate resp. lower P-conjugate (or: P-conjugate functions) of f with respect to the polarities Δ resp. $\tilde{\Delta}$.

In a more general manner it is possible to introduce upper and lower P-conjugate functions without the use of symmetrical polarities. The approach given above takes into consideration substantial examples of P-conjugates such as generalized FENCHEL-conjugation and Φ-conjugation (cf. Example 3.1 and Example 3.2).

Let us emphasize that, in general, we cannot conclude from the identity of two (upper or lower) P-conjugates to the identity of the used polarities; simple examples are given in [9]. But we have the assertion of

Corollary 2.1. Let Δ_1, Δ_2 be epigraphical polarities on $X \times \bar{R}$ and $f \in \bar{R}^X$. Then
$$f^{\Delta_1}(x) = f^{\Delta_2}(y) \text{ for all } y \in \bar{R}$$
implies $\Delta_1 = \Delta_2$.

A proof is given in [9] .

A corresponding proposition exists in the case of hypographic polarities and lower P-conjugates.

By Corollary 2.1 we obtain convenient representations of P-conjugate functions in the case of epigraphic resp. hypographic polarities.

Corollary 2.2. Let Δ_1 be an epigraphical polarity and let Δ_2 be a hypographical polarity on $X \times \bar{R}$. Then there exist uniquely determined functionals $e: X \times \bar{R} \times Y \to R$ and $h: X \times \bar{R} \times Y \to R$ such that

$$f^{\Delta_1}(y) = \sup_{x \in D(f)} e(x, f(x), y), \tag{2.3}$$

$$f_{\Delta_2}(y) = \inf_{x \in D(f)} h(x, f(x), y). \tag{2.4}$$

Proof. By Theorem 1.2 there exist uniquely determined functionals e,h such that $-e(x,k,y)+l$ is a generating functional of Δ_1 and $h(x,k,y)-l$ is a generating functional of Δ_2 for each $(x,k,y,l) \in X \times \bar{R} \times Y \times R$. Hence we obtain

$$\begin{aligned} f^{\Delta_1}(y) &= \inf\{l \in R \mid (y,l) \in \Delta_1(\text{graph } f)\} \\ &= \inf\{l \in R \mid \forall (x,k) \in \text{graph } f: -e(x,k,y)+l \geq 0\} \\ &= \inf\{l \in R \mid \forall (x,k) \in \text{graph } f: l \geq e(x,k,y)\} \\ &= \sup_{(x,k) \in \text{graph } f} e(x,k,y) \\ &= \sup_{x \in D(f)} e(x,f(x),y) . \end{aligned}$$

Analogously (2.4) can be proved. //

3. Examples

3.1. Generalized FENCHEL-Conjugation ([8], [11], [12], [15])

Let X,Y be nonempty sets, $\varphi: X \times Y \to \bar{R}$ a coupling functional, $\Delta^{\varphi}: 2^{X \times \bar{R}} \to 2^{Y \times \bar{R}}$ a polarity with the generating functional

$p^{\varphi}(x,k,y,l) := -\varphi(x,y) \dotplus k+l, (x,k,y,l) \in X \times \bar{R} \times Y \times R.$

By Theorem 1.2 Δ^{φ} is an epigraphical polarity. The polarity $\tilde{\Delta}^{\varphi}$ symmetrical to Δ^{φ} has the generating functional

$$p_{\varphi}(x,k,y,l) := \varphi(x,y) \dotplus -k-l \ ;$$

$\tilde{\Delta}^{\varphi}$ is hypographical by Theorem 1.2. For the sake of simplicity we write Δ_{φ} instead of $\tilde{\Delta}^{\varphi}$.

Then for the P-conjugate function generated by Δ^{φ} we have

$$\begin{aligned} f^{\Delta^{\varphi}}(y) &= \inf\{ l \in R \mid (y,l) \in \Delta^{\varphi}(\text{graph } f)\} \\ &= \inf\{ l \in R \mid \forall (x,k) \in \text{graph } f: -\varphi(x,y) \dotplus k+l \geqq 0\} \\ &= \inf\{ l \in R \mid \forall (x,k) \in \text{graph } f: l \geqq \varphi(x,y) \dotplus -k\} \\ &= \sup_{x \in D(f)} (\varphi(x,y) \dotplus -f(x)). \end{aligned}$$

Analogously we obtain

$$f_{\Delta_{\varphi}}(y) = \inf_{x \in D(f)} (\varphi(x,y) \dotplus -f(x)).$$

Remark: These results can also be derived by Corollary 2.2.

3.2. Φ-Konjugation ([2], [3], [4], [5], [6])

Let $X = Y =: V$ be a linear space, $\dim V=1$, $E := R \times V \times R$, V^* the algebraical dual space of V, $E^* = R \times V^* \times R$, $A: E \to E^*$ a linear mapping according

$$\begin{aligned} t' &= a_0 t + \langle a,x \rangle + cz \\ x' &= at + Bx + ez \\ z' &= ct + \langle e,x \rangle + bz \end{aligned}$$

where $a_0, b, c \in R$; $a, e \in V^*$; $B: V \to V^*$ a symmetrical linear mapping and $(b,e,c) \neq (0,0,0)$.

Furthermore, let

$$V^0 := \{(x,z) \in V \times R \mid c + \langle e,x \rangle + bz > 0\}.$$

Then the function

$$p^{\Phi}(x,z,x^*,z^*) := \langle A(1,x,z), (1,x^*,z^*) \rangle$$

is the generating functional of a polarity $\Delta^{\Phi}: 2^{V \times \bar{R}} \to 2^{V^* \times R}$, the symmetrical polarity $\tilde{\Delta}^{\Phi} =: \Delta_{\Phi}$ has the generating functional

$$p_{\Phi}(x,z,x^*,z^*) := -\langle A(1,x,z),(1,x^*,z^*) \rangle.$$

Generally, the polarities $\Delta^{\Phi}, \Delta_{\Phi}$ are not epigraphical resp. hypographical. But assuming for each $f \in R^V$ the condition

$$c + \langle e,x\rangle + bf(x) > 0 \quad \forall\ x \in D(f), \qquad (3.1)$$

we obtain epigraphical resp. hypographical functionals dividing by $c + \langle e,x\rangle + bf(x)$.

Exactly for such functions the Φ-conjugation was introduced by DEUMLICH/ELSTER.

If (3.1) is satisfied we obtain

$$\begin{aligned} f^{\Delta^{\Phi}}(x^*) &= \inf\{z^* \in R \mid (x^*,z^*) \in \Delta^{\Phi}(\text{graph } f)\} \\ &= \inf\{z^* \in R \mid \forall (x,z) \in \text{graph } f: p^{\Phi}(x,z,x^*,z^*) \geq 0\} \\ &= \inf\{z^* \in R \mid \forall (x,z) \in \text{graph } f:\ a_0 + \langle x^*,a\rangle + \\ &\quad + \langle x^*B,x\rangle + \langle a,x\rangle + \langle x^*,e\rangle + cz + cz^* + \\ &\quad + \langle e,x\rangle z^* + bzz \geq 0\} \\ &= \inf\{z^* \in R \mid \forall (x,z) \in \text{graph } f: \\ &\quad z^* \geq -\frac{a_0 + \langle x^*,a\rangle + \langle x^*B,x\rangle + \langle a,x\rangle + \langle x^*,e\rangle + cz}{c + \langle e,x\rangle + bz}\} \\ &= \sup_{x \in D(f)} -\frac{a_0 + \langle x^*,a\rangle + \langle x^*B,x\rangle + \langle a,x\rangle + \langle x^*,e\rangle + cf(x)}{c + \langle e,x\rangle + bf(x)} \\ &= f^{\Phi}(x^*). \end{aligned}$$

Analogously we can show

$$f_{\Delta_{\Phi}}(x^*) = f_{\Phi}(x^*).$$

3.3. Level-Set-Conjugation ([13], [14])

In [14], VOLLE gave a general concept for level-set-conjugation which can be used in quasiconvex optimization. Let U and V be abitrary nonempty sets and let $\Delta: 2^U \rightarrow 2^V$ be a polarity. For a function $f \in \bar{R}^U$ the conjugate function f^{Δ} is defined according to

$$f^{\Delta}(v) := \sup\{-f(u) \mid u \notin \Delta^*(\{v\})\} \quad ,$$

where Δ^* is the dual polarity of Δ.

In [14] a coupling functional $\varphi_{\Delta}: U \times V \rightarrow \bar{R}$ was associated to Δ such that the level-set-conjugation can be represented as a special case of generalized FENCHEL-conjugation:

$$\varphi_\Delta(u,v):= \begin{cases} 0 \text{ if } u \notin \Delta^*(\{v\}), \\ -\infty \text{ if } u \in \Delta^*(\{v\}). \end{cases}$$

According to Example 3.1 we obtain a generating functional $p^N: U \times \bar{R} \times V \times \bar{R} \to R$ and a polarity $\Delta^N: 2^{U \times \bar{R}} \to 2^{V \times \bar{R}}$ for describing the level-set-conjugation as a special case of P-conjugation:

$$p^N(u,k,v,l):= -\varphi_\Delta(u,v) \dotplus k+l.$$

3.4. F-Conjugation ([1])

The F-conjugation was introduced by BEN TAL/BEN ISRAEL in [1].

Let $\mathfrak{F}$ be a family of continuous functions $F: X \to R$, where $X \subseteq R^n$. Moreover we assume that the functions F depend on $n+1$ parameters $(x^*,\eta) \in X^* \times Y \subseteq R^n \times R$.

Such a family $\mathfrak{F}$ is called complete if for any convergent sequence $\{x_n^*,\eta_n\} \subset X^* \times Y$ with $\{x_n^*,\eta_n\} \xrightarrow[n\to\infty]{} (x_\infty^*,\eta_\infty)$ holds

$$\lim_{n\to\infty} F(x_n^*,\eta_n,x) = \infty \quad \text{for any } x$$

or

$$F(x_\infty^*,\eta_\infty,\cdot) \in \mathfrak{F}.$$

An example for a non-complete family is the family of all non-horizontal affine functions.

A family belongs to the class C, if for any $x^* \in X^*, x \in X$ the functions $F(x,\cdot,x^*)$ are strictly decreasing.

Let $F^I(x,\cdot,x^*)$ be the inverse function of $F(x,\cdot,x^*)$.

We introduce conjugates for families belonging to the class C according

$$\mathfrak{F}^*:= \{F^*(\cdot) = F^*(x,\zeta,\cdot) \mid x \in X,\ \zeta \in Z \subseteq R\},$$

where $F^*(x,\zeta,x^*):= F^I(x,\zeta,x^*)$ for any $x^* \in X^*$.

To define the F-conjugation we consider families $\mathfrak{F}$ such that both $\mathfrak{F}$ and $\mathfrak{F}^*$ are complete.

Let $\mathfrak{F}$ be in the class C and $f: R^n \to R$ a function. Then

$$f^F(x^*):= \sup_{x \in D(f)} F^*(x,f(x),x^*)$$

is called the convex (upper) F-conjugate of f.
In a similar way (but using the infimum) we obtain the concave (lower) F-conjugate of f.
Obviously we can describe the upper F-conjugation as an upper P-conjugation using a polarity Δ^F: $2^{X\times\bar{R}} \rightarrow 2^{X^*\times R}$ with the generating functional $p^F(x,\zeta, x^*,\eta)$: $= -F^*(x,\zeta,x^*)+\eta$.
Hence, by Theorem 1.2, Δ^F is epigraphical.
The representation of the lower F-conjugation as a lower P-conjugation can be established analogously.

4. Comparison

By the P-conjugation we are enabled to compare generalized FENCHEL-conjugation, Φ-conjugation, level-set-conjugation and F-conjugation.

Theorem 4.1. Let L be a linear space, L^* the algebraical dual space of L, A: $R\times L\times R \rightarrow R\times L^*\times R$ a linear mapping as in 3.2. and Δ^Φ the corresponding polarity.
Then there exists a coupling functional $\varphi_\Phi(x,x^*)$ such that $-\varphi_\Phi(x,x^*)\dotplus z+z^*$ is the generating functional of Δ^Φ if and only if A satisfics the condition

$$e = 0,\ b = 0,\ c > 0\ .$$

Then we obtain

$$\varphi_\Phi(x,x^*) := \frac{1}{c}(a_0+\langle a,x\rangle + \langle x^*,a\rangle + \langle x^*B,x\rangle)\ . \qquad (4.1)$$

Conversely, we get only for symmetrically bi-affine functionals φ linear mappings such that the conjugations in 3.1. and 3.2. are identically.

The proof is given in [9] .
By the following theorem the relation between the generalized FENCHEL-conjugation and the F-conjugation is described.

Theorem 4.2. A generalized FENCHEL-conjugation with the coupling functional φ: $X\times Y \rightarrow R$ can be represented as F-conjugation if and only if $X=Y=R^n$ and the coupling functional φ is bijective and continuous in the variable x.

Proof. By definition of F-conjugation (cf. Example 3.4) for the upper F-conjugate of a function $f: R^n \rightarrow R$ holds

$$f^F(x^*) = \sup_{x \in D(f)} F^*(x, f(x), x^*).$$

Hence the upper F-conjugation can be represented as P-conjugation with the epigraphical generating functional

$$p^F(x, \zeta, x^*, \eta) := -F^*(x, \zeta, x^*) + \eta$$

(cf. Example 3.4).

According to Example 3.1 it is possible to represent the upper generalized FENCHEL-conjugation as upper P-conjugation with the epigraphical generating functional

$$p^{\varphi}(x, k, y, l) := -\varphi(x, y) \dotplus k + l.$$

Identifying $X = R^n$, $Y = R^n$ and correspondingly $x^* = y$, $\zeta = k$, $\eta = l$ and using Theorem 1.2 (ii) we have

$$-F^*(x, \zeta, x^*) + \eta = -\varphi(x, x^*) \dotplus \zeta + \eta, \quad (x, \zeta, x^*, \eta) \in R^n \times R \times R^n \times R .$$

Since $\zeta, \eta \in R$ we can replace the extended addition by the usual addition.

$F^*(x, \cdot, x^*)$ is defined as the inverse function of a continuous bijection $F(x, \cdot, x^*)$, hence F^* is a continuous bijection, too.

Analogously we can prove the assertion of the theorem for lower conjugations.//

By the following theorem the connection between F-conjugation and Φ-conjugation is given.

Theorem 4.3. Any F-conjugation (BEN TAL/BEN ISRAEL) can be represented as a Φ-conjugation (DEUMLICH/ELSTER) if and only if holds $L = R^n = L^*$, for the linear mapping A (cf. Example 3.2)

$$e = 0, \; b = 0, \; c \neq 0$$

and for a function $F^*: R^n \times R \times R^n \rightarrow R$ (cf. Example 3.4)

$$F^*(x, \zeta, x^*) = -\frac{1}{c}(a_0 + \langle a, x \rangle + \langle x^*, a \rangle + \langle x^* B, x \rangle + c\zeta).$$

Proof. Analogously to the proof of Theorem 4.2 we can (because of Theorem 1.2 (ii)) identify the epigraphical generating functionals p^{Φ} and p^{F} (cf. Example 3.2 and Example 3.4). One obtains under condition $R^n = L = L^*$ (correspondingly $z = \zeta$, $z^* = \eta$):

$$p^{\Phi}(x,z,x^*,z^*) = p^{F}(x,z,x^*,z^*)$$

which means

$$\frac{a_0 + \langle x^*,a\rangle + \langle x^*B,x\rangle + \langle a,x\rangle + \langle x^*,e\rangle + cz}{c + \langle e,x\rangle + bz} + z^* = -F^*(x,z,x^*) + z^*.$$

If $F^*(x,z,x^*)$ is continuous and decreasing in z (cf. Example 3.4) and is also a continuous bijection with respect to x and z, then the statement of the theorem follows for the upper conjugation.
The proof for the lower conjugation can be given analogously. //

Finally we prove an assertion concerning Φ-conjugation and level-set-conjugation.

Theorem 4.4. If $U = V = L = L^*$ and if L is a linear space with dim $L > 1$ (cf. Example 3.2) then there is one and only one level-set-conjugation (SINGER/VOLLE) which can be represented as Φ-conjugation (DEUMLICH/ELSTER). For the polarity $\Delta_V: 2^U \to 2^V$ corresponding to the level-set-conjugation (cf. Example 3.3) holds

$$\Delta_V(A) = \emptyset \text{ for all } A \subseteq L.$$

Proof. According to Example 3.3 each level-set-conjugation with respect to a polarity $\Delta_V: 2^U \to 2^V$ can be represented as generalized FENCHEL-conjugation with the coupling functional $\varphi_{\Delta_V}: U \times V \to \bar{R}$ with

$$\varphi_{\Delta_V}(u,v) := \begin{cases} 0 & \text{if } u \notin \Delta_V^*(\{v\}), \\ - & \text{if } u \in \Delta_V^*(\{v\}). \end{cases}$$

As in Example 3.1 there exists an epigraphical generating functional $p^{\varphi_{\Delta_V}}: U \times \bar{R} \times V \times \bar{R} \to R$ for the representation as upper P-conjugation.
According to Theorem 1.2 (ii) it is possible to identify

$p^{\varphi\Delta_v}$ with p^{Φ}, the epigraphical generating functional for the Φ-conjugation (cf. Example 3.2). By Theorem 4.1 for the representation as a generalized FENCHEL-conjugation the generating functional p^{Φ} has to be symmetrically bi-affine.

Under the sondition $U = V = L = L^*$, which is necessary for a comparison, the functional

$$p^{\varphi\Delta_v}(u,k,v,l) := -\varphi\Delta_v(u,v) \underset{\cdot}{+} z+z^*$$

can be symmetrically bi-affine only in that case, when it is zero on the whole domain.

For the lower conjugations the proof can be given analogously.

Remark 1: Under the assuption $U = V = L = X = X^* = R^n$ the treated relations between the conjugation concepts can be illustrated ($A \Longrightarrow B$ means: A is a special case of B):

FENCHEL-conjugation	$\Longrightarrow$	Φ-conjugation
$\Downarrow$		$\Downarrow$
generalized FENCHEL-conjugation	$\Longrightarrow$	P-conjugation
$\Uparrow$		$\Uparrow$
level-set-conjugation		F-conjugation

Remark 2: We have to take into consideration that a comparison for more restrictive classes of functions may give other assertions.

Remark 3: The generalized FENCHEL-conjugation is useful for dualities of the type f-g to $g_* - f^*$ ($f,g \in R^X$ and $g_*, f^* \in R^Y$) but not for fractional dualities such as $\frac{f}{g}$ to $\frac{g_*}{f^*}$ and others. Such types can be described by P-conjugations. By another paper we will give an example of conjugations which are convenient for fractional programming problems.

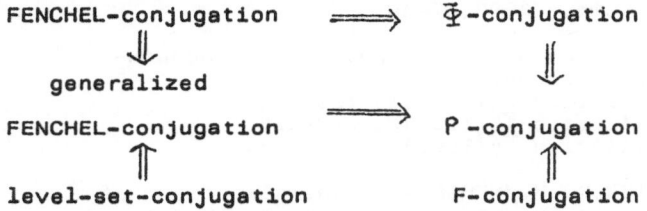

5. References

[1] BEN-TAL,A./BEN-ISRAEL,A.:F-convex Functions: Properties and Applications. In: Schaible,S./Ziemba,W.T. (eds.): Generalized Concavity in Optimization and Economics. Academie Press New York, San Francisco, London, 1981.

[2] DEUMLICH,R.: Ein geometrischer Zugang zur Optimierungstheorie auf der Grundlage der Φ-konjugierten Funktionen. Diss. B. Pädagogische Hochschule "N.K. Krupskaja" Halle 1980.

[3] DEUMLICH,R./ELSTER,K.-H.: On the theory of conjugate functions. Studies on Mathematical Programming. Akademiai Kiadó, Budapest, 1980, 19-43.

[4] DEUMLICH,R./ELSTER,K.-H.: Duality theorems and optimality conditions for nonconvex optimization problems. Math. Operationsforsch. Statist., Ser. Optimization 11(1980)2, 181-219.

[5] DEUMLICH,R./ELSTER,K.-H.: Recent results on Φ-conjugation and nonconvex optimization. In: Feichtinger,G. and P. Kall (eds.). Operations Research in Progress, D. Reidel Publ. Comp., 1982,27-40.

[6] DEUMLICH;R./ELSTER, K.-H.: Φ-Conjugation and nonconvex optimization. A survey.
Part I: optimization 14(1983)1, 125-149.
Part II: optimization 15(1984)4, 499-515.
Part III: optimization 16(1985)6, 789-803.

[7] DEUMLICH,R./ELSTER,K.-H./NEHSE,R.: Generalizations of conjugate functions. Survey of Math. Progr. (Proc. of the 9th. Intern. Math. Progr. Symp., Budapest 1976) Akademiai Kiadó, Budapest 1979, Vol. 1, 193-204.

[8] ELSTER,K.-H./NEHSE,R.: Zur Theorie der Polarfunktionale. Math. Operationsforsch. Statist., Ser. Optimization (1971), 3-21.

[9] ELSTER,K.-H./WOLF,A.: Recent Results on Generalized Conjugate Functions (To appear.).

[10] EWERS,J.J.K./VAN MAAREN,H.: Duality principles in Mathematics and their Relations to Conjugate Functions. Math. Communications. Vol. 16, No. 2, 1981, Dpt. of Applied Mathematics, Twente Univ. of Technology, Enschede, The Netherlands.

[11] MOREAU,J.J.: Fonctions convexes en dualité. Sem. Math. Fac. Sci., Montpellier, 1962.

[12] MOREAU,J.J.: Fonctionelles convexes. Seminaire sur les Equations aux Derivées Partielles II. Collège de France, Paris 1966-67.

[13] SINGER,J.: Conjugation operators. In: Hammer,G. and Pallaschke (eds.): Selected Topics in Operational Research and Mathematical Economics. Proc. 8th Symp. on OR, Karlsruhe, Aug. 22-25, 1983, Lecture Notes in Economics and Mathematical Systems. Springer Verlag, Berlin - Heidelberg - New York - Tokyo, 1984, 80-97.

[14] VOLLE,M.: Conjugation par tranches. Annali di Mathematica pura et applicata, 1985 (IV), Vol. CXXXIX, 279-312.

[15] WEISS,E.A.: Konjugierte Funktionen. Arch. Math. 20 (1969), 538-545.

OPTIMAL CONTROL WITH INITIAL STATE NOT A PRIORI GIVEN AND BOUNDARY CONDITION INVOLVING A DELAY

Adam Kowalewski
Institute of Automatic Control, University of Mining and Metallurgy, Al. Mickiewicza 30, 30-059 Cracow, Poland

1. Introduction

The purpose of this paper is to show the use of Milutin-Dubovicki's method in solving optimal control problems for distributed - parameter systems. As an example, an optimal control problem for the system described by a linear partial differential equation of parabolic type with boundary condition involving a constant time delay is considered. The initial condition of this equation is not given by a known function but it belongs to a certain set (the initial state is not a priori given). The right-hand side of this equation and the initial condition are not continuous functions usually, but they are measurable functions belonging to L^2 or L^∞ spaces. Therefore, the solution of this equation is given by a certain Sobolev space. The performance functional has the quadratic form. The control time is fixed. Finally, we impose some constraints on the control. Making use of Milutin-Dubovicki's theorem necessary and sufficient conditions of optimality with the quadratic performance functional and constrained control are derived for the Neumann problem. We also present a particular example in which the set of admissible controls and the one of initial conditions are given by means of the norm constraints. The application of the well-known projective gradient method [9] in the Hilbert space allows to obtain the numerical solution of our optimization problem.
This paper concerns the research programme P.R.I. 02/ASO-2.2.

2. Statement of optimal control problem and the optimization theorems

Now, we formulate the control problem for the system described by the following equation

$$\frac{\partial y}{\partial t} + A(t)y = u \qquad x \in \Omega,\ t \in (0,T) \tag{1}$$

$$\frac{\partial y}{\partial \eta_A} = \phi(x)\left\{ b(x,t)\ y(w(x),t-\tau) + v \right\} \qquad x \in \Gamma,\ t \in (0,T) \tag{2}$$

$$y(x,t') = g_0\ (x,t') \qquad x \in \Gamma,\ t' \in (-\tau,0) \tag{3}$$

$$y(x,0) \in K \qquad x \in \Omega \tag{4}$$

where : $\Omega \subset R^n$ - a bounded, open set with boundary Γ, which is a C^∞ manifold of dimension (n-1). Locally, Ω is totally on one side Γ.

$y \equiv y(x,t,v)$, $u \equiv u(x,t)$, $v \equiv v(x,t)$

$Q = \Omega \times (0,T)$, $\Sigma = \Gamma \times (0,T)$, $\Sigma_0^2 = \Gamma^2 \times (-\tau,0)$

ϕ - C^∞ function defined on Γ with compact support in Γ^1,

b - a real C^∞ function defined on Σ ,

w - a continuously differentiable bijection of Γ onto Γ with nonvanishing Jacobian J_w on Γ and having the following properties : $w(x) = x$ if $x \in \Gamma^2$ and $w(x) \in \Gamma^2$ if $x \in \Gamma^1$,

τ - a specified positive number representing a time delay,

K - a closed, convex subset with non-empty interior in the space $H^{1/2}(\Omega)$.

The operator A(t) has the form

$$A(t)y = -\sum_{i,j=1}^{n} \frac{\partial}{\partial x_i}\left(a_{ij}(x,t) \frac{\partial y(x,t)}{\partial x_j} \right)$$

and the functions $a_{ij}(x,t)$ satisfy the conditions

$$\sum_{i,j=1}^{n} a_{ij}(x,t)\, \xi_i\, \xi_j \geq \alpha \sum_{i=1}^{n} \xi_i^2 \ ;\ \alpha > 0,\ \forall (x,t) \in \overline{Q},\ \forall \xi_i \in R$$

where : $a_{ij}(x,t)$ - real C^∞ functions defined on $\overline{Q}$ (closure of Q)

It is easy to notice that the equations (1) ÷ (4) make the Neumann problem. The left-hand side of the Neumann boundary condition (2) is written in the following form :

$$\frac{\partial y}{\partial \eta_A} = \sum_{i,j=1}^{n} a_{ij}(x,t) \cos(n,x_i) \frac{\partial y(x,t)}{\partial x_j}$$

where : $\frac{\partial}{\partial \eta_A}$ is a normal derivative at Γ , directed towards the exterior of Ω, $\cos(n,x_i)$ is a i-th direction cosine of n, n-being the normal at Γ exterior to Ω.

The initial condition (4) is not given by a known function but it belongs to a certain set K (the initial state is not a priori given).

It is easy to notice that only $\tilde{g}_0$, the restriction of g_0 to $\Gamma^2 \times (-\tau, 0)$ is important in the mixed initial-boundary value problem (1) $\div$ (4). The equations (1) $\div$ (4) make in a linear approximation a universal mathematical model for the control of diffusion processes (e.g. plasma control) in which time-delayed feedback signals are introduced at the boundary of a system's spatial domain. Then the signal at the boundary of a system's spatial domain at any time depends on the signal which escaped earlier. This leads to boundary conditions involving time delays We also have some freedom in choosing the initial state of the controlled object. In this paper we shall consider the optimal boundary control problem i.e. $v \in L^2(\Sigma)$. Then the following result is fulfilled :

If the initial condition $y(x,0) \in K \subset H^{1/2}(\Omega)$ is an arbitrary fixed function, $u \in H^{-1/2,-1/4}(Q)$ and $\tilde{g}_0 \in L^2(\Sigma_0^2)$, then there exists a unique solution $y \in H^{3/2,3/4}(Q)$ (Theorem 2 [10]) for the mixed initial-boundary value problem (1) $\div$ (4). Moreover, $y(\cdot, j\tau) \in H^{1/2}(\Omega)$ for $j = 1,\ldots,N$. Let us denote by $Y = H^{3/2,3/4}(Q)$ the space of states and by $U = L^2(\Sigma)$ the space of controls. The control time T is fixed in our problem. The performance functional is given by

$$I(y,v) = \lambda_1 \int_Q |y(x,t,v) - z_d|^2 dx dt + \lambda_2 \int_0^T \int_{supp(\phi)} (Nv) v d\Gamma dt \qquad (5)$$

where : $\lambda_i \geq 0$ and $\lambda_1 + \lambda_2 > 0$; z_d is given element in $L^2(Q)$ and N is a strictly positive linear operator on $L^2(\Sigma)$ into $L^2(\Sigma)$.

We note from Theorem 2 [10] that for any $v \in U_{ad}$ the performance functional (5) is well defined since $y(v) \in H^{3/2,3/4}(Q)$.

We assume the following constraints on controls : $v \in U_{ad}$ is a closed, convex subset of U with non-empty interior in the space $L^2(\Sigma)$ (6)

We shall also denote by

$$y(0) \equiv y(x,0), \quad p(0) \equiv p(x,0), \quad y(T) \equiv y(x,T), \quad p(T) \equiv p(x,T)$$

The optimal control problem (1) $\div$ (6) will be solved as the optimization one in which the function v and the initial condition y(0) are the unknown functions.

It is easy to show that optimal control v^0 which suits optimal condition $y^0(0)$ gives the smallest value for the performance functional. Every other optimal control $v^{0'}$ which suits to initial condition $y(0) \in K$ gives the greater value for the performance functional. So proceeding in this way with the lack of additional information about the set K and the performance functional, we reach a certain compromise.

Making use of Milutin-Dubovicki's Theorem [3] we shall derive the necessary and sufficient conditions of optimality for the optimization problem (1) $\div$ (6). The idea of Milutin-Dubovicki's method was particularly described in [3], therefore we shall not present this method here. The solution of the stated optimal control problem is equivalent to seeking a triplet $(y^o(0), y^o, v^o) \in E = H^{1/2}(\Omega) \times H^{3/2,3/4}(Q) \times L^2(\Sigma)$ which satisfies equations (1) $\div$ (4) and minimizing the performance functional (5) with the constraints on controls (6).

We formulate the necessary and sufficient conditions of the optimality in the form of Theorem 1.

Theorem 1. The solution of the optimization problem (1) $\div$ (6) exists and it is unique with the assumptions mentioned above; the necessary and sufficient conditions of the optimality are characterized by the following system of partial differential equations and inequalities :

State equation

$$\frac{\partial y^o}{\partial t} + A(t)y^o = u \qquad x \in \Omega,\ t \in (0,T) \qquad (7)$$

$$\frac{\partial y^o}{\partial \eta_A} = \phi(x)\left\{b(x,t)y^o(w(x),t-\tau) + v^o\right\} \qquad x \in \Gamma,\ t \in (0,T) \qquad (8)$$

$$y^o(x,t') = \tilde{\Phi}_o(x,t') \qquad x \in \Gamma^2,\ t' \in (-\tau,0) \qquad (9)$$

Adjoint equation

$$-\frac{\partial p}{\partial t} + A^*(t)p = \lambda_1(y^o - z_d) \qquad x \in \Omega,\ t \in (0,T) \qquad (10)$$

$$\frac{\partial p}{\partial \eta_{A^*}} = 0 \qquad (11)$$

$$(x,t) \in ([\Gamma - w(\mathrm{supp}(\phi))] \times (0,T)) \cup (w(\mathrm{supp}(\phi)) \times (T-\tau,T))$$

$$\frac{\partial p}{\partial \eta_{A^*}} = \phi(w^{-1}(x))\, b(w^{-1}(x),t+\tau)\left|J_w(x)\right| p(w^{-1}(x),t+\tau) \qquad (12)$$

$$(x,t) \in w(\mathrm{supp}(\phi)) \times (0,T-\tau)$$

$$p(x,T) = 0 \qquad x \in \Omega \qquad (13)$$

Maximum conditions

$$\int_0^T \int_{\mathrm{supp}(\phi)} (p\phi(x) + \lambda_2 N v^o)(v - v^o)\, d\Gamma dt \geq 0 \qquad \forall v \in U_{ad} \qquad (14)$$

$$\int_\Omega p(0)\left[k - y^o(0)\right] dx \geq 0 \qquad \forall k \in K \qquad (15)$$

We can also remark that :

$$\left.\begin{aligned} \frac{\partial p}{\partial \eta_{A^*}} &= \sum_{i,j=1}^{n} a_{ij}(x,t)\cos(n,x_i)\frac{\partial p}{\partial x_j} \\ A^*(t)p &= -\sum_{i,j=1}^{n} \frac{\partial}{\partial x_j}\left(a_{ij}(x,t)\frac{\partial p}{\partial x_i}\right) \end{aligned}\right\} \tag{16}$$

Outline of the proof

According to Milutin-Dubovicki´s Theorem we approximate the set representing the inequality constraints by the regular admissible cone, the equality constraint by the regular tangent cone and the performance functional by the regular improvement cone.

a) Analysis of the equality constraint

The set Q_1 representing the equality constraint has the form

$$Q_1 = \left\{ (y(0),y,u) \in E;\ \begin{aligned} &\frac{\partial y}{\partial t} + A(t)y = u && x \in \Omega,\ t \in (0,T) \\ &\frac{\partial y}{\partial \eta_A} = \phi(x)\left\{b(x,t)y(w(x),t-\tau)+v\right\}, && x \in \Gamma,\ t \in (0,T) \\ &y(x,t') = \tilde{g}_0(x,t') && x \in \Gamma^2,\ t' \in (-\tau,0) \\ &y(x,0) = y(0) && x \in \Omega \end{aligned} \right. \tag{17}$$

We construct the regular tangent cone of the set Q_1 using the Lusternik Theorem (Theorem 9.1 [3]).

For this purpose we define the operator P in the form

$$P(y(0),y,v) = \left(\frac{\partial y}{\partial t} + Ay - u,\ \frac{\partial y}{\partial \eta_A} - \phi(x)\left\{b(x,t)y(w(x),t-\tau) + v\right\},\ y(x,t') - \tilde{g}_0(x,t'),\ y(x,0) - y(0)\right) \tag{18}$$

This operator P is the mapping from the space $H^{1/2}(\Omega) \times H^{3/2,3/4}(Q) \times L^2(\Sigma)$ into the space $H^{-1/2,-1/4}(Q) \times L^2(\Sigma) \times L^2(\Sigma_0^2) \times H^{1/2}(\Omega)$.

The Fréchet differential of the operator P can be written in the following form :

$$P'(y^0(0),y^0,v^0)(\bar{y}(0),\bar{y},\bar{v}) = \left(\frac{\partial \bar{y}}{\partial t} + A\bar{y},\ \frac{\partial \bar{y}}{\partial \eta_A} - \phi(x)\left\{b(x,t)\bar{y}(w(x),t-\tau)+\bar{v}\right\},\ \bar{y}(x,t'),\ \bar{y}(x,0) - \bar{y}(0)\right) \tag{19}$$

Really, $\frac{\partial}{\partial t}$ (Theorem 2.8 [6]), $A(t)$ (Theorem 2.1 [4]) and $\frac{\partial}{\partial \eta_A}$ (Theorem

2.1 [5] are linear and bounded mappings.
Using Theorem 2 [10] we can prove that P' is the operator "one to one" from the space $H^{1/2}(\Omega) \times H^{3/2,3/4}(Q) \times L^2(\Sigma)$ onto $H^{-1/2,-1/4}(Q) \times L^2(\Sigma) \times L^2(\Sigma_0^2) \times H^{1/2}(\Omega)$.
Considering that the assumptions of Lusternik's Theorem are fulfilled, we can write down the regular tangent cone for the set Q_1 in the point $(y^0(0),\ y^0, v^0)$ in the form

$$RTC(Q_1,(y^0(0),y^0,v^0)) = \left\{ (\bar{y}(0),\bar{y},\bar{v}) \in E;\ P'(y^0(0),y^0,v^0)(\bar{y}(0),\bar{y},\bar{v}) = 0 \right\} \tag{20}$$

It is easy to notice that it is a subspace. Therefore, using Theorem 10.1 [3] we know the form of the functional belonging to the adjoint cone

$$f_1\ (\bar{y}(0),\bar{y},\bar{v}) = 0 \quad \forall (\bar{y}(0),\bar{y},\bar{v}) \in RTC(Q_1,(y^0(0),y^0,v^0)) \tag{21}$$

b) Analysis of the constraint on controls
The set $Q_2 = K \times Y \times U_{ad}$ representing the inequality constraints is a closed and convex one with non-empty interior in the space E.
Using Theorem 10.5 [3] we find the functional belonging to the adjoint regular admissible cone, i.e.

$$f_2(\bar{y}(0),\bar{y},\bar{v}) \in \left[RAC(Q_2,(y^0(0),y^0,v^0) \right]^*$$

We can note if E_1, E_2, E_3 are three linear topological spaces, then the adjoint space to $E = E_1 \times E_2 \times E_3$ has the form

$$E^* = \left\{ f = (f_1,\ f_2,\ f_3)\ ,\ f_1 \in E_1^*\ ,\ f_2 \in E_2^*\ ,\ f_3 \in E_3^* \right\}$$

and

$$f(x) = f_1(x_1) + f_2(x_2) + f_3(x_3)$$

So we note the functional $f_2(\bar{y}(0),\bar{y},\bar{v})$ as follows

$$f_2(\bar{y}(0),\bar{y},\bar{v}) = f_1'''(\bar{y}) + f_2'(\bar{y}(0)) + f_2''(\bar{v}) \tag{22}$$

where :

$f_1'''(\bar{y}) = 0, \quad \forall y \in Y$ (Theorem 10.1 [3])

$f_2'(\bar{y}(0))$, $f_2''(\bar{v})$ are the support functionals to the sets K and U_{ad} in the points $y^0(0)$ and v^0, respectively (Theorem 10.5 [3]).

c) Analysis of the performance functional
Using Theorem 7.5 [3] we find the regular improvement cone of the performance functional (5)

$$RFC(I,(y^o(0),y^o,v^o)) = \left\{ (\bar{y}(0),\bar{y},\bar{v}) \in E,\ I'(y^o(0),y^o,v^o)(\bar{y}(0),\bar{y},\bar{v}) < 0 \right\} \tag{23}$$

where : $I'(y^o(0),y^o,v^o)(\bar{y}(0),\bar{y},\bar{v})$ is the Fréchet differential of the performance functional (5) and it can be written as

$$I'(y^o(0),,y^o,v^o)(\bar{y}(0),\bar{y},\bar{v}) = 2\lambda_1 \int_Q (y^o - z_d)\bar{y}\,dxdt + 2\lambda_2 \int_0^T \int_{supp(\phi)} (Nv^o)\bar{v}\,d\Gamma dt$$

On the basis of Theorem 10.2 [3] we find the functional belonging to the adjoint regular improvement cone, which has the form

$$f_3(\bar{y}(0),\bar{y},\bar{v}) = -\lambda_o\lambda_1 \int_Q (y^o - z_d)\bar{y}\,dxdt - \lambda_o\lambda_2 \int_0^T \int_{supp(\phi)} (Nv^o)\bar{v}\,d\Gamma\,dt \tag{24}$$

where : $\lambda_o > 0$

d) Analysis of Euler - Lagrange´s equation

The Euler - Lagrange equation for our optimization problem has the form

$$\sum_{i=1}^{3} f_i = 0 \tag{25}$$

Let $p(x,t)$ be the solution of (10) ÷ (13) for $v^o, y^o(0), y^o$ and denote by $\bar{y}$ the solution of $P'(\bar{y}(0),\bar{y},\bar{v}) = 0$ for any fixed $\bar{y}(0)$ and $\bar{v}$. Then taking into account (21),(22) and (24) we can express (25) in the form

$$f_2'(\bar{y}(0)) + f_2''(\bar{v}) = \lambda_o\lambda_1 \int_Q (y^o - z_d)\bar{y}\,dxdt + \lambda_o\lambda_2 \int_0^T \int_{supp(\phi)} (Nv^o)\bar{v}\,d\Gamma dt \tag{26}$$

$$\forall (\bar{y}(0),\bar{y},\bar{v}) \in RTC(Q_1(\bar{y}(0),\bar{y},\bar{v}))$$

We transform the first component of the right-hand side of (26) introducing the adjoint variable by equation (10) and using formulas (11),(12),(13),(19) and (20).

In turn , we get

$$\lambda_o\lambda_1 \int_Q (y^o - z_d)\bar{y}\,dxdt = \lambda_o \int_Q \left(-\frac{\partial p}{\partial t} + A^*p\right)\bar{y}\,dxdt =$$

$$= \lambda_o \int_\Omega p(0)\bar{y}(0)\,dx + \lambda_o \int_Q p\,\frac{\partial \bar{y}}{\partial t}\,dxdt + \lambda_o \int_Q A^*p\bar{y}\,dxdt \tag{27}$$

The last term in (27), in view of Green´s formula, can be rewritten as

$$\lambda_o \int_Q A^*p\bar{y}\,dxdt = \lambda_o \int_Q pA\bar{y}\,dxdt + \lambda_o \int_0^T \int_\Gamma p\,\frac{\partial \bar{y}}{\partial \eta_A}\,d\Gamma dt +$$

$$- \lambda_0 \int_0^T \int_\Gamma \frac{\partial p}{\partial \eta_{A^*}} \bar{y} \, d\Gamma \, dt \tag{28}$$

Using the boundary condition (2) , the second integral in the right-hand side of (28) can be expressed as

$$\lambda_0 \int_0^T \int_\Gamma p \frac{\partial \bar{y}}{\partial \eta_A} d\Gamma \, dt = \lambda_0 \int_0^T \int_{supp(\phi)} p(x,t)\phi(x)\left[b(x,t)\bar{y}(w(x),t-\tau)+\bar{v}\right] d\Gamma \, dt =$$

$$= \lambda_0 \int_{-\tau}^{T-\tau} \int_{supp(\phi)} p(x,t'+\tau)\phi(x)b(x,t'+\tau)\bar{y}(w(x),t')d\Gamma \, dt' + \lambda_0 \int_0^T \int_{supp(\phi)} p\phi(x)\bar{v} d\Gamma dt =$$

$$= \lambda_0 \int_{-\tau}^{T-\tau} \int_{w(supp(\phi))} p(w^{-1}(x),t'+\tau)\phi(w^{-1}(x))b(w^{-1}(x),t'+\tau)\bar{y}(x,t')\left|J_w(x)\right| d\Gamma dt' +$$

$$+ \lambda_0 \int_0^T \int_{supp(\phi)} p\phi(x)\bar{v} d\Gamma \, dt \tag{29}$$

The last term (28) can be written as

$$\lambda_0 \int_0^T \int_\Gamma \frac{\partial p}{\partial \eta_{A^*}} \bar{y} d\Gamma dt = \lambda_0 \int_0^T \int_{w(supp(\phi))} \frac{\partial p}{\partial \eta_{A^*}} \bar{y} d\Gamma dt + \lambda_0 \int_0^T \int_{\Gamma - w(supp(\phi))} \frac{\partial p}{\partial \eta_{A^*}} \bar{y} d\Gamma dt =$$

$$= \lambda_0 \int_0^{T-\tau} \int_{w(supp(\phi))} \frac{\partial p}{\partial \eta_{A^*}} \bar{y} d\Gamma dt + \lambda_0 \int_{T-\tau}^{T} \int_{w(supp(\phi))} \frac{\partial p}{\partial \eta_{A^*}} \bar{y} d\Gamma dt + \lambda_0 \int_0^T \int_{\Gamma - w(supp(\phi))} \frac{\partial p}{\partial \eta_{A^*}} \bar{y} d\Gamma dt \tag{30}$$

Substituting (29),(30) into (28) and then the result into (27) we have

$$\lambda_0 \lambda_1 \int_Q (y^0 - z_d)\bar{y} dx dt = \lambda_0 \int_\Omega p(0)\bar{y}(0)dx + \lambda_0 \int_Q p \frac{\partial \bar{y}}{\partial t} dx dt + \lambda_0 \int_Q pA\bar{y} dx dt +$$

$$- \lambda_0 \int_0^T \int_{\Gamma - w(supp(\phi))} \frac{\partial p}{\partial \eta_{A^*}} \bar{y} d\Gamma dt - \lambda_0 \int_{T-\tau}^{T} \int_{w(supp(\phi))} \frac{\partial p}{\partial \eta_{A^*}} \bar{y} d\Gamma dt - \lambda_0 \int_0^{T-\tau} \int_{w(supp(\phi))} \frac{\partial p}{\partial \eta_{A^*}} \bar{y} d\Gamma dt +$$

$$+ \lambda_0 \int_0^{T-\tau} \int_{w(supp(\phi))} \phi(w^{-1}(x))\left|J_w(x)\right| b(w^{-1}(x),t+\tau)p(w^{-1}(x),t+\tau)\bar{y}(x,t) d\Gamma dt +$$

$$+\lambda_0 \int_{-\tau}^{0} \int_{w(\mathrm{supp}(\phi))} \phi(w^{-1}(x)) \left| J_w(x) \right| b(w^{-1}(x),t+\tau) p(w^{-1}(x),t+\tau) \bar{y}(x,t) d\Gamma dt +$$

$$+\lambda_0 \int_0^T \int_{\mathrm{supp}(\phi)} p\phi(x)\bar{v} d\Gamma dt = \lambda_0 \int_\Omega p(0)\bar{y}(0)dx + \lambda_0 \int_Q p(\frac{\partial \bar{y}}{\partial t} + A\bar{y})dxdt +$$

$$+\lambda_0 \int_0^T \int_{\mathrm{supp}(\phi)} p\phi(x)\bar{v} d\Gamma dt = \lambda_0 \int_\Omega p(0)\bar{y}(0)dx + \lambda_0 \int_0^T \int_{\mathrm{supp}(\phi)} p\phi(x)\bar{v} d\Gamma dt \qquad (31)$$

Substituting (31) into (26) gives :

$$f_2'(\bar{y}(0)) + f_2''(\bar{v}) = \lambda_0 \int_\Omega p(0)\bar{y}(0)dx + \lambda_0 \int_0^T \int_{\mathrm{supp}(\phi)} (p\phi(x) + \lambda_2 Nv^o)\bar{v} d\Gamma dt \quad (32)$$

Using the definition of the support functional [3] and dividing both members of the obtained inequality by λ_0, we finally get

$$\int_\Omega p(0)\left[k_1 - y^o(0)\right] dx + \int_0^T \int_{\mathrm{supp}(\phi)} (p\phi(x) + \lambda_2 Nv^o)(v - v^o) d\Gamma dt \geq 0 \qquad (33)$$

$$\forall k \in K, \forall v \in U_{ad}$$

The last inequality is equivalent to the maximum conditions (14),(15).
In order to prove the sufficiency of the derived conditions of the optimality we use the fact that constraints and the performance functional are convex and there exists a point (Theorem 15.2 [3])

$(\tilde{y}(0),\tilde{y},\tilde{v}) \in \mathrm{int} Q_2$ such that $(\tilde{y}(0),\tilde{y},\tilde{v}) \in Q_1$ (Theorem 2 [10]).

This fact follows immediately from the existence of non-empty interior of the set Q_2 and from the existence of the solution of the equations (1) ÷ (4) as well.
This last remark finishes the proof of Theorem 1.
One may also consider analogous optimal control problem with the performance functional

$$\hat{I}(y,v) = \lambda_1 \int_\Sigma \left| y(v)\right|_\Sigma - z_{\Sigma d} \Big|^2 d\Gamma dt + \lambda_2 \int_0^T \int_{\mathrm{supp}(\phi)} (Nv)v d\Gamma dt \qquad (34)$$

where : $z_{\Sigma d}$ is a given element in $L^2(\Sigma)$

From Theorem 2 [10] and the trace theorem [5,p.9] for each $v \in L^2(\Sigma)$, there exists a unique solution $y \in H^{3/2,3/4}(Q)$ with $y|_\Sigma \in L^2(\Sigma)$. Thus $\hat{I}(y,v)$ is defined.
Then, the solution of the formulated optimal control problem is equivalent to seeking a triplet $(y^o(0),y^o,v^o) \in E = H^{1/2}(\Omega) \times H^{3/2,3/4}(Q) \times L^2(\Sigma)$ which satisfies equations (1)÷ (4) and minimizing the cost function (34) with the constraints on controls (6).
We can prove the following theorem :

Theorem 2.The solution of the optimization problem (1) ÷ (4),(34),(6) exists and it is unique with the assumptions mentioned above ; the necessary and sufficient conditions of the optimality are characterized by the following system of partial differential equations and inequalities: state equation (7) ÷ (9), adjoint equation

$$-\frac{\partial p}{\partial t} + A^*(t)p = 0 \qquad x \in \Omega \ , \ t \in (0,T) \tag{35}$$

$$\frac{\partial p}{\partial \eta_{A^*}} = \lambda_1(y^o|_{\Sigma} - z_{\Sigma d}) \tag{36}$$

$$(x,t) \in ([\Gamma - w(\mathrm{supp}(\phi))] \times (0,T)) \cup (w(\mathrm{supp}(\phi)) \times (T-\tau,T))$$

$$\frac{\partial p}{\partial \eta_{A^*}} = \phi(w^{-1}(x))h(w^{-1}(x),t+\tau)\left|J_w(x)\right|p(w^{-1}(x),t+\tau) + \lambda_1(y^o|_{\Sigma} - z_{\Sigma d})$$

$$(x,t) \in w(\mathrm{supp}(\phi)) \times (0,T-\tau) \tag{37}$$

$$p(x,T) = 0 \qquad x \in \Omega \tag{38}$$

Maximum conditions

$$\int_0^T \int_{\mathrm{supp}(\phi)} (p\phi(x) + \lambda_2 N v^o)(v - v^o)\, d\Gamma dt \geqslant 0 \qquad \forall v \in U_{ad} \tag{39}$$

$$\int_{\Omega} p(0)\left[k - y^o(0)\right] dx \geqslant 0 \qquad \forall k \in K \tag{40}$$

The idea of the proof of the Theorem 2 is the same as in the case of the Theorem 1.

We must notice that the optimal conditions derived above (Theorems 1,2) allow us to obtain an analytical formula for the optimal control in particular cases only (e.g. there are no constraints on controls). It results from the following : the determining of the function $p(x,t)$ in the maximum condition is possible from the adjoint equation, if and only if, we know that $y^o(x,t)$ will suit the control $v^o(x,t)$. These mutual connections make the practical usage of the derived optimization formulas difficult. Thus we resign from the exact determining of the optimal control and we use approximation methods.

In the case of non-coercive quadratic performance functional (in formulas (5) and (34), $N = 0$) the optimal control problem reduces to the minimizing of the functional on a closed and convex subset in a Hilbert space. Then the optimization problem is equivalent to a quadratic programming one which can be solved by the use of the well-known algorithms e.g. Gilbert's [2] , Barr's [1] or Nahi-Wheeler's [7] one. These problems will

be discussed elsewhere.
The application of the well-known projective gradient method [9] in the Hilbert space allows us to obtain the numerical solutions for particular cases of the optimization problems mentioned in the paper.

3. Projective gradient method

Let us assume that : V is a closed, convex and bounded subset in a Hilbert space

$I : V \to R^1$ is a functional belonging to $C^1(V)$.

We must find the point $q^0 \in V$, so that :

$$I(q^0) = \inf_{q \in V} I(q) \tag{41}$$

For this purpose we shall construct the sequence $\{q_n\}$ $(n = 0,1,....)$ according to the formula

$$q_{n+1} = P_V(q_n - \beta_n I'(q_n)) \tag{42}$$

where : P_V denotes the projective operator on the set V, $\beta_n > 0$ can be chosen using one of the method given in [9].
The proof of the weak convergence of the projective gradient method is given in [9]. To obtain the strong convergence one can use the method of regularization [9].

4. Certain example of optimal control problem

To illustrate practical applications of the method mentioned above we shall formulate the following control problem as an example.
We consider the parabolic equation (in which u = 0) describing the dynamics of a controlled system

$$\frac{\partial y}{\partial t} + A(t)y = 0 \qquad x \in \Omega,\ t \in (0,T) \tag{43}$$

$$\frac{\partial y}{\partial \eta_A} = y(x,t-\tau) + v \qquad x \in \Gamma,\ t \in (0,T) \tag{44}$$

$$y(x,t') = g_0(x,t') \qquad x \in \Gamma,\ t' \in (-\tau,0) \tag{45}$$

$$y(x,0) \in K \qquad x \in \Omega \tag{46}$$

It is easy to notice that the Neumann boundary condition (44) constitutes the simplest form of the more complex boundary condition (2); i.e. Γ^2 - is empty, $\phi(x) \equiv 1$ and $w(x) = x$ on Γ.
Then the performance functional (5) with $\lambda_1 = \lambda_2 = 1$ and $N = I$ can be written as

$$I(y,v) = \int_Q |y(x,t,v) - z_d|^2 dxdt + \int_\Sigma v^2 \, d\Gamma dt \longrightarrow \min \qquad (47)$$

Let us consider particular cases of the sets K and U_{ad} in which

$$K = \left\{ y(0) \in H^{1/2}(\Omega); \quad \int_\Omega y^2(0)dx \leq C_1^2 \, , \; C_1 \text{ is constant} \right\} \qquad (48)$$

$$U_{ad} = \left\{ v \in L^2(\Sigma); \quad \int_\Sigma v^2(x,t) \, d\Gamma dt \leq C_2^2 \, , \; C_2 \text{ is constant} \right\} \qquad (49)$$

Then the optimal control problem (43) ÷ (49) constitutes a particular case of the optimization problem (1) ÷ (6).
As it is possible to obtain the evident form of projective operator for K and U_{ad} given above, to get numerical solution of the optimal control problem (43) ÷ (49) one can use the well-known projective gradient method.
Let us denote

$$V = \left\{ (y(0),v) \in H^{1/2}(\Omega) \times L^2(\Sigma); \int_\Omega y^2(0)dx \leq C_1^2, \; \int_\Sigma v^2(x,t)d\Gamma dt \leq C_2^2 \right\} \qquad (50)$$

It is known that the space $H^{1/2}(\Omega) \times L^2(\Sigma)$ is a Hilbert one. Performing the same calculations as in the proof of the Theorem 1 we can see that

$$I'(y(0),v) = (p(0),p+v) \qquad (51)$$

Admitting that in the n - th iteration the control v is equal v_n, we get y_n as the solution of the following equation

$$\frac{\partial y_n}{\partial t} + A(t)y_n = 0 \qquad x \in \Omega, \; t \in (0,T) \qquad (52)$$

$$\frac{\partial y_n}{\partial \eta_A} = y_n(x,t-\tau) + v_n \qquad x \in \Gamma, \; t \in (0,T) \qquad (53)$$

$$y_n(x,t') = g_{no}(x,t') \qquad x \in \Gamma, \; t' \in (-\tau,0) \qquad (54)$$

with the initial condition $y_n(0)$
As $\phi(x) \equiv 1$, $w(x) = x$ and $|J_w(x)| = 1$, the adjoint equation has the form

$$-\frac{\partial p_n}{\partial t} + A^*(t)p_n = y_n - z_d \qquad x \in \Omega, \; t \in (0,T) \qquad (55)$$

$$\frac{\partial p_n}{\partial \eta_{A^*}} = 0 \qquad x \in \Gamma,\ t \in (T-\tau, T) \tag{56}$$

$$\frac{\partial p_n}{\partial \eta_{A^*}} = p_n(x, t+\tau) \qquad x \in \Gamma,\ t \in (0, T-\tau) \tag{57}$$

$$p_n(x,T) = 0 \qquad x \in \Omega \tag{58}$$

Knowing the n- th approximation $q_n = (y_n(0), v_n)$ we can find $q_{n+1} = (y_{n+1}(0), v_{n+1})$ using the projective gradient method. Taking into account the form of projective operator P_V on the set V [8], we get

$$y_{n+1}(0) = \begin{cases} y_n(0) - \beta_n p_n(0) & \text{if } \int_\Omega [y_n(0) - \beta_n p_n(0)]^2 dx \leq c_1^2 \\ \dfrac{c_1[y_n(0) - \beta_n p_n(0)]}{\left[\int_\Omega [y_n(0) - \beta_n p_n(0)]^2 dx\right]^{1/2}} & \text{otherwise} \end{cases} \tag{59}$$

$$v_{n+1} = \begin{cases} v_n - \beta_n(p_n + v_n) & \text{if } \int_\Sigma [v_n - \beta_n(p_n+v_n)]^2 d\Gamma dt \leq c_2^2 \\ \dfrac{c_2[v_n - \beta_n(p_n+v_n)]}{\left[\int_\Sigma [v_n - \beta_n(p_n+v_n)]^2 d\Gamma\, dt\right]^{1/2}} & \text{otherwise} \end{cases} \tag{60}$$

where : $\beta_n > 0$ can be chosen on the basis of [9] for instance

1) β_n may be calculated from the condition

$$f_n(\beta_n) = \inf_{\beta > 0} f_n(\beta)$$

$$f_n(\beta) = I\left[P_V((y_n(0), v_n) - \beta I'(y_n(0), v_n))\right]$$

2) β_n can be arbitrarily given as the sequence such that

$$\sum_{n=0}^{\infty} \beta_n = \infty, \qquad \sum_{n=0}^{\infty} \beta_n^2 < \infty$$

To solve equations (52) ÷ (58) one can use the convergent and stable difference method.

Finally, using the projective gradient method one may obtain the numerical solutions for particular case of the optimal control problem (1)÷÷(4)(6),(34).

5. Conclusions and perspectives

The obtained results are generalizations of the optimization theorems proved by Wang [10] with the initial condition given by a known function $y_0(x)$. Also the derived conditions of optimality are original from the point of view of applications Milutin-Dubovicki's Theorem [3] in solving optimal control problems for distributed - parameter systems. Making use of Milutin - Dubovicki's method the similar conditions of the optimality may be derived for the Dirichlet boundary condition involving a time delay.

The obtained optimization theorems demand the assumption dealing with the non-empty interior of the set Q_2 representing the inequality constraints. Therefore we approximate the set Q_2 by the regular admissible cone,(if int $Q_2 = \phi$ then this cone does not exist).

It is worth mentioning that the obtained results can be reinforced by omitting the asumption concerning the non-empty interior of the set Q_2 and utilizing the fact that the equality constraints in the form of the state equations are "decoupling" (i.e. we can define the linear, continuous mapping $R : U \rightarrow Y$). The optimal control problem reduces to seeking a couple $(y^0(0), v^0) \in Q_2'$ and minimizing the performance index $I(v)$. Then we approximate the set Q_2' representing the inequality constraints by the regular tangent cone and for the performance index $I(v)$ we construct the regular imorovement cone.

One may also derive the necessary and sufficient conditions of optimality for parabolic system with more complex boundary conditions involving multiple time delays.

We may also obtain estimates and a sufficient condition for the boundedness of solutions for parabolic systems with specified forms of feedback controls [10].

Finally, one may consider more complex optimization problems with non-differentiable and non-continuous performance functionals.

According to the author similar optimal control problems can be solved for hyperbolic and parabolic-hyperbolic systems.

The ideas mentioned above will be considered in the next papers.

References

[1] Barr, R. O., "On Efficient Computational Procedure for a Generalized Quadratic Programming Problem", SIAM J. Control, Vol. 7, 1969, pp. 415-429.

[2] Gilbert, E.S., "An Iterative Procedure for Computing the Minimum of a Quadratic Form on a Convex Set", SIAM J. Control, Vol. 4, 1966, pp. 61-80.

[3] Girsanov, I. V., "Lectures on the Mathematical Theory of Extremal Problems", Publication University of Moscow, Moscow 1970 (this monograph is written in English too).

[4] Lions, J. L., "Optimal Control of Systems Governed by Partial Differential Equations", Springer-Verlag, Berlin 1971.

[5] Lions, J. L. and Magenes, E., "Non-Homogeneous Boundary Value Problems and Applications", Springer-Verlag, Berlin, Vol. 2, 1972.

[6] Maslov, V. P., "Operators Methods", Moscow 1973 (in Russian).

[7] Nahi, N. E. and Wheeler, L. A., "Optimal Terminal Control of Continuous System via Successive Approximation of the Reachable Set", IEEE Trans. Automat. Control, Vol. AC - 12, 1967, pp. 515-521.

[8] Vasiljev, F.P., "On Gradient Methods for Solving Optimal Control Problems for Systems Described by Parabolic Equations", Moscow 1978 (in Russian).

[9] Vasiljev, F.P., "Methods of Solving Extremal Problems", Moscow 1981 (in Russian).

[10] Wang, P.K.C., "Optimal Control of Parabolic Systems with Boundary Conditions Involving Time Delays", SIAM J. Control, Vol. 13, 1975, pp. 274-293.

SENSITIVITY AND OPTIMAL CONTROL OF ELASTIC STRUCTURES WITH DISTRIBUTED PARAMETERS

Gwidon Szefer
Institute of Structures Mechanics, Technological University of Cracow
ul. Warszawska 24
31-155 Kraków, Poland

1. Introduction

The development of structural design and optimal control shows, that the sensitivity analysis constitutes an important part of investigations in this fiels. This fact follows from two reasons: firstly-the sensitivity operators play a decisive role in construction of optimality conditions, secondly - in design of structures /specially by CAD/ the sensitivity analysis leads always to more rational solutions /in comparison with the standard procedures/. There is a lot of results of optimization and sensitivity obtained in the last years for linear systems /see e.g. [1], [2], [4], [5] /. Few results only are known up to now for nonlinear structures [6], [7].
In the present paper a unified variational approach to sensitivity and optimal control of distributed parameter elastic structures will be given. A peculiar attention we devote to non linear systems loaded by configuration dependent forces. Small motion superimposed on finite /large/ deformation is considered too. To find the sensitivity operators, the adjoint system has been used.
Effective determination of the functional - state - and eigenvalue - sensitivity operators, constitutes the main result of the paper.
Owing to the variational formulation of the boundary value problem /BVP/ the optimal control can be find in a weaker class of regularity. The numerical advantage of the presented approach follows from the fact, that both - the primal as well the adjoint BVP may be solved by means of a one common computer program /using e.g. the finite element technique/.
The following denotations and nomenclature will be used:

Ω_R reference configuration of the body, $\Omega_R \subset R^n$ /n=1,2,3/
Γ_R boundary of Ω_R, $\Gamma_R = \Gamma_G \cup \Gamma_u$, $\Gamma_G \cap \Gamma_u = \phi$
X^K Lagrangean coordinates of the particle $\underset{\sim}{X} \in \Omega_R$

$\underset{\sim}{u} = (u_i)$	displacement
$\nabla \underset{\sim}{u} = (u_{i,K})$	displacement gradient
$\underset{\sim}{v} = (v_i)$	virtual displacemnet,
$\underset{\sim}{F} = (F_{iK})$	deformation gradient
$\underset{\sim}{T} = (T_{Ki})$	I Piola-Kirchhoff stress tensor
ϱ_R	mass density
s	control
$\underset{\sim}{b} = (b_i)$	external body force
$\underset{\sim}{p} = (p_i)$	surface traction
$\mathcal{U}$	set of controls
V	set of kinematical admissible displacements
$(,)$	derivative.

2. Statement of the problem

Let us consider an elastic body which in initial configuration occupies the region Ω_R . The body is loaded by configuration dependent forces $\underset{\sim}{b}(\underset{\sim}{X}, \underset{\sim}{u}, \nabla\underset{\sim}{u})$, $\underset{\sim}{p}(\underset{\sim}{X}, \underset{\sim}{u}, \nabla\underset{\sim}{u})$ and satisfies the kinematical boundary conditions $u_i(\underset{\sim}{X}) = g_i(\underset{\sim}{X})$ for $\underset{\sim}{X} \in \Gamma_u$. Let the constitutive equation of the material will be of the form

$$T_{Ki} = \mathcal{F}_{Ki}(s, \underset{\sim}{X}, \underset{\sim}{F}) \qquad /1/$$

Then, taking into consideration the relations

$$F_{iK} = \delta_{iK} + u_{i;K} \qquad /2/$$

$$T_{Ki}\, F_{jK} = T_{Kj}\, F_{iK} \qquad /3/$$

the variational formulation of the BVP for the body has the form

$$\int_{\Omega_R} T_{Ki}(s, \nabla\underset{\sim}{u})\, v_{i;K}\, d\Omega = \int_{\Omega_R} \varrho_R\, b_i(\underset{\sim}{X}, \underset{\sim}{u}, \nabla\underset{\sim}{u})\, v_i\, d\Omega + \int_{\Gamma_R} p_i(\underset{\sim}{X}, \underset{\sim}{u}, \nabla\underset{\sim}{u})\, v_i\, d\Gamma\, , \quad \forall \underset{\sim}{v} \in V_0 \qquad /4/$$

Here

$$V_0 = \left\{ \underset{\sim}{v} : \; v_i = 0 \quad \text{on } \Gamma_u \right\}$$

δ_{iK} means the Kroneckers symbol and the summation convention holds.

Formula /4/ expresses the well known principle of virtual displacements which in mechanical sense is equivalent to the equilibrium equations

$$T_{Ki,K} + \varrho_R b_i = 0 \qquad \text{in } \Omega_R$$
$$T_{Ki} N_K = p_i \qquad \text{on } \Gamma_R \qquad /5/$$

/N_K - means coordinates of the unit outside normal rector of the boundary Γ_R /

Introducing for simplicity the forms

$$B(s, \underline{u}, \underline{v}) = \int_{\Omega_R} T_{Ki} v_{i,K} \, d\Omega \qquad /6/$$

$$L(\underline{u}, \underline{v}) = \int_{\Omega_R} \varrho_R b_i v_i \, d\Omega + \int_{\Gamma_R} p_i v_i \, d\Gamma \qquad /7/$$

one can write /4/ shortly

$$B(s, \underline{u}, \underline{v}) = L(\underline{u}, \underline{v}) \qquad \forall \; \underline{v} \in V_0 \qquad /8/$$

which constitutes the state equation of the elastic distributed parameter system.

Constrains and the objective function of the system is described by functionals of the type

$$J_i(s, \underline{u}) = \int_{\Omega_R} \Phi_i(\underline{X}, s, \underline{u}, \nabla \underline{u}) \, d\Omega + \int_{\Gamma_R} f_i(\underline{X}, s, \underline{u}, \nabla \underline{u}) \, d\Gamma \qquad /9/$$

$$\xi = \xi(s) \qquad /10/$$

ξ - eigenvalue of the system

which represents all most interesting quantities met in mechanics /volume, energy, norm of displacements and stresses, cost, eigenvibration frequency, buckling force etc./

The functional /10/ is not explicitly given; it can be discussed by the analysis of the eigenvalue problem /EVP/ of the system, only. This problem will be formulate later.

Now, the problem of sensitivity is as follows:

find operators $A_u : \mathcal{U} \rightarrow V, \; A_i : \mathcal{U} \rightarrow \mathbb{R}, \; A_\xi : \mathcal{U} \rightarrow \mathbb{R}$

such that

$$\delta \underset{\sim}{u} = A_u \, \delta s$$
$$\delta J_i = A_i \, \delta s \qquad /11/$$
$$\delta \xi = A_\xi \, \delta s$$

where $\delta(\ )$ denotes variation of the given quantity.
A general optimal control problem for nonlinear statics has the form:
find $s \in \mathcal{U}$, $\underset{\sim}{u} \in V$ subject to
- the state equation /8/
- the constrains

$$J_i(s,\underset{\sim}{u}) = c_i \qquad i = 1,2,\dots,r \qquad c_i \in \mathbb{R}$$
$$J_i(s,\underset{\sim}{u}) \leq c_i \qquad i = r+1,\dots,m \qquad /12/$$
$$\xi \leq \xi_0$$

to achieve the minimum value of a given objective function

$$J_0(s,\underset{\sim}{u}) \longrightarrow \underset{s}{\mathrm{Min}} \qquad /13/$$

This problem as well the sensitivity analysis of the system can be solved by means of any version of the steepest descent procedure /see e.g. [3]/.
Therefore, the effective determination of the sensitivity operators /11/ is of decisive meaning.

3. Sensitivity operators

To find the sensitivity operators let us consider the change of control $s \rightarrow s+\delta s$ and their influence on J_i . Then assuming that the functional J_i is Gateaux differentiable it will be

$$\delta J_i = \int_{\Omega_R} \left(\frac{\partial \Phi_i}{\partial \underset{\sim}{u}} \delta \underset{\sim}{u}_s + \frac{\partial \Phi_i}{\partial \nabla \underset{\sim}{u}} \delta \nabla \underset{\sim}{u}_s \right) d\Omega + \int_{\Omega_R} \frac{\partial \Phi_i}{\partial s} \delta s \, d\Omega \; +$$
$$+ \int_{\Gamma_R} \left(\frac{\partial f_i}{\partial \underset{\sim}{u}} \delta \underset{\sim}{u}_s + \frac{\partial f_i}{\partial \nabla \underset{\sim}{u}} \delta \nabla \underset{\sim}{u}_s \right) d\Gamma + \int_{\Gamma_R} \frac{\partial f_i}{\partial s} \delta s \, d\Gamma \; = \qquad /14/$$
$$= \left\langle \frac{\delta J_i}{\delta s}, \delta s \right\rangle + \left\langle \frac{\delta J_i}{\delta \underset{\sim}{u}}, \delta \underset{\sim}{u}_s \right\rangle$$

where for simplicity the scalar product $\langle \cdot\cdot, \cdot\cdot \rangle$ in spaces U and V is introduced. Variation $\delta \underset{\sim}{u}_s$ means this increment of the state $\underset{\sim}{u}$ which is caused by the variation δs . Variation of /8/ yields

$$\left\langle \frac{\delta B(s,\underset{\sim}{u},\underset{\sim}{v})}{\delta s}, \delta s \right\rangle + \left\langle \frac{\delta B(s,\underset{\sim}{u},\underset{\sim}{v})}{\delta \underset{\sim}{u}}, \delta \underset{\sim}{u}_s \right\rangle = \left\langle \frac{\delta L(\underset{\sim}{u},\underset{\sim}{v})}{\delta \underset{\sim}{u}}, \delta \underset{\sim}{u}_s \right\rangle \qquad /15/$$

Defining the adjoint state $\underset{\sim}{v}$ by the equation

$$\underset{\sim}{v}: \left\langle \frac{\delta B(s,\underset{\sim}{u},\underset{\sim}{v})}{\delta \underset{\sim}{u}}, \delta \underset{\sim}{u}_s \right\rangle - \left\langle \frac{\delta L(\underset{\sim}{u},\underset{\sim}{v})}{\delta \underset{\sim}{u}}, \delta \underset{\sim}{u}_s \right\rangle = -\left\langle \frac{\delta J_i(s,\underset{\sim}{u})}{\delta \underset{\sim}{u}}, \delta \underset{\sim}{u}_s \right\rangle, \forall \delta \underset{\sim}{u}_s \qquad /16/$$

we obtain from /14/ and /15/

$$\delta J_i = \left\langle \frac{\delta J_i}{\delta s}, \delta s \right\rangle + \left\langle \frac{\delta B(s,\underset{\sim}{u},\underset{\sim}{v})}{\delta s}, \delta s \right\rangle =$$

$$= \int_{\Omega_R} \frac{\partial \Phi_i}{\partial s} \delta s \, d\Omega + \int_{\Gamma_R} \frac{\partial f_i}{\partial s} \delta s \, d\Gamma + \int_{\Omega_R} \frac{\partial T_{Ki}(s,\nabla\underset{\sim}{u})}{\partial s} v_{i,K} \delta s \, d\Omega \qquad /17/$$

Formula /17/ determines the sensitivity operator A_i.
Equation /16/ admits the explicite form

$$\int_{\Omega_R} \frac{\partial T_{Ki}(s,\nabla\underset{\sim}{u})}{\partial u_{j,L}} v_{i,K} \delta u^s_{j,L} \, d\Omega = \int_{\Omega_R} \left(\varrho_R \frac{\partial b_i}{\partial u_j} v_i \delta u^s_j + \varrho_R \frac{\partial b_i}{\partial u_{j,K}} v_i \delta u^s_{j,K} \right) d\Omega +$$

$$+ \int_{\Gamma_R} \left(\frac{\partial p_i}{\partial u_j} v_i \delta u^s_j + \frac{\partial p_i}{\partial u_{j,K}} v_i \delta u^s_{j,K} \right) d\Gamma + \int_{\Omega_R} \left(\frac{\partial \Phi_i}{\partial u_j} \delta u^s_j + \right.$$

$$\left. + \frac{\partial \Phi_i}{\partial u_{j,L}} \delta u^s_{j,L} \right) d\Omega + \int_{\Gamma_R} \left(\frac{\partial f_i}{\partial u_j} \delta u^s_j + \frac{\partial f_i}{\partial u_{j,L}} \delta u^s_{j,L} \right) d\Gamma$$

Denoting by

$$T^*_{Lj}(s, \nabla\underset{\sim}{u}, \underset{\sim}{v}) = \frac{\partial T_{Ki}}{\partial u_{j,L}} v_{i,K}$$

$$b^*_j(s, \underset{\sim}{u}, \nabla\underset{\sim}{u}, \underset{\sim}{v}) = \varrho_R \frac{\partial b_i}{\partial u_j} v_i - \frac{\partial \Phi_i}{\partial u_j}$$

$$
\begin{aligned}
T^{W}_{Lj}(s, \underline{u}, \nabla\underline{u}, \underline{v}) &= \rho_R \frac{\partial b_l}{\partial u_{j,L}} v_l - \frac{\partial \Phi_{\cdot}}{\partial u_{j,L}} \\
p^{*}_{j}(s, \underline{u}, \nabla\underline{u}, \underline{v}) &= \frac{\partial p_l}{\partial u_j} v_l - \frac{\partial f_{\cdot}}{\partial u_j} \\
p^{W}_{jL}(s, \underline{u}, \nabla\underline{u}, \underline{v}) &= \frac{\partial p_l}{\partial u_{j,L}} v_l - \frac{\partial f_{\cdot}}{\partial u_{j,L}}
\end{aligned} \qquad /18/
$$

one can write

$$
\begin{aligned}
\int_{\Omega_R} T^{*}_{Lj}(s, \nabla\underline{u}, \underline{v})\, \delta u^{s}_{j,L}\, d\Omega &= \int_{\Omega_R} \left[b^{*}_{j}(s, \underline{u}, \nabla\underline{u}, \underline{v})\, \delta u^{s}_{j} + T^{W}_{Lj}(s, \underline{u}, \nabla\underline{u}, \underline{v})\, \delta u^{s}_{j,L} \right] d\Omega \; + \qquad /19/ \\
&+ \int_{\Gamma_R} \left[p^{*}_{j}(s, \underline{u}, \nabla\underline{u}, \underline{v})\, \delta u^{s}_{j} + p^{W}_{jL}(s, \underline{u}, \nabla\underline{u}, \underline{v})\, \delta u^{s}_{j,L} \right] d\Gamma \\
&\forall\, \delta \underline{u}_s \in V_0
\end{aligned}
$$

The last term on the right hand side can be transformed into a more convenient form. To do this let us perform a coordinate mapping $\{X^K\} \longrightarrow \{\theta^{\alpha}\}$, $\alpha = \bar{1}, \bar{2}, \bar{3}$ where the orthogonal system $\{\theta^{\alpha}\}$ is introduced in such a way, that lines $\theta^{\bar{1}}, \theta^{\bar{2}}$ parametrizy the surface Γ_R whereas $\theta^{\bar{3}}$ coincides with the normal N /fig.1a/

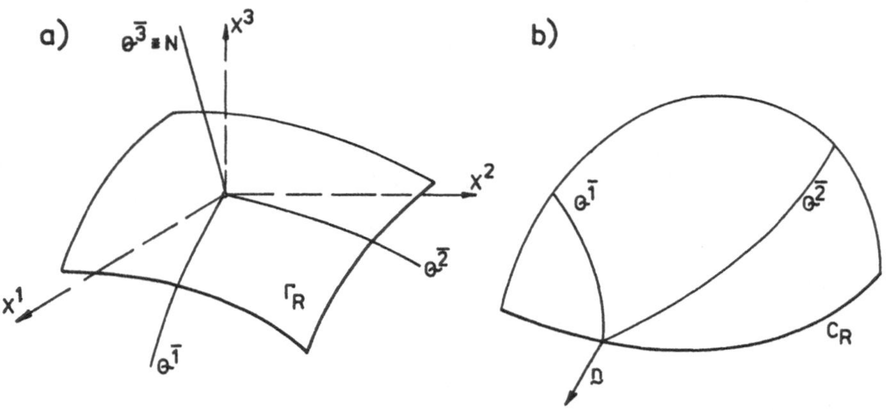

Fig.1

Then it will be

$$I = \int_{\Gamma_R} p^w_{jL} \, \delta u^s_{j,L} \, d\Gamma = \int_{\Gamma_R} G^{\alpha\beta} p^w_{j\alpha} \, \delta u_{j,\beta} \, d\Gamma = \int_{\Gamma_R} G^{\alpha\beta} p^w_{j\alpha} \underset{\alpha,\beta=\bar{1},\bar{2}}{\delta u_{j,\beta} \, d\Gamma} + \int_{\Gamma_R} G^{\bar{3}\bar{3}} p^w_{j\bar{3}} \, \delta u_{j,\bar{3}} \, d\Gamma \qquad /20/$$

where $G^{\alpha\beta}$ - means the metric tensor of $\{\theta^\alpha\}$.
Taking into account the assumption

$$p^w_{j\bar{3}} = \frac{\partial p_i}{\partial u_{j,\bar{3}}} v_i - \frac{\partial f_i}{\partial u_{j,\bar{3}}} = 0 \qquad /21/$$

/what means, that the loads $\tilde{p}$ as well the function f_i can not depend on the gradients $u_{j,\bar{3}}$ normal to Γ_R /
we have

$$I = \int_{\Gamma_R} G^{\alpha\beta} p^w_{j\alpha} \, \delta u^s_{j,\beta} \, d\Gamma = \int_{\Gamma_R} (G^{\alpha\beta} p^w_{j\alpha} \, \delta u^s_j)_{,\beta} \, d\Gamma - \int_{\Gamma_R} (G^{\alpha\beta} p^w_{j\alpha})_{,\beta} \, \delta u^s_j \, d\Gamma$$

$$j = 1,2,3 \qquad \alpha,\beta = \bar{1},\bar{2}$$

Using the Green - Gauss theorem we obtain /fig.1b/

$$I = \int_{C_R} G^{\alpha\beta} p^w_{j\alpha} n_\beta \, \delta u^s_j \, ds - \int_{\Gamma_R} (G^{\alpha\beta} p^w_{j\alpha})_{,\beta} \, \delta u^s_j \, d\Gamma$$

For a fixed contour C_R as well for the surface Γ_G which restricts the whole region Ω_R - the integral over C_R vanishes and there is

$$I = - \int_{\Gamma_R} (G^{\alpha\beta} p^w_{j\alpha})_{,\beta} \, \delta u^s_j \, d\Gamma \qquad /22/$$

Denoting further the quantity

$$\sigma^\beta_j = G^{\alpha\beta} p^w_{j\alpha} \qquad j = 1,2,3 \qquad \alpha,\beta = \bar{1},\bar{2} \qquad /23/$$

one can write /19/ in the final form

$$\int_{\Omega_R} T^*_{Lj} \, \delta u^s_{j,L} \, d\Omega = \int_{\Omega_R} b^*_j \, \delta u^s_j \, d\Omega + \int_{\Omega_R} T^w_{Lj} \, \delta u^s_{j,L} \, d\Omega + \qquad /24/$$

$$+\int_{\Gamma_R} (p_j^* - \sigma_{j,\beta}^{\beta})\, \delta u_j^s \, d\Gamma$$

or for similarity to /8/ shortly

$$B^*(s, \underset{\sim}{u}, \underset{\sim}{v}; \delta \underset{\sim}{u}_s) = L^*(s, \underset{\sim}{u}, \underset{\sim}{v}; \delta \underset{\sim}{u}_s) \qquad \forall\, \delta \underset{\sim}{u}_s \in V_0 \qquad /25/$$

where the forms B^* and L^* mean the left and the right hand side of eq./24/ respectively. It is easy to show, that the adjoint equation /24/ /or /25// is equivalent to the following ones

$$\begin{aligned} T^*_{Lj,L} - T^W_{Lj,L} + b^*_j &= 0 \qquad \text{in } \Omega_R \\ T^*_{Lj} N_L = p^*_j - \sigma^{\beta}_{j,\beta} + T^W_{Lj} N_L & \qquad \text{on } \Gamma_R \end{aligned} \qquad /26/$$

This result allows to state a simple and clear physical interpretation of the adjoint system, namely: the adjoint variable $\underset{\sim}{v} \in V_0$ determined by the variational BVP /24/ /virtual displacement principle for the adjoint system/ describes such a state of the system which is defined by a fictitious constitutive law

$$T^*_{Lj} = \frac{\partial T_{Ki}(s, \nabla \underset{\sim}{u})}{\partial u_{j,L}}\, v_{i,K}$$

subjected to fictitious loads $\underset{\sim}{b}^*$, $\underset{\sim}{p}^*$, initial stresses $\underset{\sim}{T}^W$ and additional boundary forces div $\underset{\sim}{\sigma}$ /see fig.2/

Having $\underset{\sim}{v}$, we effectively calculate the variation of J_i from /17/. To find the sensitivity operator A_u for the state $\underset{\sim}{u} \in V$ we can write

$$\underset{\sim}{u}: \qquad u_i(\underset{\sim}{X}) = J_i = \int_{\Omega_R} u_i(\underset{\sim}{Y})\, \delta(\underset{\sim}{X} - \underset{\sim}{Y})\, d\Omega_Y \qquad /27/$$

where $\delta(\underset{\sim}{X} - \underset{\sim}{Y})$ - is the Dirac delta - funtion

Hence

$$\Phi_i(\underset{\sim}{X}, \underset{\sim}{u}) = \delta(\underset{\sim}{X} - \underset{\sim}{Y})\, u_i(\underset{\sim}{Y}), \qquad f_i = 0$$

$$\frac{\partial \Phi_i}{\partial s} = 0, \quad \frac{\partial \Phi_i}{\partial u_i} = \delta(\underset{\sim}{X} - \underset{\sim}{Y})\, \delta u_i^s$$

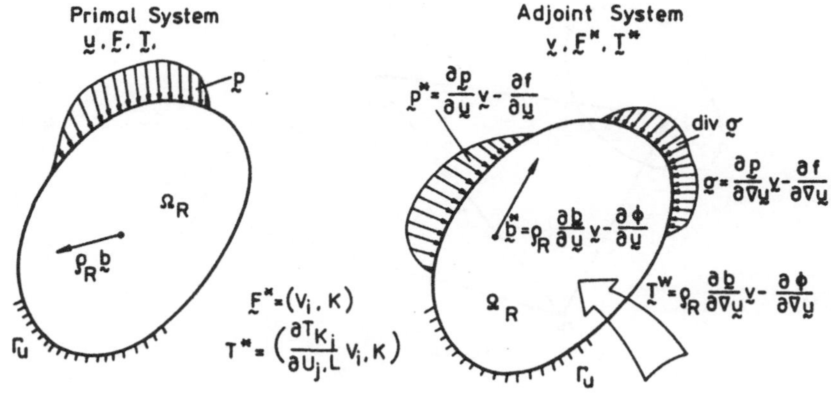

Fig.2

and the previons results can be applied. The adjoint equation for this case has the form

$$\underset{\sim}{v}: \quad \int_{\Omega_R} \frac{\partial T_{Ki}}{\partial u_{j,L}} v_{i,K}\, \delta u^s_{j,L}\, d\Omega = -\int_{\Omega_R} \delta(\underset{\sim}{X}-\underset{\sim}{Y})\, \delta u^s_i\, d\Omega \qquad /28/$$

From /17/ follows the variation of the state

$$\delta u_i(\underset{\sim}{X}) = \int_{\Omega_R} \frac{\partial T_{Kj}}{\partial s} v_{j,K}\, \delta s\, d\Omega \qquad /29/$$

It is worth noting that the adjoint equation /24/ is allways linear with respect to $\underset{\sim}{v}$.

4.Eigenvalue sensitivity

Many results of eigenvalue sensitivity analysis were obtained for linear distributed parameter systems. Eigenvalue sensitivity and optimal control problems in terms of finite deformations /nonlinear systems/ constitute a still open problems. In the present chapter we would like to discuss this kind of problem, limiting our considerations to the case of small vibrations superimposed on finite elastic deformation /fig.3/ and to simple eigenvalue only.

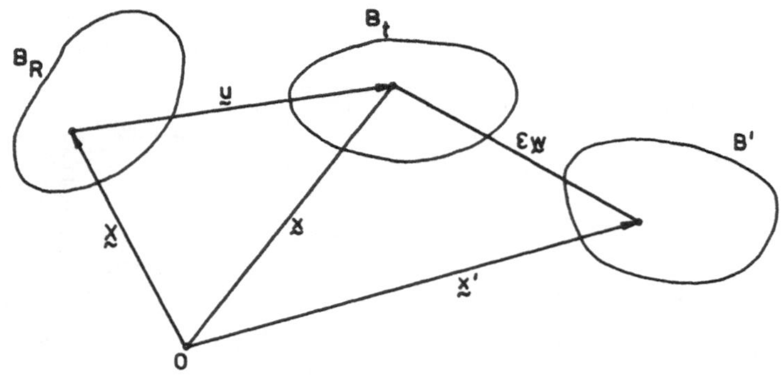

Fig.3

For such a situation the variational formulation of the EVP has the form

$$\int_{\Omega_R} T'_{Ki}(s,\nabla\underset{\sim}{u},\nabla\underset{\sim}{w})\, v_{i;K}\, d\Omega = \xi \int_{\Omega_R} k(s)\, w_i v_i\, d\Omega \tag{/30/}$$

where

$$T'_{Ki} = \frac{\partial T_{Ki}}{\partial u_{j,L}} w_{j,L} = A_{KijL}(s,\nabla\underset{\sim}{u})\, w_{j,L} \tag{/31/}$$

Denoting

$$\begin{aligned} \bar{B}(s,\underset{\sim}{u};\underset{\sim}{w},\underset{\sim}{v}) &= \int_{\Omega_R} T'_{Ki}\, v_{i;K}\, d\Omega \\ \bar{b}(s,\underset{\sim}{w},\underset{\sim}{v}) &= \int_{\Omega_R} k(s)\, w_i v_i\, d\Omega \end{aligned} \tag{/32/}$$

equation /30/ yields

$$\bar{B}(s,\underset{\sim}{u};\underset{\sim}{w},\underset{\sim}{v}) = \xi\, \bar{b}(s,\underset{\sim}{w},\underset{\sim}{v}) \qquad \forall\, \underset{\sim}{v} \in V_0 \tag{/33/}$$

To calculate the variation $\delta\xi$ we take $\underset{\sim}{v} = \underset{\sim}{w}$ and differentiate /33/. Then we have

$$< \frac{\delta \bar{B}(s,\underline{u};\underline{w},\underline{w})}{\delta s}, \delta s > + < \frac{\delta \bar{B}(s,\underline{u};\underline{w},\underline{w})}{\delta \underline{u}}, \delta \underline{u}_s > +$$

$$+ 2\bar{B}(s,\underline{u};\underline{w},\delta \underline{w}_s) = \xi \left[< \frac{\delta \bar{b}(s,\underline{w},\underline{w})}{\delta s}, \delta s > + 2\bar{b}(s,\underline{w},\delta \underline{w}_s) \right] +$$

$$+ \bar{b}(s,\underline{w},\underline{w})\, \delta \xi$$

and hence

$$\delta \xi = - \frac{1}{\bar{b}(s,\underline{w},\underline{w})} \left\{ < \frac{\delta \bar{B}(s,\underline{u};\underline{w},\underline{w})}{\delta s}, \delta s > + < \frac{\delta \bar{B}(s,\underline{u};\underline{w},\underline{w})}{\delta \underline{u}}, \delta \underline{u}_s > + \right.$$

$$\left. + \xi < \frac{\delta \bar{b}(s,\underline{w},\underline{w})}{\delta s}, \delta s > + 2 \left[\bar{B}(s,\underline{u};\underline{w},\delta \underline{w}_s) - \xi \bar{b}(s,\underline{w},\delta \underline{w}_s) \right] \right\}$$

The expression in the square brackets vanishes since /33/ must be satisfied. So it will be

$$\delta \xi = - \frac{1}{\bar{b}(s,\underline{w},\underline{w})} \left[< \frac{\delta \bar{B}(s,\underline{u};\underline{w},\underline{w})}{\delta s}, \delta s > + \xi < \frac{\delta \bar{b}(s,\underline{w},\underline{w})}{\delta s}, \delta s > + \right.$$

$$\left. + < \frac{\delta \bar{B}(s,\underline{u};\underline{w},\underline{w})}{\delta \underline{u}}, \delta \underline{u}_s > \right] \qquad /34/$$

To find the variation $\delta \underline{u}_s$, the finite deformation state B_t should be considered. Then it is

$$B(s,\underline{u},\underline{v}) = L(\underline{v}) \qquad \forall\, \underline{v} \in V_0 \qquad /35/$$

from where after variation results

$$< \frac{\delta B(s,\underline{u},\underline{v})}{\delta s}, \delta s > + < \frac{\delta B(s,\underline{u},\underline{v})}{\delta \underline{u}}, \delta \underline{u}_s > = 0 \qquad /36/$$

Definig now the adjoint state

$$\underline{v}: \quad < \frac{\delta B(s,\underline{u},\underline{v})}{\delta \underline{u}}, \delta \underline{u}_s > = < \frac{\delta \bar{B}(s,\underline{u};\underline{w},\underline{w})}{\delta \underline{u}}, \delta \underline{u}_s > \qquad /37/$$

we obtain from /36/

$$\left\langle \frac{\delta \overline{B}(s, \underset{\sim}{u}; \underset{\sim}{w}, \underset{\sim}{w})}{\delta \underset{\sim}{u}}, \delta \underset{\sim}{u}_s \right\rangle = - \left\langle \frac{\delta B(s, \underset{\sim}{u}, \underset{\sim}{v})}{\delta s}, \delta s \right\rangle$$

Hence, return to /34/ it will be finally

$$\delta \xi = \frac{1}{\overline{b}(s, \underset{\sim}{w}, \underset{\sim}{w})} \left[\left\langle \frac{\delta B(s, \underset{\sim}{u}, \underset{\sim}{v})}{\delta s}, \delta s \right\rangle - \left\langle \frac{\delta \overline{B}(s, \underset{\sim}{u}; \underset{\sim}{w}, \underset{\sim}{w})}{\delta s}, \delta s \right\rangle - \right.$$

$$\left. - \xi \left\langle \frac{\delta \overline{b}(s, \underset{\sim}{w}, \underset{\sim}{w})}{\delta s}, \delta s \right\rangle \right] =$$

$$= \frac{1}{\overline{b}(s, \underset{\sim}{w}, \underset{\sim}{w})} \int_{\Omega_R} \left[\frac{\partial T_{Ki}(s, \nabla \underset{\sim}{u})}{\partial s} v_{i,K} - \frac{\partial T'_{Ki}(s, \nabla \underset{\sim}{u}, \nabla \underset{\sim}{w})}{\partial s} w_{i,K} - \right. \qquad /38/$$

$$\left. - \xi \frac{dk}{ds} w_i w_i \right] \delta s \, d\Omega$$

This formula determines the sensitivity operator A_ξ provided that $\underset{\sim}{u}, \underset{\sim}{v}, \underset{\sim}{w}$ are known from /35/, /33/ and /37/ respectively.

5. Conclusions

In the paper a variational method of sensitivity analysis for distributed parameter structures was presented. Elastic systems described by nonlinear, nonpotential operators were taken into account. Using the notion of the adjoint system the sensitivity operators for constraints - and objective functionals as well for a simple eigenvalue of the system, were derived effectively. A clear and useful physical interpretation of the adjoint system in terms of configuration dependent loads was established. By means of finite dimensional approximation /FEM and steepest - descent procedure/ several optimal control problems for nonlinear elastic structures can be solved. One of them is given in [6]. Results obtained in the paper extend the possibilities and applications of the optimal computer aidded design /OCAD/.

References

1 Haug, E.J. „Review of Literature on Distributed Parameter Optimization Theory Related to Elastic Structural Optimization" Techn.Report N° 35, Materials Division, College of Engineering, The University of Iowa, Aug. 1977.

2 Haug, E.J., Arora, J.S. „Applied Optimal Design. Mechanical and Structural Systems" J.Wiley, New York, 1979.

3 Haug,E.J., Arora, J., Matsui,K. „A Steepest - Descent Method for Optimization of Mechanical Systems", JOTA, vol.19, Nº 3 July 1976.

4 Haug, E., Cea,J. /ed./„Optimization of Distributed Parameter Structures" Sijthoff-Noordhoff, E50,1981, vol.1,2.

5 Mróz, Z., Mironov,A.„Optimal Design for Global Mechanical Contraints" Arch.Mech. 32,4,1980.

6 Szefer, G., Demkowicz, L.„Optimal Design of von Karman Plates" J.Struct.Mech. 12,1,1984.

7 Szefer,G., Mróz,Z., Demkowicz, L.„Variational Approach to Sensitivity Analysis in Non-Linear Elasticity" /submitted to Arch. Mech./.

Computational Strategies for the Tension Parameters of the Exponential Spline

P. Rentrop, U. Wever
Fachbereich Mathematik, Universität Kaiserslautern
D - 6750 Kaiserslautern

Summary

Three different strategies to determine the tension parameters p_i of the exponential spline (or spline under tension) are discussed. A first heuristic strategy is based on the knowledge of the interpolating cubic spline and the p_i-values are proposed in order to eliminate undesired inflection points. Convexity or monotonicity of the interpolant cannot be guaranteed. A convexity preserving C^2 - interpolant (if possible) is constructed by solving a constrained nonlinear optimization problem for the tension parameters p_i. This second strategy characterizes an 'optimal' set of p_i-values. The optimization problem is the base to derive 'a priori' estimates for the p_i in a third strategy. The convexity arguments are supplemented by monotonicity constraints. The performance of all strategies is demonstrated in several examples.

1. Introduction

For the interpolation of given data points x_i, y_i, $i=0,\dots,n$ spline functions are a useful tool, see Ahlberg, Nilson, Walsh [1], Böhmer [3], Bulirsch, Rutishauser [4], De Boor [6]. However, in practical interpolation using cubic splines undesired oscillations can occur. In physical terms a cubic spline is interpreted as a tie rod, which links the given data points and is of minimal bending energy

$$\min. \int_{x_0}^{x_n} y''(x)^2 \, dx. \tag{1.1}$$

In order to avoid the undesired oscillations, one can apply tensile forces to the tie rod. Following Timoshenko [17], this leads to the minimal bending energy

$$\min. \int_{x_0}^{x_n} \left(y''(x)^2 + P(x)^2 y'(x)^2 \right) dx, \tag{1.2}$$

where $P(x)$ is proportional to the tensile forces in each interval: $P(x) = p_i$ for $x_i \leq x < x_{i+1}$, $i=0,\dots,n-1$.

The solution of the variational problem (1.2) is called the exponential spline or spline under tension, which was studied in Cline [5], Schweikert [14], Späth [15], Stoer, Bulirsch [16] and in [13]. The approximation properties of the exponential spline - a global C^2 - interpolant - are similar to those of the cubic spline and were discussed in Pruess [12].

Compared with the cubic spline the exponential spline has the important advantage that a 'good choice' of the tension parameters p_i leads to a shape preserving interpolant. This follows from the limit cases of the functional (1.2). For all p_i equal zero, one achieves the cubic spline; if all p_i are tending to infinity, one gets straight lines. In the latter case the C^2 - property is lost. An example with convex data points, which allows only a C^1 - interpolant is studied in Heß, Schmidt [9].

An open question is, how to choose the tension parameters p_i, so that the exponential spline is shape preserving. Values of p_i too large are unsuitable because they result in sharp edges of the interpolant. Three different techniques to estimate the tension parameters p_i are discussed in part 3. The numerical results and the plots are presented in part 4. In part 2 the necessary computational formulas of the exponential spline are prepared.

2. Computation of the Exponential Spline

The associated Euler-Lagrange differential equations of the variational problem (1.2) are:

$$\left(\frac{d^4}{dx^4} - p_i^2 \frac{d^2}{dx^2}\right) y(x) = 0 \quad \text{for } x_i \leq x \leq x_{i+1},\ i=0,\dots,n-1. \tag{2.1}$$

The fundamental system $\{1, x, \exp(p_i x), \exp(-p_i x)\}$ leads in each interval to the local representation of the exponential spline:

$$E(x) = y_{i+1}t + y_i(1-t) + \frac{d_{i+1}}{p_i^2}\left(\frac{\sinh(u_i t)}{\sinh(u_i)} - t\right) + \frac{d_i}{p_i^2}\left(\frac{\sinh(u_i(1-t))}{\sinh(u_i)} - (1-t)\right) \tag{2.2}$$

where: $x_i \leq x \leq x_{i+1}$, $h_i = x_{i+1} - x_i$, $t = \frac{1}{h_i}(x - x_i)$, $u_i = p_i h_i$, $i=0,\dots,n-1$

$y_i = E(x_i)$ the given data points

$d_i = E''(x_i)$ the unknown second derivatives

For given boundary conditions, i.e. $d_o = d_n = 0$, the second derivatives d_i are uniquely determined by the solution of the symmetric, positive definite tridiagonal system:

$$T\,d = b \tag{2.3}$$

where: $d = (d_1,\dots,d_{n-1})^T$, $b = (b_1,\dots,b_{n-1})^T$,

$$b_i = \frac{y_{i+1} - y_i}{h_i} - \frac{y_i - y_{i-1}}{h_{i-1}}, \quad u_i = p_i h_i,$$

$$q_i = \frac{u_i \cosh(u_i) - \sinh(u_i)}{u_i^2 \sinh(u_i)} h_i, \qquad r_{i+1} = \frac{\sinh(u_i) - u_i}{u_i^2 \sinh(u_i)} h_i,$$

$$T = \begin{bmatrix} q_1+q_0 & r_2 & & \\ r_2 & \ddots & \ddots & \\ & \ddots & \ddots & r_{n-1} \\ & & r_{n-1} & q_{n-1}+q_{n-2} \end{bmatrix} \quad \text{and } 0 < 2\, r_{i+1} < q_i .$$

The representation (2.2), (2.3) is numerically unstable in the limit cases $p_i \longrightarrow 0$ and $p_i \longrightarrow \infty$. A stable algorithm, which allows different types of boundary conditions is presented in [13].

3. Estimation of the Tension Parameters

Three different strategies for the estimation of the tension parameters p_i have been studied. In a first step the investigations are restricted to convex sets of data points.

First Strategy: Heuristic p_i - generator

This strategy was applied in Späth [15] and in [13]. In a first test run the cubic spline - that means all p_i are chosen as zero - is computed and the sign of the second derivative d_i of the cubic spline at x_i is compared with the sign of the second difference quotient - essentially the b_i of (2.3) - of the data. Different signs indicate undesired oscillations and a tension parameter $p_i \neq 0$ is chosen. From heuristic studies we have:

$$p_i h_i = \begin{cases} 15 & , y_i = y_{i+1} \\ 4 + \dfrac{1}{0.1 + \dfrac{|y_{i+1}-y_i|}{\max(|y_{i+1}|,|y_i|)}} & , \text{else} \end{cases} \tag{3.1}$$

Formula (3.1) reflects the observation, that the cubic spline leads to oscillations, if $y_i \simeq y_{i+1}$. If $y_i = y_{i+1}$, a reasonable choice seems to be $p_i h_i = 15$, in all other cases (3.1) gives values $4 < p_i h_i < 14$. For $p_i h_i \gg 15$ the interpolation curve may have sharp edges, for $0 < p_i h_i < 4$ no significant difference between the cubic and the exponential spline is noticed.

Strategy (3.1) suffers under the large amount of computation time, because both, the cubic spline and the exponential spline must be computed.

Second Strategy: Optimization Strategy

A convexity preserving interpolant is obtained, if the signs of the b_i and d_i in (2.3) are the same. In order to avoid sharp edges the tension parameters p_i should be as small as possible. These properties are reproduced by the minimization problem:

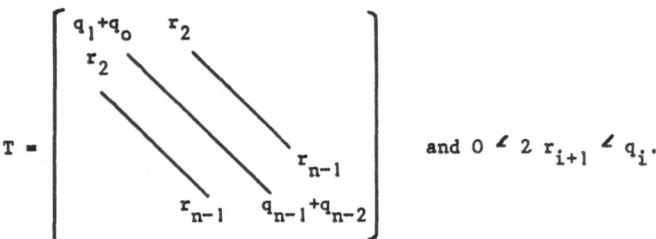

$$\min. (p_o,\dots,p_{n-1}) \text{ subject to}$$

$$T(p_o,\dots,p_{n-1})\, d = b \qquad \text{from (2.3)} \tag{3.2}$$

$$d_i b_i \geq 0 \qquad i=1,\dots,n-1.$$

Introducing the Euclidian norm of the tension parameters gives a scalar objective function. The standard nonlinear optimization problem with constraints is:

$$\min. \sum_{i=0}^{n-1} p_i^{\,2} \qquad \text{subject to}$$

$$T(p_o,\dots,p_{n-1})\, d = b \tag{3.3}$$

$$\text{if } b_i \neq 0 \text{ then } b_i d_i \geq 0$$

$$\text{else } -\text{eps} \leq d_i \leq \text{eps}.$$

Problem (3.3) is solved via the Sequential Quadratic Programming (SQP) method due to Powell [11]. The algorithm E04VDF from the NAG - Library [10], and Gill et al. [8] was used. Usually the algorithm E04VDF terminated successfully after 6 to 8 iterations. In the tested examples the algorithm was not sensitive to the initial guesses p_i.

Remark: The choice of the objective function in (3.3) is not a 'natural' one. In a forthcoming paper of Wever the tension parameters p_i in functional (1.2) are interpreted as control variables. Transformation of the functional (1.2) leads to the optimization problem:

$$\min. \Big\{ y_n' y_n'' - y_o' y_o'' + y_n \,(p_{n-1}^2 y_n' - y_n'''^{-}) - y_o \,(p_o^2 y_o' - y_o'''^{+}) + + \sum_{i=?}^{n-1} y_i \,(y_i' (p_{i-1}^2 - p_i^2) + y_i'''^{+} - y_i'''^{-}) \Big\}$$

subject to

$$\text{if } b_i \neq 0 \text{ then } b_i y_i'' \geq 0$$

$$\text{else } -\text{eps} \leq y_i'' \leq \text{eps}\ ,$$

where

$y_i \quad i=0,\dots,n$ the given data,

$$y_i' = \frac{y_{i+1} - y_i}{h_i} - d_i q_i - d_{i+1} r_{i+1} \qquad i=0,\dots,n-1,$$

$$y_n' = \frac{y_n - y_{n-1}}{h_{n-1}} + d_{n-1} r_n + d_n q_{n-1},$$

$y_i'' = d_i \quad i=0,\dots,n$ the unknown second derivatives,

$$y_i'''^{+} = \frac{d_{i+1} - d_i}{h_i} - p_i^{\,2}(d_{i+1} r_{i+1} + d_i q_i),$$

$$y_i'''^{-} = \frac{d_i - d_{i-1}}{h_{i-1}} + p_i^{\,2}(d_i q_{i-1} + d_{i-1} r_i).$$

For the solution of this nonlinear optimization problem, Wever uses again the SQP - method. The gradient of the objective function and the Jacobian of the constraints are calculated numerically. An optimization problem of this kind allows additional constraints of the interpolant like positivity. In our case the solutions of both optimization problems do not differ too much.

Third Strategy: Estimation Strategy

In practice the solution of an optimization problem (3.3) for large n is very expensive. Therefore some rough estimates for the p_i are derived.

The weaker requirement $d^T b \geq 0$ in (3.3) leads to a necessary condition for the matrix T of (2.3):

$T\,d = b \Longrightarrow d^T T\, b = d^T b \geq 0 \Longrightarrow$ T positive semidefinite;

The necessary condition for T is satisfied, because of $0 \leq 2r_{i+1} \leq q_i$.

Sufficient for $d_i b_i \geq 0$ $i=1,\ldots,n-1$ would be an increasing main diagonal of T, which can be achieved by increasing values of p_i.

An equivalent representation of r_{i+1} and q_i from (2.3) is:

$$q_i = \frac{1-(1-\exp(-2u_i))/u_i + \exp(-2u_i)}{u_i(1-\exp(-2u_i))}\, h_i, \quad u_i = p_i h_i,$$

$$r_{i+1} = \frac{(1-\exp(-2u_i))/u_i - 2\exp(-2u_i)}{u_i(1-\exp(-2u_i))}\, h_i, \quad i=0,\ldots,n-1. \tag{3.4}$$

For $u_i \geq 3$ holds $0 \leq \exp(-2u_i) \leq 2.5 * 10^{-3}$ and one may replace q_i by

$\hat{q}_i = \frac{u_i-1}{u_i^2} h_i$ and r_{i+1} by $\hat{r}_{i+1} = \frac{1}{u_i^2} h_i$. In the simplified tridiagonal system:

$$\hat{T}\,\hat{d} = b \tag{3.5}$$

holds $\hat{T}$ is positive definite , because

$2\hat{r}_{i+1} < \hat{q}_i \iff 2 < u_i - 1 \iff u_i > 3.$

The i-th. row of the simplified system (3.5) is:

$$\frac{h_{i-1}}{u_{i-1}^2}\hat{d}_{i-1} + \left(\frac{u_i-1}{u_i^2} h_i + \frac{u_{i-1}-1}{u_{i-1}^2} h_{i-1}\right)\hat{d}_i + \frac{h_i}{u_i^2}\hat{d}_{i+1} = b_i$$

One has $\mathrm{sign}(\hat{d}_i) = \mathrm{sign}(b_i)$ - as a sufficient condition - if:

$(u_i-1)|\hat{d}_i| \geq |\hat{d}_{i+1}|$ and $(u_{i-1}-1)|\hat{d}_i| \geq |\hat{d}_{i-1}|$.

Combining these inequalities from the i-th. and the (i+1)-th. row give:

$$\frac{|\hat{d}_{i+1}|}{u_i-1} \leq |\hat{d}_i| \leq (u_i-1)|\hat{d}_{i+1}|. \tag{3.6}$$

(3.6) essentially splits the three term recursion for the unknown $\hat{d}_i$ into a two term recursion. If the tension parameters p_i, respectively u_i, are large enough, the tri-diagonal system degenerates to a diagonal system. One has

$$\hat{d}_i \sim \frac{b_i}{\frac{u_i - 1}{u_i^2} h_i + \frac{u_{i-1} - 1}{u_{i-1}^2} h_{i-1}} , \qquad (3.7)$$

and finally with $p_i \sim p_{i-1}$:

$$\hat{d}_i \sim p_i b_i . \qquad (3.8)$$

Replacing the unknown second derivatives $\hat{d}_i$ in (3.6) by (3.8) - essentially the second order difference quotients - leads to a simple and cheap estimation for the unknown tension parameters:

$$\frac{|b_{i+1}|}{u_i - 1} \le |b_i| \le (u_i - 1)|b_{i+1}| . \qquad (3.9)$$

All presented strategies must fail, when the given data are of <u>mixed convex and concave</u> type. A typical situation is shown in Sketch 1.

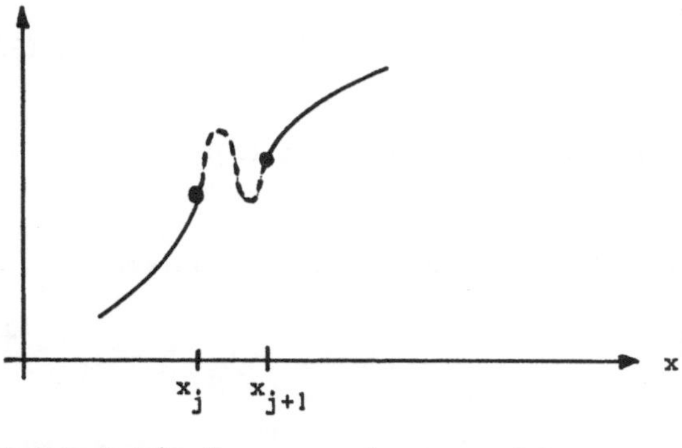

$x \le x_j$ convex data

$x_{j+1} \le x$ concave data

<u>Sketch 1:</u> Mixed convex and concave data

In practical interpolation two extremal values in the interval $x_j \le x \le x_{j+1}$ may occur, although the conditions $\text{sign}(b_j) = \text{sign}(d_j) > 0$ and $\text{sign}(b_{j+1}) = \text{sign}(d_{j+1}) < 0$ are satisfied. The first derivatives and the first order order difference quotients at x_j and x_{j+1} will give no additional information or warning to damp these oscillations. In this case one can only force the interpolant to be a monotone function.

For $u_i > 3$ a numerically stable representation of the exponential spline $E(x)$ from (2.2) is given by:

$$E(x) = y_{i+1}t + y_i(1-t) + \frac{d_{i+1}}{p_i^2}\left(\exp(-u_i(1-t)) \frac{1-\exp(-2u_i t)}{1-\exp(-2u_i)} - t \right) + \frac{d_i}{p_i^2}\left(\exp(-u_i t) \frac{1-\exp(-2u_i(1-t))}{1-\exp(-2u_i)} - (1-t) \right) \quad (3.10)$$

For $u_i \geq 3$ and in the inner interval $\left[x_i + \delta , x_{i+1} - \delta \right]$, $\delta \geq 0$ one may drop the exponential terms in the first derivative $E'(x)$ and obtain approximately:

$$E'(x) \sim \frac{y_{i+1}-y_i}{h_i} + \frac{d_i - d_{i+1}}{h_i p_i^2} \qquad x_i + \delta \leq x \leq x_{i+1} - \delta .$$

The interpolant is monotone, if the sign of $E'(x)$ is constant. This gives the additional condition:

$$\left| y_{i+1} - y_i \right| \geq \frac{1}{p_i^2} \left| d_{i+1} - d_i \right| \quad (3.11)$$

(3.11) can be used as an additional restriction in the optimization problem (3.3). In order to improve the heuristic generator (3.1) and the estimation strategy , the unknown d_i in (3.11) are replaced by (3.8). Further simplifications lead to:

$$p_i \geq \left| \frac{b_{i+1} - b_i}{y_{i+1} - y_i} \right| . \quad (3.12)$$

Remark: The heuristic generator (3.1) reflects the structure of (3.12).

As a first summary the complete 'a priori' <u>estimation procedure</u> is listed:

for $i=1,\dots,n-1$: $\quad b_i = \frac{y_{i+1}-y_i}{h_i} - \frac{y_i - y_{i-1}}{h_{i-1}}$ (3.13)

for $i=2,\dots,n-2$:

```
    if ( b_i = 0 or b_{i+1} = 0 ) then u_i = 30
    if ( b_i b_{i+1} > 0 ) then q = |b_i / b_{i+1}|
                                u_i = max ( q, 1/q ) + 1
                      elseif ( y_i = y_{i+1}) then u_i = 30
                                          else u_i = h_i abs( (b_{i+1}-b_i) / (y_{i+1}-y_i) )
    u_i = min (u_i, 30 )
    if ( u_i < 3 ) then u_i = 0
```

$u_1 = u_2$, $u_{n-1} = u_{n-2}$

4. Test Results

The three strategies are compared in typical examples, where the cubic spline interpolant works unsatisfactory. The Tables 1 till 3 show the data and the chosen tension parameters of all strategies. The plots belonging to Table 1 and Table 2 are presented in Figure 1 and Figure 2. The cubic spline and the exponential spline of all strategies are given. Table 3 includes two different data sets. Only the plots for the cubic spline and for the exponential spline with the estimation strategy are noticed in Figure 3. The two other strategies lead to comparable results.

Table 1: 8 data points, due to Späth [15].

x_i		-6	1	3	6	8	10	11	12
y_i		-2	2	3.5	3.5	2.8	-4	2.8	5
p_i	generator (3.1)	0	0	4.6	3.6	0	0	0	
	minimization (3.3)	0	0	1.8	4.0	0	0	0	
	estimation (3.13)	0	0	10	4.3	1.9	0	0	

Table 2: 11 data points, due to Akima [2], Fritsch, Carlson [7].

x_i		0	2	3	5	6	8	9	11	12	14	15
y_i		10	10	10	10	10	10	10.5	15	50	60	85
p_i	generator (3.1)	7	14	7	14	7	10	3.2	0	0	0	
	minimization (3.3)	0	0	0	0	0	0	4.9	0	5	10	
	estimation (3.13)	15	30	15	30	15	3.5	9.3	0	5	10	

Table 3: 7 data points with y(± 7) = A, due to Heß, Schmidt [9].

x_i		-7	-6	-5	0	5	6	7
y_i		A	1	0	0	0	1	A
A = 10								
p_i	generator (3.1)	4.9	4.9	3	3	4.9	4.9	
	minimization (3.3)	0	3.3	0	0	3.3	0	
	estimation (3.13)	8	8	6	6	8	8	
A = 2.1								
p_i	generator (3.1)	4.9	4.9	3	3	4.9	4.9	
	minimization (3.3)	0.2	8	20	20	8	0.2	
	estimation (3.13)	10	10	6	6	10	10	

Remark: For A = 2 only the straight line will preserve convexity of the given data.

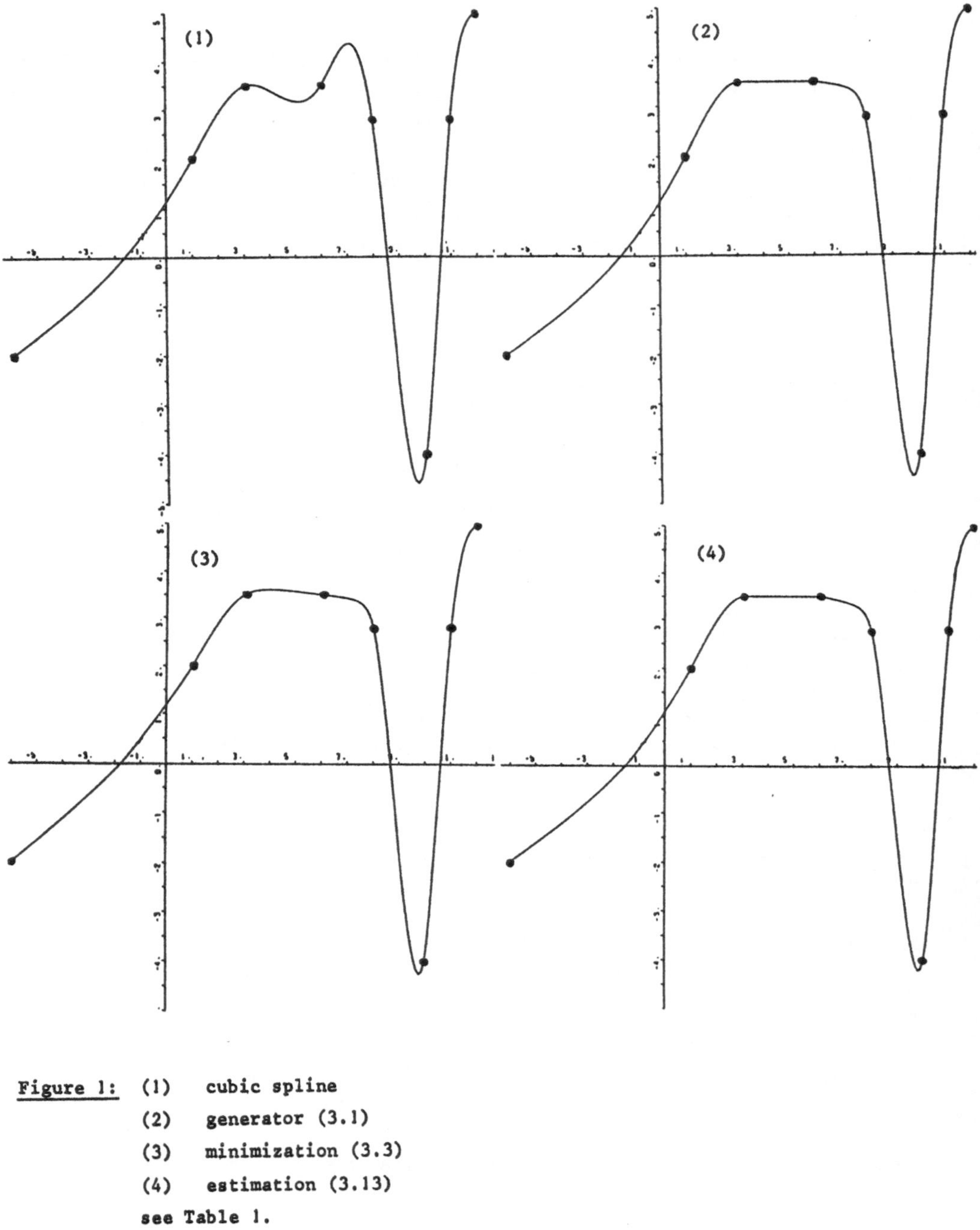

<u>Figure 1:</u> (1) cubic spline
(2) generator (3.1)
(3) minimization (3.3)
(4) estimation (3.13)
see Table 1.

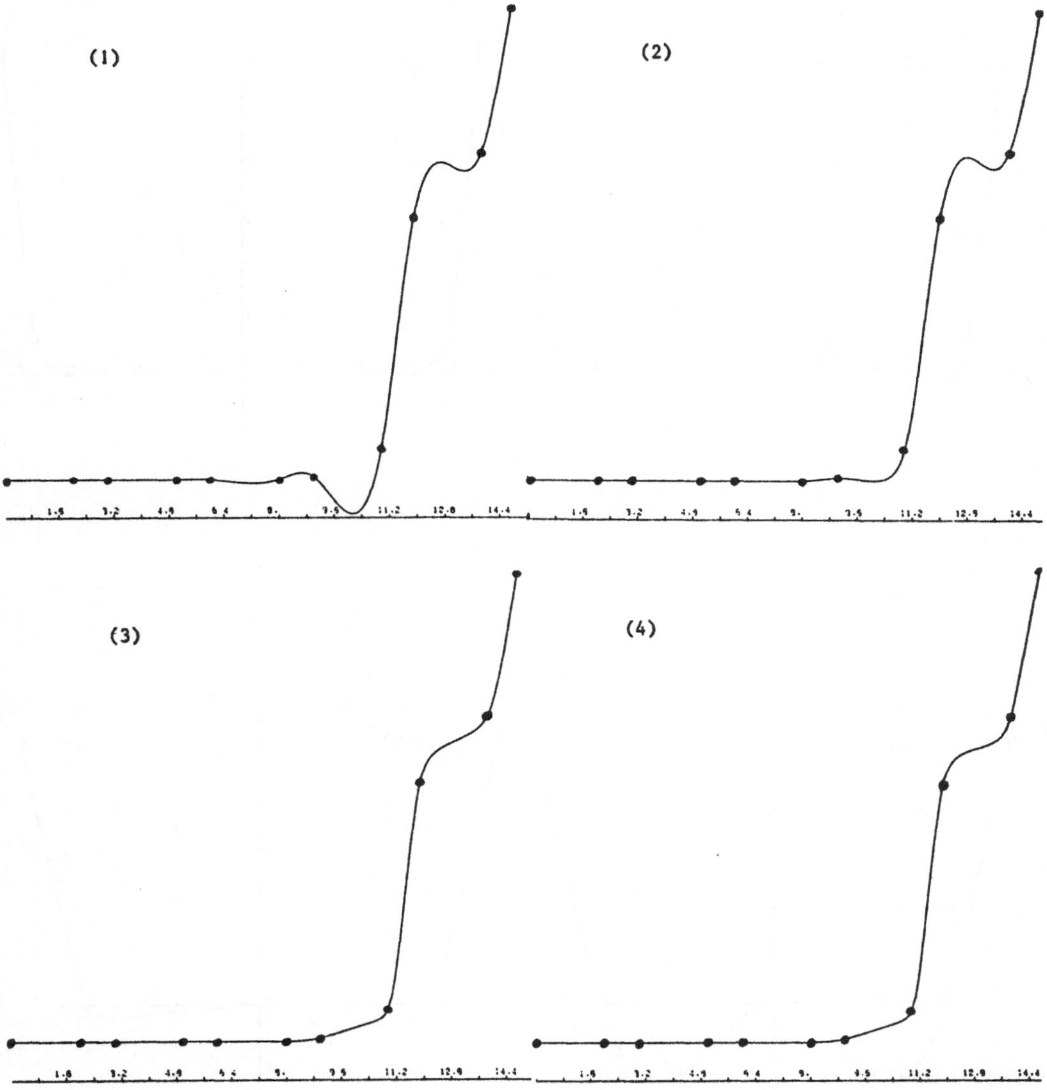

<u>Figure 2:</u> (1) cubic spline
(2) generator (3.1)
(3) minimization (3.3)
(4) estimation (3.13)
see Table 2.

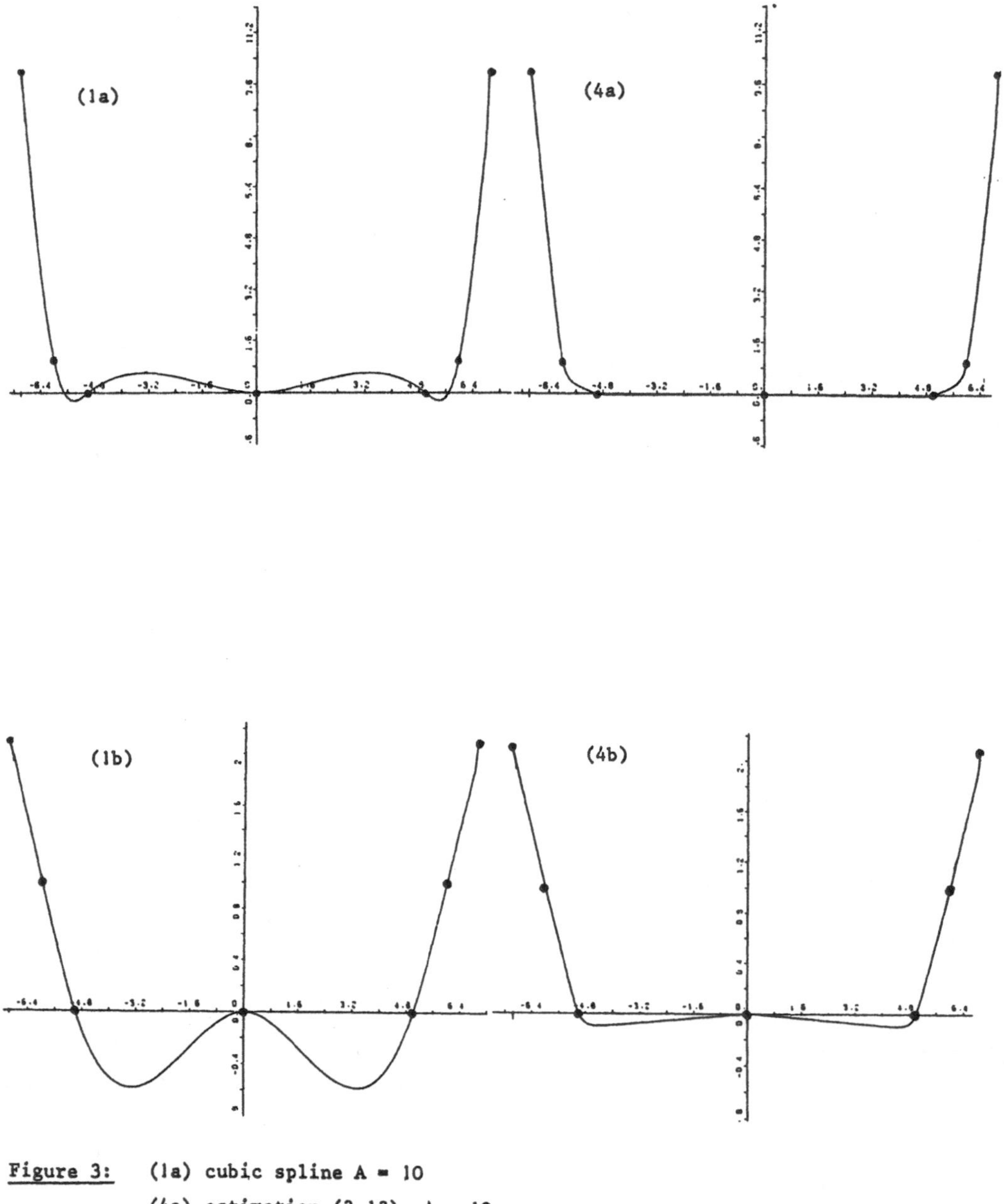

<u>Figure 3:</u> (1a) cubic spline A = 10

(4a) estimation (3.13), A = 10

(1b) cubic spline A = 2.1

(4b) estimation (3.13), A = 2.1

see Table 3.

5. Conclusion

Among the three tested strategies, the estimation strategy (3.13) is the cheapest one and works reliable. The heuristic generator strategy (3.1) enlarges the computing time by a factor two, because both, the cubic and the exponential spline must be computed. The minimization strategy (3.3) is too time consuming in general, but gives valuable hints for an 'optimal' tension parameter set. The proposed values p_i are remarkable smaller than in the other strategies.

An exponential spline algorithm coupled with a 'a priori' estimate of the tension parameters works much more satisfactory than the cubic spline and can be as easily implemented. Overall the computing time for the exponential spline is 30% larger than for the cubic spline. In our opinion, the good performance of the exponential spline justifies the additional amount of computing time.

All computations were performed in FORTRAN 77 single precision - 24 bit mantissa - on the SIEMENS 7051 of the Regionales Hochschulrechenzentrum Kaiserslautern (RHRK) under BS 2000. Usually the execution times for a data set are below one second.

Acknowledgement: The authors wish to thank R. Bulirsch for his stimulating interest. For the design of the exponential spline algorithm the procedure splex2 from the LRZ - Library, München was a valuable help. We thank C. Reinsch for the introduction to splex2. We are indebted to H. Neunzert and the Technomathematic group for their support and we thank H. Kraft, DFVLR Oberpfaffenhofen for helpful discussions.

References

[1] Ahlberg, J.H., Nilson, E.N., Walsh, J.L.: The theory of splines and their application. New York: Academic Press 1967

[2] Akima, H.: A new method for interpolation and smooth curve fitting based on local procedures. J. ACM 17 (1970) 589-602

[3] Böhmer, K.: Spline Funktionen. Stuttgart: Teubner Verlag 1974

[4] Bulirsch, R., Rutishauser, H.: Interpolation und genäherte Quadratur. In: Mathematische Hilfsmittel des Ingenieurs, Teil III, Hrsg.: R. Sauer, I.Szabo Berlin-Heidelberg-New York: Springer Verlag 1968

[5] Cline, A.: Scalar and planar-valued curve fitting in one and two-dimensional spaces using splines under tension. Comm. ACM 17 (1974) 218-223

[6] De Boor, C.: A practical guide to splines. Apl . Math. Sc. No. 27, Berlin-Heidelberg-New York: Springer 1978

[7] Fritsch, F.N., Carlson, R.E.: Monotone piecewise cubic interpolation. SIAM J. Numer. Anal. 17 (1980) 238-246

[8] Gill, P., Murray, W., Wright, M.: Practical optimization. New York: Academic Press 1981

[9] Heß, W., Schmidt, J.W.: Convexity preserving interpolation with exponential splines. Appears in Computing 1987

[10] NAG - FORTRAN Library, MK10 Febr. 1983

[11] Powell, M.J.D.: A fast algorithm for nonlinearly constrained optimization calculation. In: Numerical Analysis 144-157. Berlin-Heidelberg-New York: Springer 1978

[12] Pruess, S.: Properties of splines in tension. J. Approximation Theory 17 (1976) 86-96

[13] Rentrop, P.: An algorithm for the computation of the exponential spline. Handbook Series Approximation. Numer. Math. 35 (1980) 81-93

[14] Schweikert, D.G.: An interpolation curve using a spline in tension. J. Math. Phys. 45 (1966) 312-317

[15] Späth, H.: Exponential spline interpolation. Computing 4 (1969) 225-233

[16] Stoer, J., Bulirsch, R.: Introduction to numerical analysis. Berlin-Heidelberg-New York: Springer 1980

[17] Timoshenko, S.: Strength of materials, part II. New York: Van Nostrand 1951

AIRCRAFT TRAJECTORY CONTROL

AIRCRAFT MINIMUM TIME-TO-CLIMB MODEL COMPARISON

Bion L. Pierson and Shaw Y. Ong
Department of Aerospace Engineering
Iowa State University
Ames, Iowa 50011, U.S.A.

1. Introduction

The aircraft minimum time-to-climb problem has been treated by many investigators [1-10]. But many of these studies have been concerned primarily with the development of improved numerical techniques for solving the minimum time-to-climb problem. Often, various numerical methods are applied to a specific aircraft dynamic model, and the corresponding results are compared to reveal the relative advantages and disadvantages of the methods.

However, many modeling questions arise. For example, is the choice of a particular dynamic model a "good" one or not? What happens if another dynamic model is used? Will a two-state model produce nearly the same results as a three-state model? What are the tradeoffs between model choice and computational expense? These are just a few of the many unanswered questions. In the past, it was sufficiently difficult to obtain numerical solutions (say, with a "continuous" gradient method) that alternate model choices were often left unexamined. Now, with the aid of advanced nonlinear programming codes, it has become feasible to examine numerically a whole range of related models for each class of problems.

Ardema [6] has compared the results of three different models: energy-state, two-state, and a modified two-state. He concludes that thrust and weight influence the time-to-climb most strongly and that the modified two-state model is significantly better than the other two. However, no comparable numerical results are presented for higher-order models. Thus, it is likely that other factors may be present that influence significantly the minimum time-to-climb. Other related studies include those by Bryson, Desai and Hoffman [4] and Rader and Hull [7].

In this paper, five dynamic models are examined. The models used range from the simple energy-state model to the complete point-mass model with five state variables. The primary objective is to compare the solutions for each of these five models with regard to accuracy and computational effort. Numerical results are presented for an early represention of the F-4 fighter aircraft.

2. Problem Formulation

We want to find the flight path of a supersonic aircraft for a minimum-time climb from a given initial state to a given final state. Five dynamic models are examined. They range from a 5-state point-mass model to a single energy-state model. Since it can be shown that these minimum-time trajectories occur for maximum thrust, each model involves only a scalar "steering" control function. The maximum thrust magnitude $T(V,h)$ is a specified function of speed V and altitude h and therefore is not regarded as a control function. In addition, we assume that the thrust force is directed along the (zero-lift) longitudinal axis of the aircraft.

2.1 Model 1

This model consists of the usual 4th-order, point-mass, translational equations of motion for planar flight in still air over a flat earth plus a mass rate equation to account for the consumption of fuel [4]. No rotational dynamics are present. The control function is angle of attack α, and the five state variables are: speed V, flight path angle γ, altitude h, horizontal range x, and mass m. The optimal control problem for this model can be stated as:

Find the angle of attack time history $\alpha(t)$, $0 \le t \le t_f$, which minimizes

$$J = t_f \tag{1}$$

subject to

$$m\dot{V} = T(V,h)\cos\alpha - D(V,h,\alpha) - m g \sin\gamma \tag{2}$$

$$m V \dot{\gamma} = T(V,h)\sin\alpha + L(V,h,\alpha) - m g \cos\gamma \tag{3}$$

$$\dot{h} = V \sin\gamma \tag{4}$$

$$\dot{x} = V \cos\gamma \tag{5}$$

$$\dot{m} = - f(V,h) \tag{6}$$

and the boundary conditions $V(0) = V_o$, $\gamma(0) = \gamma_o$, $h(0) = h_o$, $x(0) = 0$, $m(0) = m_o$, $V(t_f) = V_f$, $h(t_f) = h_f$.

We assume that the fuel consumption rate f is proportional to thrust so that

$$f(V,h) = T(V,h)/cg \tag{7}$$

where $c = 1600$ sec and $g = 32.174$ ft/sec^2. Also, the aerodynamic lift and drag forces are given by

$$L = q S C_L \quad \text{and} \quad D = q S C_D \tag{8}$$

respectively, where $q = \rho V^2/2$ is the dynamic pressure, ρ is the atmospheric density, S is the aircraft planform area, and C_L and C_D are the nondimensional lift and drag coefficients, respectively. We will adopt the usual "linear lift curve"

$$C_L = C_{L_\alpha} \alpha \tag{9}$$

and "parabolic drag polar"

$$C_D = C_{D_0} + \beta C_L^2 \tag{10}$$

for these coefficients. In general, the parameters C_{L_α}, C_{D_0}, and $\eta = \beta C_{L_\alpha}$ depend on Mach number, $M = V/a$, where a is the local speed of sound. Each of these three empirical aerodynamic parameters is assumed to be available in the form of an analytical function of M. Finally, both ρ and a depend on the altitude h. A standard "exponential atmosphere"

$$\rho(h) = \rho_0 e^{-h/h_1} \tag{11}$$

is used for the density model, and the function

$$a(h) = \begin{cases} (k_1 - k_2 h)^{1/2}, & h \le 36{,}000 \text{ ft} \\ 968.1 \text{ ft/sec}, & h > 36{,}000 \text{ ft} \end{cases} \tag{12}$$

is used for the speed of sound where $\rho_0 = 2.54(10^{-3})$ slug/ft^3, $h_1 = 2.73(10^4)$ ft, $k_1 = 1.244(10^6)$ ft^2/sec^2, and $k_2 = 8.57$ ft/sec^2 [11].

2.2 Model 2

In this and the remaining models, range is used to replace time as the independent variable. The range equation (5) can be used to modify both the performance index and the remaining state equations. The result is a point-mass model with only four states. Thus, we have the following minimum-time problem.

Find the angle of attack range history $\alpha(x)$, $0 \le x \le x_f$, which minimizes

$$J = \int_0^{x_f} \frac{1}{V \cos \gamma} \, dx \tag{13}$$

subject to

$$V' = \frac{T(V,h) \cos \alpha - D(V,h,\alpha) - m g \sin \gamma}{m V \cos \gamma} \tag{14}$$

$$\gamma' = \frac{T(V,h) \sin \alpha + L(V,h,\alpha) - m g \cos \gamma}{m V^2 \cos \gamma} \tag{15}$$

$$h' = \tan \gamma \tag{16}$$

$$m' = - \frac{f(V,h)}{V \cos \gamma} \tag{17}$$

and $V(0) = V_0$, $\gamma(0) = \gamma_0$, $h(0) = h_0$, $m(0) = m_0$, $V(x_f) = V_f$, $h(x_f) = h_f$.

The primes denote derivatives taken with respect to range. Models 1 and 2 are basically the same. The only difference is that Model 1 uses time rather than range as the independent variable.

2.3 Model 3

The previous four-state model is now reduced to a three-state model by neglecting the mass differential equation and by instead approximating the mass history with a linear function of range.

$$m(x) = c_1 x + m_0 \tag{18}$$

Here, the slope c_1 is computed from the solution values for x_f and $m(x_f)$ obtained from Model 1. The problem now becomes:

Find the angle of attack history $\alpha(x)$, $0 \le x \le x_f$, which minimizes (13)

subject to

$$V' = \frac{T(V,h)\cos\alpha - D(V,h,\alpha) - m(x)\,g\sin\gamma}{m(x)\,V\cos\gamma} \tag{19}$$

$$\gamma' = \frac{T(V,h)\sin\alpha + L(V,h,\alpha) - m(x)\,g\cos\gamma}{m(x)\,V^2\cos\gamma} \tag{20}$$

$$h' = \tan\gamma \tag{21}$$

and $V(0) = V_o$, $\gamma(0) = \gamma_o$, $h(0) = h_o$, $V(x_f) = V_f$, $h(x_f) = h_f$.

2.4 Model 4

We now assume: (i) small angle of attack so that $\sin\alpha \cong 0$ and $\cos\alpha \cong 1$, and (ii) small flight path angle range rate γ'. Therefore, equation (20) reduces to

$$0 = L - m(x)\,g\cos\gamma \tag{22}$$

By combining (8), (10), and (22), we obtain the following expression for the drag.

$$D(V,h,x,\gamma) = q\,S\left\{ C_{D_O} + \beta\left[\frac{m(x)\,g\cos\gamma}{qS}\right]^2 \right\} \tag{23}$$

Flight path angle now plays the role of the control and the problem becomes:

Find the flight path angle history $\gamma(x)$, $0 \le x \le x_f$, which minimizes (13)

subject to

$$V' = \frac{T(V,h) - D(V,h,x,\gamma) - m(x)\,g\sin\gamma}{m(x)\,V\cos\gamma} \tag{24}$$

$$h' = \tan\gamma \tag{25}$$

and $V(0) = V_o$, $h(0) = h_o$, $V(x_f) = V_f$, $h(x_f) = h_f$.

2.5 Model 5

The last model is the well-known energy-state model [4-6,8,10,12]. In this approximation, the specific energy

$$E = \frac{1}{2}V^2 + gh \tag{26}$$

is the only state variable. By differentiating (26), we obtain

$$E' = VV' + gh' \tag{27}$$

Using (24) and (25) to eliminate V' and h', we get

$$E' = \frac{T(V,h) - D(V,h,x,\gamma)}{m(x)\cos\gamma} \tag{28}$$

An additional assumption made in the energy-state approximation is that the flight path angle is small so that $\cos\gamma \cong 1$ and thus

$$E' = [T(V,h) - D(V,h,x)]/m(x) \tag{29}$$

where $D(V,h,x) = D(V,h(E,V),x) = qS[C_{D_0} + \beta\{m(x)g/(qS)\}^2]$. (30)

The problem now becomes:

Find the velocity history $V(x)$, $0 \le x \le x_f$, which minimizes

$$J = \int_0^{x_f} \frac{1}{V}\,dx \tag{31}$$

subject to (29) and $E(0) = E_0$ and $E(x_f) = E_f$.

Note that velocity now plays the role of the control. Also, h is a function of E and V as given in (26).

3. Numerical Method Used

All of these optimal control problems are "variable end-time" problems. That is, neither t_f nor x_f is specified. We choose to convert each problem to a "fixed end-time" problem via a simple linear transformation of the independent variable and an added control parameter. For example, if we let

$$x = \sigma\,\tau, \qquad 0 \le x \le x_f, \qquad 0 \le \tau \le x_{f_e} \tag{32}$$

where x_{f_e} is a specified estimate of the optimal x_f, the resulting problem statement for Model 5 becomes:

Find the velocity history $V(\tau)$, $0 \le \tau \le x_{f_e}$, and the parameter σ which minimize

$$J = \int_0^{x_{f_e}} V^{-1}\,d\tau \tag{33}$$

subject to $\quad dE/d\tau = \sigma[T(V,h) - D(V,h,\sigma\tau)]/m(\sigma\tau)$ (34)

and $\quad E(0) = E_0$ and $E(x_{f_e}) = E_f$.

These problems, however, will not be solved as optimal control problems. Instead, they will be solved as parameterized optimization problems. In particular, we replace each control history with a piecewise linear function obtained by interpolation among equally-spaced "control points", and minimization takes place over this set of control points, rather than over the entire control history.

The solution of our minimum time-to-climb problems involves numerical integration of differential equations. The computational cost associated with doing this can be exceedingly high, especially if the differential equations are very complex and highly nonlinear. Since we want to make a model comparison between five dynamic models, the solutions to each model must be accurate. Hence, we desire a method of solution that is both relatively accurate and inexpensive. In addition, the method must be flexible so as to accommodate changes in the dynamic model and performance index with relatively little reprogramming. For these reasons, the method of sequential quadratic programming [13-16] has been chosen.

Sequential quadratic programming is a constrained Quasi-Newton method which exhibits superlinear convergence. It consists of four essential steps:

i) For an initial guess of the control parameters and an initial (positive definite) estimate of the Hessian matrix, compute the required first partial derivatives via numerical integration and finite-difference approximation and solve a quadratic programming problem for the corrections to the control parameter vector and the associated Lagrange multipliers.

ii) Perform a one-dimensional search along the direction of search vector obtained in step (i) by minimizing an auxiliary performance index. This step-size selection procedure is used to enhance convergence from poor initial control parameter estimates.

iii) Update the control parameter vector and test for convergence.

iv) If convergence is not achieved, update the Hessian matrix estimate by a variable-metric formula and repeat from step (i).

This method, which solves a sequence of approximating quadratic programming problems, has proven to be very useful for problems with expensive function and gradient evaluations. The specific algorithm used here is due to Pouliot [17].

4. Numerical Results and Model Comparison

All numerical computations for these problems have been performed on the Iowa State University NAS/9160 computer using FORTRAN 77 with double precision arithmetic. Simpson's rule is used to evaluate the integral performance indices, and a fourth-order, fixed-step, Runge-Kutta, numerical integration scheme is used to integrate the differential constraints. One hundred integration steps are used.

Data for the maximum thrust, T(V,h), and the aerodynamic coefficients are taken from an early representation of the F-4 fighter aircraft [4]. We are using analytical representations of these data prepared by Ong [18]. The thrust model is a two-variable, fourth-order, polynomial, least-squares fit; the aerodynamic coefficients are modeled by segmented cubic polynomials in Mach number M.

We will present comparative results for a climb from h_o = 20000 ft and V_o = 829.8 ft/sec (M(0) = 0.8) to h_f = 50000 ft and V_f = 1162.2 ft/sec ($M(x_f)$ = 1.2). The initial mass m_o is 1305 slugs, and we start from level flight so that $\gamma_0 = 0$. The final flight path angle is free.

One of the pleasant surprises of using a nonlinear programming approach is that satisfactory optimal controls and state trajectories are often obtained with a relatively small number of control points. In Fig. 1 we show the optimal angle of attack histories for Model 2 with 6, 11, and 21 control points. All three plots are qualitatively similar. Since the use of 21 rather than 11 control points provides only modest improvement, as shown in Fig. 1 and Table 1, we will use 11 control points for most of our comparisons.

The theoretical equivalence of Models 1 and 2 is substantiated by the numerical results in Table 1. The minimum t_f values differ by only 0.5%, and the curves for the optimal trajectories in Fig. 3 are indistinguishable. The CPU time per iteration, however, is about 1.5% less for Model 2.

The slope of the linear mass loss function (18), $c_1 = - 3.98(10^{-4})$ slug/ft, has been computed to match the optimal $m(x_f)$ obtained from the Model 1 solution. Since (18) is a good approximation of the optimal mass history as shown in Fig. 2, the optimal trajectories for Models 2 and 3 in Fig. 3 not surprisingly are also quite close. Model 3 produces a slightly lower minimum t_f.

In Model 4, the flight path angle dynamics are no longer present. Although the optimal trajectory shown in Fig. 3 retains the same general features of the more realistic solutions, the flight path angle history shown in Fig. 4 exhibits a very oscillatory behavior. The last γ control point has even reached the imposed upper limit of 80°. This solution has the appearance of a discrete approximation to a lower bound/singular arc/upper bound optimal control history similar to those discussed by Ardema [6] and Breakwell [19] for different but related models. We have not yet attempted a theoretical analysis of the optimal control structure for this problem, but similar numerical results have been obtained with more control points. The Model 4 solution with 11 control points yields optimal t_f and x_f values which are 13.6% and 22.0%, respectively, below those for Model 1.

The comparisons between Models 1-3 and Model 4 are not "fair" in the sense that the initial γ is not specified in Model 4 since γ is the control. Therefore, we have obtained additional Model 2 and 3 solutions with γ_0 treated as an additional control parameter. Although the initial portion of each new optimal trajectory is slightly different, the optimal t_f and x_f values listed in Table 1 are nearly unchanged from their former values.

As expected, the energy-state Model 5 yields the most overly optimistic minimum time, a value 17.1% less than that for Model 1. The optimal final range, however, lies midway between the results from Models 3 and 4. The Model 5 optimal trajectory and h-V diagram are shown in Figs. 3 and 5, respectively. The slope discontinuities are due to the piecewise-linear nature of the control V. Note also that Model 5 is unable to meet the boundary conditions on h and V except by instantaneous shifts of the control V at the initial and final range. These shifts take place along the curves of constant energy shown in Fig. 5.

5. Conclusions and Discussion

Models 4 and 5 provide easily solved but optimistic results for the minimum time-to-climb. However, the results are qualitatively correct and can be used to generate starting data for higher-order models.

Although the addition of flight path angle dynamics in Models 1-3 increases the solution cost, its addition significantly influences the minimum time-to-climb. For accurate prediction of the minimum flight time and the corresponding trajectory, the flight path angle dynamics must be included. Also, tests with γ_o as an added control parameter indicate that the influence of the flight path angle dynamics are much more important than the flight path angle boundary values.

The Model 3 results show that, instead of integrating the m' equation, a simple linear mass function can be used without significant loss in accuracy. This reduces the model order by one.

Obviously, the Model 4 solution is of intrinsic interest and deserves further study. Other (h,V) boundary values have not produced a highly oscillatory γ history similar to that presented here. The boundary values used here do not result in a pronounced accelerating dive which is characteristic of the "zoom climbs" usually obtained. Thus, other combinations of $(h_o,V_o;h_f,V_f)$ should be tried.

Many extensions are possible. Higher-order models including rotational dynamics should be compared. It would also be useful to study the sensitivity to the thrust and aerodynamic models used and to examine alternate aircraft types.

References

[1] Miele, A. "On the Non-Steady Climb of Turbojet Aircraft," J. Aeronautical Sciences, 21(11), 781-783, 1954.

[2] Kelley, H.J. "An Investigation of Optimal Zoom-Climb Techniques," J. Aeronautical Sciences, 26(12), 794-802, 1959.

[3] Bryson, A.E. and Denham, W.F. "A Steepest-Ascent Method for Solving Optimum Programming Problems." J. Applied Mechanics, 29(2), 247-257, 1962.

[4] Bryson, A.E., Desai, M.N. and Hoffman, W.C. "Energy State Approximation in Performance Optimization of Supersonic Aircraft," J. Aircraft, 6(6), 481-488, 1969.

[5] Schultz, R.L. and Zagalsky, N.R. "Aircraft Performance Optimization," J. Aircraft, 9(2), 108-114, 1972.

[6] Ardema, M.D. "Approximations in the Minimum Time-To-Climb Problem," NASA TM X-62292, Aug. 1973.

[7] Rader, J.E. and Hull, D.G. "Computation of Optimal Aircraft Trajectories Using Parameter Optimization Methods," J. Aircraft, 12(11), 864-866, 1975.

[8] Parsons, M.G., Bryson, A.E. and Hoffman, W.C. "Long-Range Energy-State Maneuvers for Minimum Time to Specified Terminal Conditions," J. Optimization Theory and Applications, 17(5/6), 447-463, 1975.

[9] Ardema, M.D. "Solution of the Minimum Time-to-Climb Problem by Matched Asymtotic Expansions," AIAA Journal, 14(7), 843-850, 1976.

[10] Merrit, S.R., Cliff, E.M. and Kelley, H.J. "Energy-Modelled Climb and Climb-Dash - The Kaiser Technique," Automatica, 21, 319-321, 1985.

[11] Bryson, A.E. and Hoffman, W.C. "A Study of Techniques for Real-Time On-Line Flight Path Control-Minimum Time Turns to a Specified Track," Rept. ASI-TR-4, Aerospace Systems Inc., Burlington, Mass., Sept. 1971.

[12] Kelley, H.J., Cliff, E.M. and Weston, A.R. "Energy State Revisited," Optimal Control Applications and Methods, 7(2), 195-200, 1986.

[13] Han, S.P. "Superlinearly Convergent Variable Metric Algorithms for General Nonlinear Programming Problems," Mathematical Programming, 11, 263-282, 1976.

[14] Han, S.P. "A Globally Convergent Method for Nonlinear Programming," J. Optimization Theory and Applications, 22(3), 297-309, 1977.

[15] Powell, M.J.D. "A Fast Algorithm for Nonlinearly Constrained Optimization Calculations," G.A. Watson (Ed.), Numerical Analysis. Springer-Verlag, Berlin, 144-157, 1978.

[16] Pouliot, M.R., Pierson, B.L. and Brusch, R.G. "Recursive Quadratic Programming Solutions to Minimum-Time Aircraft Trajectory Problems," K.H. Well (Ed.), Collection of Papers, Second IFAC Workshop on Control Applications of Nonlinear Programming and Optimization, DFVLR, Oberpfaffenhofen, West Germany, 253-261, 1980.

[17] Pouliot, M.R. "CONOPT2: A Rapidly Convergent Constrained Trajectory Optimization Program for TRAJEX," Rept. GDC-SP-82-008, General Dynamics, Convair Division, San Diego, California, 1982.

[18] Ong, S.Y. "A Model Comparison of a Supersonic Aircraft Minimum Time-to-Climb Problem," M.S. Thesis, Iowa State University, Ames, Iowa, May 1986.

[19] Breakwell, J.V. "Optimal Flight-Path-Angle Transitions in Minimum-Time Airplane Climbs," J. Aircraft, 14(8), 782-786, 1977.

Table 1. Minimum time, optimal final range, and computing time comparison.

Model	NU	γ_0	t_f (sec)	x_f (ft)	CPU time per iteration (sec)
1	11	0	165.56	211299	0.40875
2	6	0	165.77	214106	0.23082
2	11	0	164.70	211300	0.40243
2	21	0	164.63	211193	0.95525
2	11	free	164.35	210558	0.44074
3	11	0	160.97	205213	0.32460
3	11	free	160.62	204279	0.42500
4	11	0	143.07	164862	0.35846
5	11	0	137.25	180725	0.28068

NU: number of control points

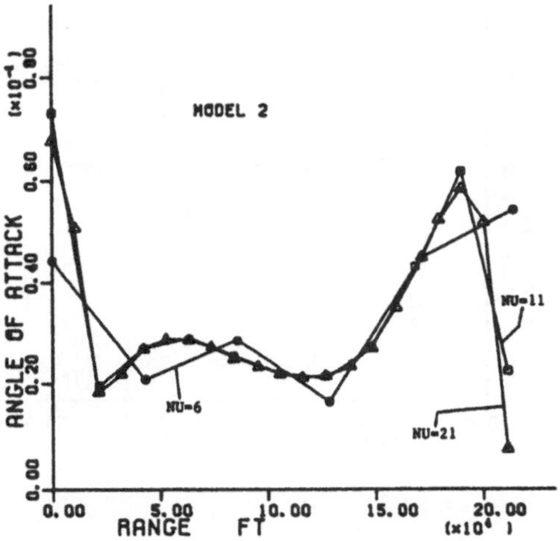

Fig. 1. Optimal angle of attack range histories for Model 2: the effect of the number of control points, NU.

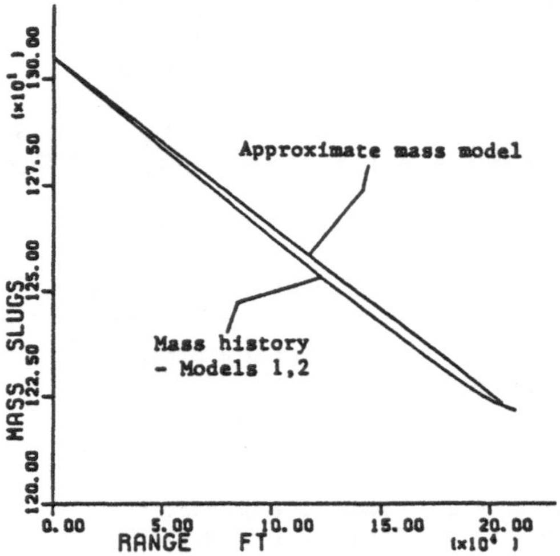

Fig. 2. Optimal and approximate mass histories.

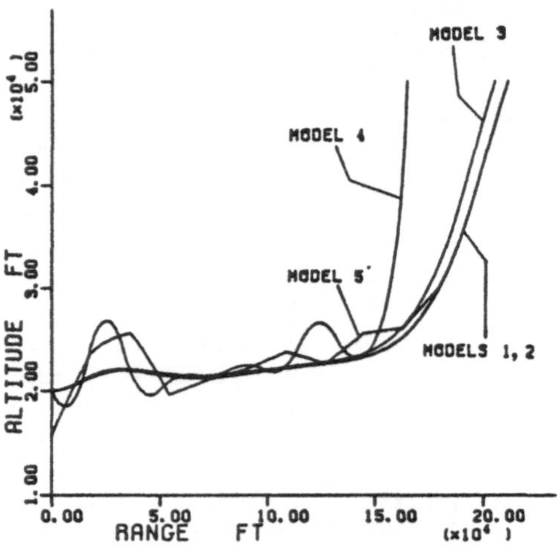

Fig. 3. Optimal trajectories: 11 control points and $\gamma_0 = 0$ (Models 1-3).

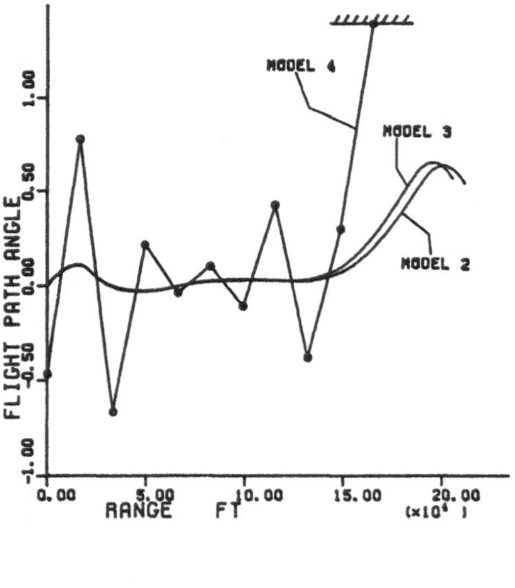

Fig. 4. Optimal flight path angle histories for 11 control points.

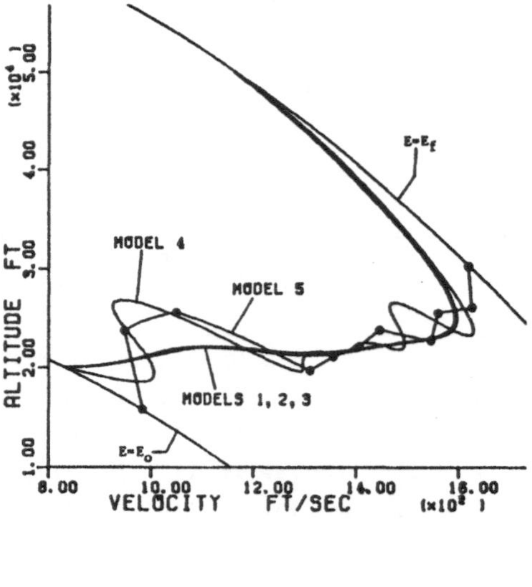

Fig. 5. Optimal (h,V)-diagrams for 11 control points.

AIRCRAFT TRAJECTORY OPTIMIZATION BY CURVATURE CONTROL

Rainer Walden
University Paderborn
D-4790 Paderborn, West Germany

1. Introduction

There is a long history in the endeavor of optimizing trajectories of vehicles flying in the atmosphere. The numerous results obtained depend on the mathematical model and the payoff function used in the calculations. This paper concerns the modelling of an aircraft in 3-dimensional space over a flat earth. We confine our considerations to the point mass model which has proven sufficient and useful for operational studies. In the usual point mass model there are the three control functions angle of attack (or lift), bank angle and power setting. These functions of time together with the initial conditions determine the flight path of the model aircraft.

We would prefer to treat the optimization problems analytically and to avoid numerical calculations. But the more the mathematical model becomes realistic the sooner we fail because of the complexity of the mathematical model and the fact that there are involved not only constant design parameters but also complicated functions like thrust, lift and drag which are at least functions of hight and mach number. So it is hard to obtain general results analytically and in most cases we withdraw to numerical computations.

In this paper we introduce a new point mass model with reduced complexity. The adventage of this model is its usefulness for analytical investigations. The difference between this model and the usual one concerns mainly two points. The first is that we do not care about the angle of attack, which is the input parameter for the calculation of the lift via the lift coefficients. So these coefficients are not involved. Drag coefficients are involved. We use a quadratic drag law. The second point is that we do not use the bank angle but the bank angle rate as a control function. This has a remarkable consequence: at any time the state of aircraft is independent of the choise of the controls at this instant. This is not the case in the usual model: the bank angle determines the normal vector of the flight path. There are some more technical assumptions and simplifications to be described later.

The main idea is to look at the flight path as a curve in 3-dimensional space and to describe it by its curvature and torsion, two functions known from elementary differential geometry and known as the genuine controls of curves in 3-space. This two functions substitute in our model for lift and bank angle. The parametrisation of the curve is given by the speed of the aircraft. The resulting differential equations are the FRENET equations. The whole dynamic of the flight process is described by one differential equation for the speed v, which now of course is very complex. The simplification mentioned above concerns this formula, but it hopefully will not spoil the usefulness of the model for operational studies. This introductory paper shows how this model works and gives some examples. We confine ourselfes to the important case of time optimal trajectories.

2. The mathematical model

We do not use the usual differential equations (see e.g. [2]) for a point mass aircraft. Instead we regard the aircraft's trajectory as a curve in 3-dimensional space and describe it by its FRENET equations. The control functions are now curvature κ, torsion τ (e.g. [3], p. 16) and the power setting δ. We ignore the physical control functions lift angle α and bank angle μ, involved in the technical control process. The stall of the aircraft results in a force orthogonal to the flight path tangent, described by (see [2])

(1) $$F(h,Ma,\delta)\ \sin(\alpha+\alpha_0) + 1/2\ \rho(h)\ v^2 S c_\alpha(Ma,\alpha).$$

We have to add to this force the component of the gravity. The resulting force has the direction of the normal to the flight path and is proportinal to the normal acceleration n. Since

$$n = v^2\kappa\ ,$$

we may use as well κ as a control function. In doing so we assume implicitly that the aircraft is capable of flying a path with this curvature. So we have to include a constraint on the minimal speed.

The force in the direction of the flight path is given by

(2) $$F(h,Ma,\delta)\ \cos(\alpha+\alpha_0) - 1/2\ \rho(h)\ v^2 S c_w(Ma,\alpha).$$

Again the component of the gravitational force has to be added. We avoid the problem to deduce the formula of the correct tangential acceleration and use the following simplification

(3) $$dv/dt = -g\ t_3 + (\delta F - W)/m.$$

Here t_3 is the third component of the flight path tangent T ($t_3 = \langle T,E_3\rangle$, $\{E_1,E_2,E_3\}$ standard basis in $\mathbb{R}^3$) and $W = 1/2\ \rho v^2 c_w$ is the drag. δ, $0 \le \delta \le 1$, is the power setting control. We assume for c_w the quadratic drag law

$$c_w = c_{wo} + k\ c_a^{\ 2}\ ,$$

which leads to

$$c_w = c_{wo} + Cv^2\kappa^2 \qquad \text{with} \qquad C=2mk/(\rho S)$$

if we express c_a as function of κ. This gives the following system of differential equations describing the motion of the model aircraft

$$\begin{aligned} & dX/dt = v\,T \\ & dT/dt = v\,\kappa\,N \\ (4a) \quad & dN/dt = -v\,\kappa\,T + v\,\tau\,B \\ & dB/dt = -v\,\tau\,N \\ & dv/dt = -g\,t_3 + f(\delta,h,v,\kappa) \end{aligned}$$

$$\begin{aligned} & d\lambda_X/dt = -\partial f/\partial X\ \lambda_v \\ & d\lambda_T/dt = -v\,\lambda_X + v\,\kappa\,\lambda_N + g\,E_3\,\lambda_v \\ (4b) \quad & d\lambda_N/dt = -v\,\kappa\,\lambda_T + v\,\tau\,\lambda_B \\ & d\lambda_B/dt = -v\,\tau\,\lambda_N \\ & d\lambda_v/dt = -\lambda_X T - \kappa(\lambda_T N-\lambda_N T) - \tau(\lambda_N B-\lambda_B N) - \partial f/\partial v\ \lambda_v . \end{aligned}$$

We abbreviate $Y=(X,T,N,B,v)$. $X(t)=(x(t),y(t),h(t))$ is the position vector of the aircraft at time t, $T(t)$ the tangent of the flight path, $N(t)$ the normal orthogonal to T and $B(t)$ the binormale to T and N in positive orientation, $v(t) = \| dX/dt \|$ the speed, $\kappa(t)$ die curvature, $\tau(t)$ the torsion, $f(\delta,h,v,\kappa) = (\delta F-W)/m = f_o - Cv^2\kappa^2$ the tangential acceleration (without gravitational acceleration), where $f_o = (\delta F-W_o)/m$ is the acceleration in a straight line flight path. The variable $\Lambda = (\lambda_X,\lambda_T,\lambda_N,\lambda_B,\lambda_v)$ is the adjoint variable and the system of differential equations (4b) is the adjoint system. Notations like $\lambda_T N$ or $\langle\lambda_T,N\rangle$ denote the scalar product between (the row vector) λ_T and (the column vector) N. We always assume $\kappa > 0$. In simplified models, e.g. if speed is constant or gravitation is ignored, the case $\kappa = 0$ may occur, but we exclude this in our more realistic consideration.

There are some differences between the usual model and the one described here which should be mentioned.

In the usual model the directional position of the aircraft in 3-space located at X(t) is given by the velocity vektor dX/dt and the bank angle β. So β provides two informations at the same time: it informs us about the pilot's normal direction and it is a control function. This coupling is not very favorable to theoretical considerations. There is no such coupling in our model.

In our model the formula for the tangential acceleration dv/dt is not exact. The drag W depends only on κ, but not on the bank angle β, which would be case if we transform the equations in [2] exactly. We ignore this here. We also ignore the dependece of F on κ. But this is of minor importance since in most cases we have boundary control for the power setting.

3. Optimality conditions

The structure of optimal control depends heavily on the state and control constraints (this is quite clear since the interior necessary conditions derived from maximum principle are the same for all types of time optimal problems). We mention the most important one's.

(5)
(a) $v^2\kappa \leq n_{max}$
(b) $\kappa \leq \kappa_{max}(v,h)$
(c) $q \leq q_{max}$
(d) $v_{min}(h) \leq v$
(e) $h \geq h_{min}$
(f) $F_{min}(v,h) \leq F \leq F_{max}(v,h)$
(g) $|\tau| \leq \tau_{max}$

For simplification let us assume here that (a) - (e) are not activ. Then it is possible to show that $F=F_{max}$. We calculate the other controls using the maximum principle.

The Hamiltonian for the time-optimal problem is

$$H = \langle\Lambda, Y\rangle - 1.$$

We add no constraints. The optimal control functions $(\kappa^*,\tau^*,\delta^*)$ maximize the Hamiltonain. This gives

$$\partial H/\partial\kappa = 0 , \quad \partial H/\partial\tau = 0, \quad \delta^* = +1.$$

Moreover we have

$$H = 0,$$

since H depends not explicitely on time. We use this to transform the last equation (4b) into the form

(6) $$d\lambda_v/dt = ([f-gt_3]/v - \partial f/\partial v)\lambda_v - 1/v.$$

To calculate the optimal controls κ^* and τ^* the following equations are available

(7) $$0 = \partial H/\partial\kappa = v(\lambda_T N-\lambda_N T) + \partial f/\partial\kappa\ \lambda_v = v(\lambda_T N-\lambda_N T) - 2Cv^2\kappa\lambda_v,$$

(8) $$0 = \partial H/\partial\tau = v(\lambda_N B-\lambda_B N) ,$$

(9) $$0 = H = v\lambda_x T+v\kappa(\lambda_T N-\lambda_N T)+v\tau(\lambda_N B-\lambda_B N)+[f - gt_3]\lambda_v-1$$

Since τ is linear in H we have to take into consideration boundary and singular controls. Form the visual point of view it seem apparent that long trajectories have to include singular parts. We easily calculate the sigular control. Since $\lambda_N B-\lambda_B N=0$ we have

$$0 = d/dt(\lambda_N B-\lambda_B N) = -v\kappa(\lambda_T B-\lambda_B T).$$

Assuming $v\kappa \neq 0$ we have $\lambda_T B-\lambda_B T=0$ and

$$0 = d/dt(\lambda_T B-\lambda_B T) = -v\lambda_x B + g\langle E_3,B\rangle\lambda_v - v\tau(\lambda_T N-\lambda_N T),$$

such that

(10) $$v\kappa^* = (\lambda_T N-\lambda_N T)/(2C\lambda_v),$$

(11) $$v\kappa^* = (-v\lambda_x B + g\langle E_3,B\rangle\lambda_v)/(\lambda_T N-\lambda_N T)$$

In (11) the gravitation plays an important role. If $g = 0$ we have $\tau = 0$ and a piecewise planar motion (see [1]).

To calculate $\kappa^*$and τ^*we need not know all adjoint functions. It is sufficient in in principle to know the function λ_v.

The first two components of $\lambda_x = (\lambda_x,\lambda_y,\lambda_h(t))$ are constant, $\lambda_h(t)$ is the result of integrating

$$d\lambda_h/dt = - \partial f/\partial h \ \lambda_v.$$

$\lambda_T,\lambda_N,\lambda_B$ are the solutions of the following linear differential equations

$$d\begin{pmatrix}\lambda_T\\ \lambda_N\\ \lambda_B\end{pmatrix}/dt = \begin{pmatrix}0 & v\kappa & 0\\ -v\kappa & 0 & v\tau\\ 0 & -v\tau & 0\end{pmatrix}\begin{pmatrix}\lambda_T\\ \lambda_N\\ \lambda_B\end{pmatrix} + \begin{pmatrix}-v\lambda_x+gE_3\lambda_v\\ 0\\ 0\end{pmatrix}$$

The homogeneous solution in known ([1]):

(12) $$\lambda_Z(t) = \langle T_f,Z\rangle\lambda_{T_f} + \langle N_f,Z\rangle\lambda_{N_f} + \langle B_f,Z\rangle \lambda_{B_f},$$

where $Z \in \{T,N,B\}$. The index "f" denotes evaluation at final time t_f. U(t) denotes the fundamental matrix of (12). A particular solution is

(13) $$\lambda_Z(t) = U(t) \int_{t_f}^{t}\langle Z_f,T(\rho)\rangle(-v\lambda_x+gE_3\lambda_v)(\rho)\,d\rho$$
$$= \int_{t_f}^{t}\langle Z(t), T(\rho)\rangle (-v\lambda_x+gE_3\lambda_v)(\rho)\,d\rho.$$

We obtain the solution as a sum of (12) and (13). Analytical solutions with no further special assumption are hardly to obtain. The following examples show that we gain important insight to the behaviour of time optial trajectories if we allow some simplifications. Even if the results are known widely their derivation becomes quite clear and we may obtain simply related results by useing this model.

4. Examples

We give three examples with increasing model sophistication. The first two are resticted to a plane, which is simply expressed by $\tau=0$. The main difference in the three cases concerns the equation of dv/dt in (4a).

Example 1: The brachistochrone

Let us look at the famous example of the brachistochrone (see [4] p. 139, [5] pp. 81, 119, 142). The problem is, given an initial point, to reach a usually lower target point in the vertical plane in shortest time. The motion is influenced only by gravitation, without any friction so that the sum of kinetic and potential energy is constant. The optimal trajectory is known as a piece of a cycloid. We look for a feedback law, i.e. a control function $\kappa=\kappa(Y)$, depending only on the states (here we have $\delta \equiv 1$, $\tau = 0$ .and κ has no sign restriction). The last equation in (4a) now reads

$$dv/dt = -g\ t_3.$$

The maximum principle allows a boundary control $\kappa^*=\pm\kappa_{max}$ (we assume κ_{max} to be constant) or a singular control, to be calculated from $\partial H/\partial\kappa=0$ and $H=0$. This gives

$$\lambda_T N-\lambda_N T = 0$$

and

(14) $$\lambda_v = (v\lambda_x T-1)/gt_3.$$

On the other hand a consequence of $0 = d(\lambda_T N-\lambda_N T)/dt$ is

(15) $$\lambda_v = v\lambda_x N/gn_3.$$

Comparing (14) and (15) we ontain $n_3(v\lambda_x T-1) = vt_3\lambda_x N$.

Another differentiation of $\lambda_N T-\lambda_T N$ gives

$$0 = d^2(\lambda_N T-\lambda_T N)/dt^2 = v\kappa + g(t_3\lambda_x N-n_3\lambda_x T).$$

So we obtain the following feedback control law

(16) $$\kappa^* = g/v^2\ n_3.$$

This is indeed the curvature of a cycloid. The speed v and the coordinate y satisfy $v^2=2gy$. The result reveals a problem in computing feedback laws analytically. The target point is not involved in formula (16). The result is true only if the state vector Y lies already on the desired singular controlled trajectory passing through the target point. If this is not the case, e.g. when tangent vector T has a wrong direction, we need first a boundary control ($|\kappa|=\kappa_{max}$) until the right direction is obtained and then we apply the singular control. The problem is to know when to switch from one type of control to the other. We happyly are able to solve this problem in our example because we know the whole family of time optimal singular trajectories analytically.

The possibility to eliminate the adjoint variables useing equations derived from the maximum principle is investigated systematically in [10]. Usually differential equations for the feedback control are obtained.

Example 2: The horizontal intercept maneuver

We confine the motion to a horizontal plane. The last equation in (4a) is

$$dv/dt = f(\delta,v,\kappa).$$

We assume $f_0 > 0$ and regard only the constraint (5a). The problem is to reach in shortest time a circle with some radius $R \geq 0$ the center of which is a given fixed traget point X_z. The following facts are taken from [1]. The condiition $\partial H/\partial\kappa = 0$ gives

$$(\lambda_T N - \lambda_N T) = 2Cv\lambda_v\kappa^* .$$

Let $J: \mathbb{E}^2 \rightarrow \mathbb{E}^2$ denote a 90°-rotation in the plane. Then

$$(\lambda_T N - \lambda_N T) = \lambda_x J(X_f-X).$$

κ^* depends only on the two adjoint variables λ_v and λ_x. X_f is the final point of the trajectory on the circle of radius R with venter X_z. λ_x is a vector with direction X_f-X_z. We assume $\lambda_x J(X_f - X) \neq 0$ (otherwise we have a straight line trajectory). λ_x is constant and λ_v can be eliminated from $H=0$. The result is

$$|\kappa^*| = (r - (r^2-s)^{1/2})/v, \tag{17}$$

where

$$r = (1-v\lambda_x T)/|\lambda_x J(X_f-X)|, \qquad s = f_0/C.$$

In the final point we have

$$\lambda_x T_f = 1/v_f \quad \text{and} \quad \lambda_x N_f = 2C\kappa_f/v_f$$

such that

$$\kappa^*_f = 1/2C\ \lambda_x N_f/\lambda_x T_f.$$

We see that κ^*_f depends not on $\|\lambda_x\|$. This is true for $\kappa^*(t)$ at any time t as seen from

(18) $r = \lambda_x(T_f v_f - Tv)/|\lambda_x J(X_f - X)|$.

So κ^* depends only on the direction φ of λ_x. Let φ be the angle between T_f and λ_x.

We now have the following procedure to calculate time optimal trajectories: choose some value for φ and integrate backwards the system (4a) starting with given initial conditions $Y_f=(X_f,T_f,N_f,v_f)$ and $\Lambda_f= (\lambda_x,0,0,0)$. The two parameter family (parameters are v_f and φ) of such trajectories is investigated in [7], [8] , denoted as "extemal trajectory map" (ETM).

Observe that the assumption $f_0 > 0$ includes a constraint on the final speed v_f,

$$v_f < v_\infty := [2F/(\rho S c_{wo})]^{1/2}$$

We easily see from (17), that $v=v_\infty$ is equivalent to $\kappa^*=0$ and this is equivalent to $\lambda_x N=0$, i.e. $\lambda_x=T/v$ or $\varphi=0$. This is the limit case of a straight line trajectory. We will see at once that $|\varphi|$ is restricted to an interval $[0,\varphi_s]$.

(17) shows that $|\kappa^*|$ is bounded by r/v. If this constraint is activ we have $r^2=s$, which implies

(19) $|\kappa^*| \leq \kappa^*_s := 1/v \ [f_0/C]^{1/2}$.

$v\kappa^*_s$ is the stationary turn rate at speed v. If the backwards integration leads to a point where $|\kappa^*|=\kappa_s$ then the only possibility to continue the trajectory further backwards is to apply a switch of κ^* to the boundary $\pm\kappa_{max}$. The two equationy H=0 and $\partial H/\partial\kappa=0$ are no longer simultaneously valid. So we have to drop the last one. These two equations simulataneously from an "interior control condition" which always has to be checked. Integrating further backwards the control function is determined by the activ constraint (5a). This phenomenon comes not unexpected since we have a similar situation in the well known case f=0 when we have a drag free motion with constant speed. It is apparent that more generally in the case k=0, the time optimal trajectories consist of pieces of circles and straight lines (see [1], [6]). The possibility of such interior jumps of κ should be observed in numerical calculations. The numerical results themselfes do not reveal this switching structure ([9]).

The equality case in (19) provides us with an upper bound for $|\varphi|$. Let be t_s the time where the jump occurs. With increasing $|\varphi|$ t_f-t_s gets smaller and in the limit we have $t_s=t_f$. This means that then

$$[f_0/C]^{1/2}/v = 1/2C \ \lambda_x N/\lambda_x T,$$

and

(20) $|\varphi| \le \varphi_s := \arctan (2/v\ [f_o C]^{1/2})$.

The time optimal trajectory reaches the target circle without applying maximal curvature if the hit angle φ (between T_f and λ_x) lies in the interval $(-\varphi_s, +\varphi_s)$. Trajectories dropping in steeper are controlled ba boundary control.

(17), and (18) show that we know the optimal control function κ^* leading to the final point X_f, if v_f, T_f and λ_x are known. In this case we have a feedback law for $\kappa^*(t)$. It is possible to estimate these quantities to obtain useful feedback laws ([12]).

Example 3: Minimum-time 180° turns of aircraft.

The following example has only narrative character but is included here because it involves trajectories in 3-space which are not plane. So the torsion formula (11) applies.

[11] reports the results of numerical calculations concerning the following problem. At initial time t=0 we prescrible X_o, T_o = (1,0,0) and v_o. At free final time t_f we prescribe $T_f = -T_o$. We seek the time optimal trajectory satisfying this boundary conditions. N_o is not prescribed but included in the optimization procedure (in accordance with the fact that in the model used in [11] N is no component of the state vector but used as a control function). The results can be interpreted by formula (11). If $N_o = \pm(0,0,1)$ then $\tau \equiv 0$ which means that we have a plane mation (this is also clear from symmetry considerations). If $N_o \neq \pm(0,0,1)$,

$$v\tau^* = (-v\lambda_h + g\lambda_v)b_3/(\lambda_T N - \lambda_N T)$$

vanisher no longer. [11] shows that for sufficient large or small initial speeds $N_o = \pm(0,0,1)$. The trajectory is a half loop up or down. For initial speeds between $b_3 \neq 0$ and the motion is no longer plane. This is an example how analytical investigations help to interpret numerical results.

5. Concluding remarks

The mathematical model introduced here is suitable for analytical investigations of operational problems. One advantage is that as far as possible a co-ordinate free terminology is used. So the structure of the problem is revealed as far as possible. Using the FRENET equations and the genuine control functions κ and τ the specific data of the aircraft only appear in the differential equation of the speed (and in the constraints). Tests for usefulness in numerical computations are at the beginning and seem promissing.

References

[1] R. Walden: Aircraft Trajectory Optimization in Homogeneous Space. MBB-Bericht Nr. Z 65/86ö

[2] G.-Ch. Shau: Transformationsbeziehungen bei der Optimierung dreidimensionaler Flugbahnen zur Umgebung von Singularitäten. Z. Flugwiss. Weltraumforsch. 6 (1982), Nr. 2, p. 90-98.

[3] M.P. do Carmo: Differentialgeometrie von Kurven und Flächen. Vieweg, Braunschweig 1983.

[4] L. Cesari: Optimation - Theory and applications. Springer, New-York 1983

[5] A.E. Bryson, Y.-C. Ho: Applied Optimal Control. Hemisphere PC, Washington, D.C. 1975.

[6] L.E. Dubins: On curvature of minimal lenght with a constraint on average curvature, and with precribed initial and terminal positions and tangents. Am. J. Math. 79, 497-516 (1957).

[7] N. Rajan, M.D. Ardema: Barriers and Dispersal Surfaces in Minimum-Time Interception. JOTA 42, p. 201-228 (1984).

[8] N. Rajan, U.R. Prasad, N.J. Rao: Planar Pursuit-Evasion with Variable Speed, Part 1 & Part 2, JOTA 33, p. 401-432 (1981).

[9] B.S.A. Järmark, A.W. Merz, J.V. Breakwell: The Variable-Speed Tail-Chase Aerial Combat Problem. J. Guidance and Control 4, p. 323-328 (1981).

[10] M. Fliess, H. Bourdache-Sigerdidjyne: Quelques remarques elementaires sur le calcul des lois de bouclage en commande optimale non lineaire. Proc. 6th Int. Conf. Analysis Optimiz. System, Nice June 1984.

[11] K.H. Well und E. Berger: Minimum-Time 180^{0} Turns of Aircraft. JOTA 38, p. 86-96 (1982).

[12] R. Walden: Das Frenet-Modell und seine Anwendung zur Berechnung von Optimalsteuerungen von Flugzeugen und Flugkörpern. MBB Technical Report 1986.

OSCILLATORY CRUISE - A PERSPECTIVE

John V. Breakwell
Professor, Dept. of Aero/Astro
Stanford University
Stanford, CA 94301

Abstract

The problem of minimum fuel/km is examined using various simplifications, introduced in the last three decades, leading to order reduction. In particular, the second variation about steady cruise, ignoring only the slow change in total weight, reveals two separate physical reasons for the possible advantage of oscillatory cruise. A linear analysis with quadratic payoff (the second variation) and with bounds on thrust variation yields quite good agreement with an exact solution, published recently, for the optimal periodic cruise of a particular airplane. Finally, an explanation is given for the much more substantial percentage saving obtainable by oscillatory maneuvers in the "endurance" problem: minimum fuel/hour.

1. Introduction: The Problem

If V, γ, h, x and m denote, respectively, velocity, flight-path angle, altitude, range and mass of fuel consumed, we may write:

$$\begin{aligned}
\dot{V} &= g\left(\frac{T-D(V,h,L)}{W} - \sin\gamma\right) \\
V\dot{\gamma} &= \left(\frac{L}{W} - \cos\gamma\right) \\
\dot{h} &= V\sin\gamma \\
\dot{x} &= V\cos\gamma \\
\dot{m} &= \sigma(V,h)T
\end{aligned} \tag{1}$$

where g will be assumed constant, T is the thrust, L the lift, W the weight, $D(V,h,L)$ the drag and $\sigma(V,h)$ an engine fuel-rate factor.

The throttle-setting is $\eta = T/T_{max}$ where $T_{max} = T_{max}(V,h)$, is the engine's maximum thrust. It has been tacitly assumed that fuel-rate is proportional to the throttle-setting; this may, in fact, be essentially true only above a certain lower bound, η_{min}, to useful settings.

If we treat W as constant, ignoring the weight of fuel consumed (for the duration of a typical cruise oscillation this will turn out to be unimportant), the optimal *steady* cruise ($\gamma = 0$) occurs at speed V_c and altitude h_c obtained by minimizing $\frac{dm}{dx} = \frac{\sigma(V,h)D(V,h,W)}{V}$, subject, of course, to $D(V,h,W) \leq T_{max}(V,h)$.

From here onward we shall assume that this occurs with D strictly less than T_{max}, so that oscillations in T above and below the cruise value are possible.

In the following sections we shall investigate the reduction in $(\frac{dm}{dx})_{ave}$ obtainable from such oscillations.

2. The Lowest Order Model: The Energy State

This model has been used by Kelley et al. in many papers (see for example [1]). Introducing the energy state:

$$E = \frac{V^2}{2} + gh \quad , \tag{2}$$

and supposing that γ remains small enough so that we may replace $\cos\gamma$ by 1 and $\dot{\gamma}$ by 0, the appropriate drag function is $D^*(V,E) = D(V,h,W)$, and the typical "maneuverability domain" for the range rates of change of the qualities $\frac{W}{g}E$ and m, regarding now V as well as T as a control, is shown in Fig. 1, evaluated for E equal to its cruise value E_c. The left boundary of the domain is concave at the cruise point, as we shall show, and it is clear that an oscillation between $T = 0$, $V = V_1$ and $T = T_{max}$, $V = V_2$, where $V_2 > V_1$ and $h_2 < h_1$, yields smaller fuel expenditure $(\frac{dm}{dx})_{ave}$ than steady cruise.

In general the oscillatory value for $(\frac{dm}{dx})_{ave}$ is minimized at an energy level E^* different from E_c, and the optimal solution, using this lowest order model, requires an infinitely fast "chatter" in thrust and altitude, so as not to deviate from the optimal E^*. This solution, of course, violates the tacit assumption of small γ used in evaluating D^*. The nonoptimality of steady cruise, then, remains unproven.

Incidentally, the first suggestion that a rapid oscillation in thrust and altitude could reduce fuel/km was made by Edelbaum [2], again treating γ as small in the calculation of drag, but restricting himself to oscillations in the neighborhood of cruise energy E_c.

As suggested to the present author by both J. Speyer and P.J.K. Menon, we now restrict ourselves to *small* oscillations along the left boundary about the cruise point in Fig. 1. The change in fuel expenditures is clearly of 2nd order, and is expressible as:

$$\frac{\delta^2 m}{\text{cycle}} = \int_0^{x_1} \left[\frac{1}{2}\frac{\partial^2 X}{\partial Y^2}(\delta Y)^2 + (\frac{\sigma D^*}{V})_{EV}\delta E\,\delta V + \frac{1}{2}(\frac{\sigma D^*}{V})_{EE}(\delta E)^2 \right] dx \quad , \tag{3}$$

where x_1 is the range of the oscillation, and the 2nd and 3rd terms in the integrand would be negligible if the oscillation were performed rapidly enough.

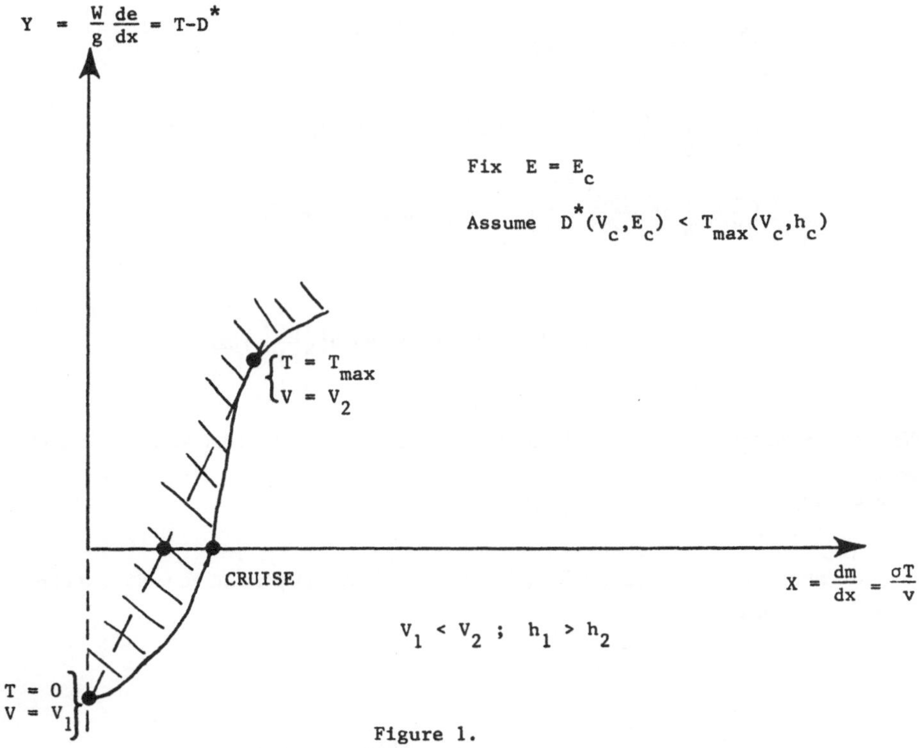

Figure 1.

Now the left boundary is given by $X = \min\limits_{V}[\frac{\sigma}{V}(Y + D^*)]$, with E and Y held fixed, so that $Y = -(\frac{\sigma D^*}{V})_V/(\frac{\sigma}{V})_V$, and in the neighborhood of $Y = 0$, where $V = V_c$ and $(\frac{\sigma D^*}{V})_V = 0$, we find that $\frac{dY}{dV} = -(\frac{\sigma D^*}{V})_{VV}/(\frac{\sigma}{V})_V$. Also $\frac{dX}{dY} = \frac{\sigma}{V} + [Y(\frac{\sigma}{V})_V + (\frac{\sigma D^*}{V})_V]\,\frac{dV}{dY}$ so that, in the neighborhood of $Y = 0$:

$$\frac{d^2X}{dY^2} = 2(\frac{\sigma}{V})_V \frac{dV}{dY} + (\frac{\sigma D^*}{V})_{VV}(\frac{dY}{dV})^2 = -[(\frac{\sigma}{V})_V]^2/(\frac{\sigma D^*}{V})_{VV} \; ,$$

which is always negative, as promised earlier, under the assumption made at the end of Section 1, unless $(\frac{\sigma}{V})_V$ happens to vanish.

The first term in the integrand in (3) may now be rewritten as $-\frac{1}{2}(\frac{\sigma D^*}{V})_{VV}(\delta V)^2$, and the fuel expenditure expressed in the form:

$$\frac{\delta^2 m}{\text{cycle}} = \int_0^{x_1} [-(\frac{\sigma D^*}{V})_{VV}(\delta V)^2 + \delta^2(\frac{\sigma D^*}{V})]dx \qquad (4)$$

The first term in (4) may be transformed as follows:

$$-\left(\frac{\sigma D^*}{V}\right)_{VV}\int_0^{x_1}(\delta V)^2dx = +\left(\frac{\sigma}{V}\right)_V\frac{W}{g}\int_0^{x_1}\frac{d(\delta E)}{dx}\,\delta V dx$$

$$= +\frac{W}{g}\left(\frac{\sigma}{V}\right)_V\int_0^{x_1}\frac{d(\delta E)}{dx}\left(\frac{\delta E}{E}-\frac{g}{V}\,\delta h\right)dx = -\frac{W}{V}\left(\frac{\sigma}{V}\right)_V\int_0^{x_1}\frac{d(\delta E)}{dx}\,\delta h dx$$

$$= +\frac{W}{V}\left(\frac{\sigma}{V}\right)_V\int_0^{x_1}\delta E\,\frac{d(\delta h)}{dx}\,dx = +\frac{W}{V}\left(\frac{\sigma}{V}\right)_V\int_0^{x_1}(VdV+g\,\delta h)\delta\gamma dx \quad ,$$

so that

$$\frac{\delta^2 m}{\text{cycle}} = \int_0^{x_1}\left[+W\left(\frac{\sigma}{V}\right)_V\delta V\delta\gamma+\delta^2\left(\frac{\sigma D^*}{V}\right)\right]dx \tag{5}$$

We shall obtain this expression in the next section, as part of the second variation in a higher order model.

3. The "Full Order" Model (But W = Const)

Rewriting the equations of motion (1) with range X as independent variable, and eliminating the thrust T, we obtain:

$$\begin{cases} \dfrac{dh}{dx} = \tan\gamma \\[2ex] \dfrac{dm}{dx} = \sigma\left\{\dfrac{W}{g}\dfrac{dV}{dx}+\dfrac{D(V,h,L)}{V\cos\gamma}+W\,\dfrac{\tan\gamma}{V}\right\} \end{cases} \tag{6}$$

The first-order variation from cruise ($\gamma=0, V=V_c, h=h_c$) is given by

$$\frac{\delta m}{\text{cycle}} = \int_0^{x_1}\left\{\frac{\sigma W}{g}\frac{d(\delta V)}{dx}+\frac{\sigma W}{V}\,\delta\gamma+\delta\left(\frac{\sigma D^*}{V}\right)+\frac{\sigma}{V}\,D_L\delta L\right\}dx \quad , \tag{7}$$

which vanishes over the cycle since $\delta\left(\frac{\sigma D^*}{V}\right)=0$ at cruise, while to 1st order: $\delta\gamma=\frac{d(\delta h)}{dx}$ and $\delta L = \frac{WV^2}{g}\frac{d(\delta\gamma)}{dx}$. The vanishing of the first variation is more usually exhibited by constructing a Hamiltonian; the present argument is perhaps more physically appealing, and it can be carried to 2nd order.

Two mechanisms are immediately apparent from (6) for a second-order reduction in $\left(\frac{dm}{dx}\right)_{\text{ave}}$:

(i) Ignoring for the moment any variation in σ, the integral of $\frac{\sigma W}{V}\tan\gamma$ can be made negative, to 2nd order, by varying the throttle so that δV has the same sign as $\delta\gamma$. This "gravity effect" can be thought of as resulting from a negative time average (to 2nd order) for $\tan\gamma$.

(ii) The fact that induced drag is always an increasing function of altitude implies that D_{Lh} is positive and hence the 2nd order term $\frac{\sigma}{V}D_{Lh}\delta L\,\delta h$ has a negative average over the cycle, δL being 180° out of phase with δh. This "induced drag effect" can be thought of as a consequence of reducing the lift at the top of the altitude cycle where its effect on drag is highest, while increasing it where the effect is lowest.

Using

$$\frac{\delta L}{W} = \frac{V^2}{g}\frac{d(\delta\gamma)}{dx} - \frac{1}{2}(\delta\gamma)^2 + 2\,\frac{\delta L}{W}\,\frac{\delta V}{V} \tag{8}$$

(to 2nd order) and dropping out integrals of all exact derivatives while integrating by parts to express $\int_0^{x_1} \delta h \,\frac{d(\delta V)}{dx}\,dx$ as $-\int \delta\gamma\delta V dx$, we obtain the complete 2nd variation as:

$$\frac{\delta^2 m}{\text{cycle}} = \int_0^{x_1}\Bigg\{\delta^2\Big(\frac{\sigma D^*}{V}\Big) + \frac{\sigma}{2V}(D - WD_L)(\delta\gamma)^2 + \frac{\sigma}{2V}D_{LL}(\delta L)^2 \tag{9}$$

$$+ \left[\frac{\sigma}{V}D_{Lh} + \frac{\sigma_h}{V}D_L\right]\delta L\,\delta h \;-\; \frac{W}{V^2}\Big(\sigma - V\sigma_V + \frac{V^2}{g}\sigma_h\Big)\delta V\,\delta\gamma$$

$$+ \left[\frac{\sigma D_L}{V^2} + \frac{\sigma D_{LV}}{V} + \frac{\sigma_V D_L}{V}\right]\delta V\,\delta L\Bigg\}dx$$

An earlier version, by the author and H. Shoaee [3], contained an error in the coefficient of $\delta V\,\delta L$.

The first term in the integrand, $\delta^2\big(\frac{\sigma D^*}{V}\big)$ is now to be understood as a positive definite quadratic form in $\delta V, \delta h$:

$$\delta^2\Big(\frac{\sigma D^*}{V}\Big) = \frac{1}{2}\Big(\frac{\sigma D^*}{V}\Big)_{VV}(\delta V)^2 + \Big(\frac{\sigma D^*}{V}\Big)_{Vh}\delta V\,\delta h + \frac{1}{2}\Big(\frac{\sigma D^*}{V}\Big)_{hh}(\delta h)^2\,,$$

partial derivatives w.r.t. V being now taken with h held constant. The term outlined by the heavy rectangle is seen to agree with that obtained in (5) and includes the gravity effect (i). The term outlined by the dotted rectangle, includes, in addition to the induced drag effect (ii), an effect due to variation of σ with altitude.

4. An Intermediate Model

If, in (9), we ignore all terms with δL as well as the term with $(\delta\gamma)^2$, steady cruise becomes a "doubly singular arc" with $\delta\gamma$ and δT as controls. As shown by Speyer, [4], this arc *fails* the Robbins test, [5], for optimality: $\frac{\partial}{\partial u_i}\left[\frac{d}{dx}\left[\frac{\partial H}{\partial u_j}\right]\right] = 0$. As a consequence, see [5], an improvement can be obtained over steady cruise by a rapid pulse in one control, say δT, accompanied by a pulse derivative in the other control, $\delta\gamma$. If this short pattern is repeated with opposite signs we get a cycle in the $\delta V - \delta h$ plane, as sketched in Fig. 2, the direction being chosen so that δV is in phase with $\delta\gamma = \frac{d(\delta h)}{dx}$. This is just the mechanism (i) described in the previous section.

However, if this cycle is performed rapidly enough, the ignored term in (9) with $(\delta\gamma)^2$ may dominate. It is interesting that the coefficient in this term actually vanishes for certain simple drag models, but not in general. In a numerical example to be investigated in Section 6 it turns out to the

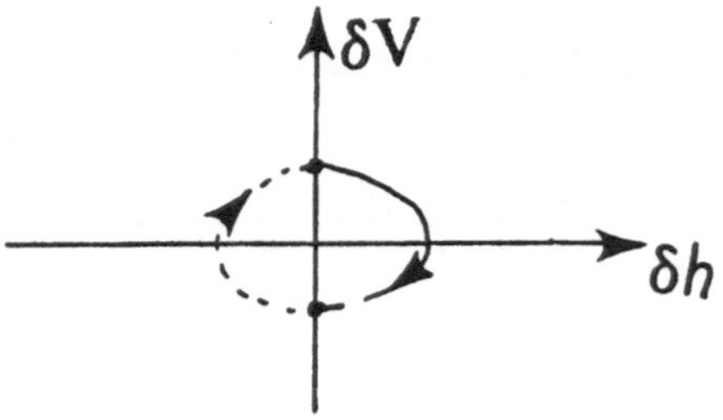

Figure 2.

rather small but positive. The failure, then of the Robbins test for this intermediate model, with $(\delta\gamma)^2$ ignored, is not sufficient to prove the nonoptimality of steady cruise.

5. Sinusoidal Variations

Following a procedure used by Speyer, [6], the second variation (9) may be minimized w.r.t. δV as a function of δh, $\delta\gamma$, δL:

$$\delta V = C_1 \delta h + C_2 \delta\gamma + C_3 \delta L \quad ,$$

where

$$C_1 = -\left(\frac{\sigma D^*}{V}\right)_{Vh} / \left(\frac{\sigma D^*}{V}\right)_{VV}$$

$$C_2 = \frac{W}{V^2}\left(\sigma - V\sigma_V + \frac{V^2}{g}\sigma_h\right) / \left(\frac{\sigma D^*}{V}\right)_{VV} \tag{11}$$

$$C_3 = -\left[\frac{\sigma D_L}{V^2} + \frac{\sigma D_{LV}}{V} + \frac{\sigma_V D_L}{V}\right] / \left(\frac{\sigma D^*}{V}\right)_{VV}$$

Note that, if for example σ is constant, $C_2 > 0$ in accordance with our discussion of the gravity effect (i).

This choice of δV leads to

$$\frac{\delta^2 m}{m} = \text{ave/cycle}\left\{Q\left(\delta h, \delta\gamma, \delta L/W\right)\right\} \quad , \tag{12}$$

where Q is a quadratic form in its arguments:

$$Q = \frac{1}{2}Q_{11}(\delta h)^2 + \frac{1}{2}Q_{22}(\delta\gamma)^2 + \frac{1}{2}Q_{33}\left[\frac{\delta L}{W}\right]^2 + Q_{13}\delta h \frac{\delta L}{W} \tag{13}$$

The coefficient Q_{22}, in typical examples, is negative, arising mainly from the gravity effect. The other coefficients are positive, the coefficients Q_{13} arising mainly from the induced drag effect.

Following the usual treatment of the 2nd variation we shall investigate the second order relative change (12) in fuel expenditure, retaining only the *linearized* form of the dynamics:

$$\frac{d(\delta h)}{dx} = \delta\gamma \ , \quad \frac{d(\delta\gamma)}{dx} = \frac{g}{V^2}\frac{\delta L}{W} \tag{14}$$

Note that the second equation (14) is the linearized form of (8).

Following an idea first introduced by Bittanti et al. [7] and applied to this problem by Speyer in [6], but with time rather than range as independent variable, we now enquire whether the change in fuel expenditure can be made negative by *sinusoidal* variations in altitude, i.e., $\delta h \sim \cos \omega x$, for some wavelength $2\pi/\omega$. The calculation is straightforward, the answer being:

$$\text{Yes, If } \frac{V^4}{g^2} Q_{33}\omega^4 - (|Q_{22}| + 2\frac{V^2}{g} Q_{13})\omega^2 + Q_{11} < 0 \tag{15}$$

We have tacitly assumed a sinusoidal variation in δh with mean zero; Speyer assumed, instead, that δh and $\delta\gamma$ both vanished at the ends of the cycle, i.e. $\delta h \sim (-1 + \cos \omega x)$. In this case the last term in (15) becomes $3Q_{11}$.

It is immediately clear from (15) that no reduction in fuel expenditure is obtained if the wavelength is too short, i.e., ω too high. The increase in drag due to lift makes rapid oscillations uneconomical. If, on the other hand, the wavelength is too large, the term Q_{11} predominates i.e., the path wanders slowly back and forth about the conditions for optimal steady cruise and fuel expenditure is increased.

It is perhaps interesting to pose a classical question in connection with (12) and (14): Is there a "conjugate point" to an initial point $x = 0$ on steady cruise? The answer, after a rather lengthy analysis, turns out to be:

$$\text{Yes, If } \left[\frac{1}{2}|Q_{22}| + \frac{V^2}{g} Q_{13}\right]^2 > \frac{V^4}{g^2} Q_{33} Q_{11} ,$$

which is just the condition that expenditure can be reduced for *some* wavelength.

6. An Example

We turn now to a particular F4-type aircraft studied by Grimm, Oberle and Well [8]. Steady cruise occurs at altitude $h_c = 11.2\ km$, velocity $V_c = 251\ m/s$ and throttle-setting η_c almost exactly 5/8. If the minimum throttle-setting η_{min} is assumed to be 1/4 the allowed throttle variations are symmetric: $-3/5 \leq \Delta\eta/\eta_c \leq 3/5$.

The coefficients Q_{ij} have been evaluated from numerical partial differentiation of D and σ in the neighborhood of cruise. The answer (15) is found here to be:

$$\text{Yes, If } 16\ km < 2\pi/\omega < 132\ km$$

Incidentally, the (damped) phugoid has a wavelength of approximately 26 km, lying in the favorable range, but this is irrelevant!

To determine how much fuel reduction is possible, we will choose the amplitude and wavelength of the altitude variation in the following way: Firstly, the amplitude will be adjusted as a function of the wavelength so that the resulting sinusoidal relative fluctuation in thrust has amplitude 3/5. Note that this relative fluctuation is expressible as:

$$\frac{\delta T}{T} = \frac{\delta L}{W} + \frac{W}{D}\left[\delta\gamma + \frac{V^2}{g}\frac{d}{dx}\left(\frac{\delta V}{V}\right)\right] + \frac{\delta V}{V} \tag{16}$$

Secondly, the wavelength is then adjusted to minimize (12). The result is: $2\pi/\omega = 58\ km$,

$\delta^2 m/m = -0.39\%$. The sinusoidal variations in V, h and T are indicated in Fig. 3. The wavelength obtained here, by our "local" analysis with sinusoidal variations, agrees rather well with the optimal wavelength, 52.8 km, obtained in [8] for the exact optimal "periodic cruise," but with $\eta_{min} = 0$ rather 1/4. The fuel saving, however, is too low.

The sinusoidal pattern for δT, on the other hand, is clearly nonoptimal. Both of the physical effects discussed in Section 3 would be amplified by choosing a square-wave pattern for δT with the same bounds. This will be examined in the next section.

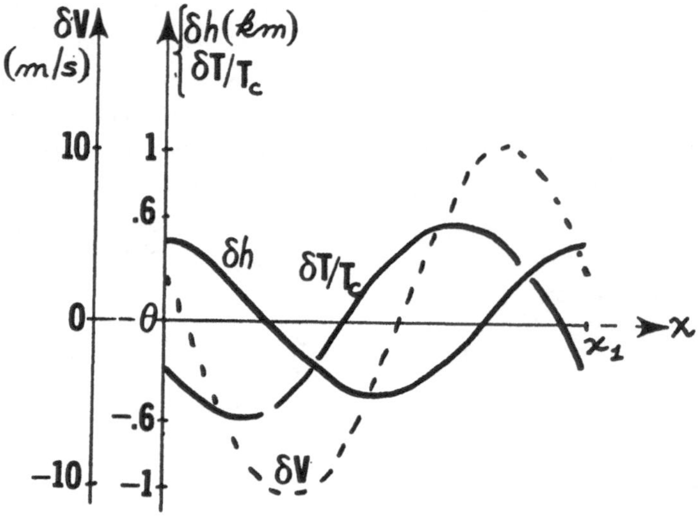

Figure 3.

7. Limited Thrust Variations

With limits imposed on δT, say

$$-\Delta_2 \le \frac{\delta T}{T} \le \Delta_1 ,$$

δV becomes an additional state, no longer satisfying (10), the total state being $X = [\delta V, \delta h, \delta\gamma]^T$, and the controls $\delta T/T$ and $\delta L/W$.

We seek the minimization of $\delta^2 m/m$, obtainable from (9) in the form:

$$J = \frac{\delta^2 m}{m} = \frac{1}{x_1}\int_0^{x_1} \left\{ \frac{1}{2} X^T A X + X^T B \frac{\delta L}{W} + \frac{1}{2} c \left(\frac{\delta L}{W}\right)^2 \right\} dx \tag{17}$$

subject to the linear equations (14), together with

$$\frac{d}{dx}(\delta V) = \frac{g}{V}\left\{ -\delta\gamma + \frac{D}{W}\left(\frac{\delta V}{V} + \frac{\delta L}{W} - \frac{\delta T}{T}\right)\right\} , \tag{18}$$

expressible in the form

$$\frac{dX}{dx} = F X + G_1 \frac{\delta L}{W} + G_2 \frac{\delta T}{T} \tag{19}$$

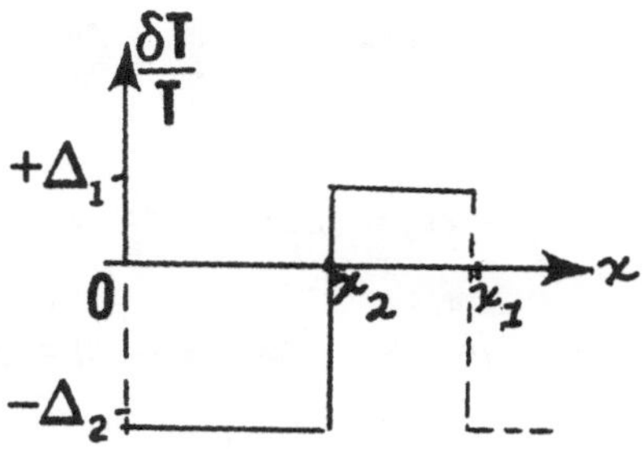

Figure 4.

This leads to the usual coupled state and adjoint equations, after elimination of $\delta L/W$, in the form:

$$\frac{d}{dx}\binom{X}{\lambda} = \overset{6\times6}{F_A}\binom{X}{\lambda} + \overset{6\times1}{G_A}\,\frac{\delta T}{T} \tag{20}$$

Assuming now the square wave pattern for $\delta T/T$ shown in Fig. 4, λ_V must vanish at the switch-points $x=0,\ x_2, x_1, \ldots$

The boundary conditions, for the optimal periodic solution of wavelength x_1, are

$$X(x_1) = X(0)\ ,\ \ \lambda(x_1) = \lambda(0) \tag{21}$$

Note that constancy of the associated Hamiltonian, namely

$$H = \frac{1}{2}X^TA\,X + X^TB\,\frac{\delta L}{W} + \frac{1}{2}c\left(\frac{\delta L}{W}\right)^2 + \lambda^T\frac{dX}{dx}\ , \tag{22}$$

implies that only 5 of the 6 boundary conditions (21) are independent.

The state and adjoint variables at $x=x_2, x_1$ are related to their values at $x=0$ by

$$\begin{bmatrix} X(x_2) \\ \lambda(x_2) \end{bmatrix} = \Phi_A(x_2)\begin{bmatrix} X(0) \\ \lambda(0) \end{bmatrix} + \Gamma_A(x_2)(-\Delta_2)$$

$$\begin{bmatrix} X(x_1) \\ \lambda(x_1) \end{bmatrix} = \Phi_A(x_1-x_2)\begin{bmatrix} X(x_2) \\ \lambda(x_2) \end{bmatrix} + \Gamma_A(x_1-x_2)(\Delta_1) \tag{23}$$

where the matrix functions $\Phi_A(x), \Gamma_A(x)$ are obtained by numerical integrations associated with the linear system (20). Eliminating $X(x_2), \lambda(x_2)$, we obtain

$$\begin{bmatrix} X(x_1) \\ \lambda(x_1) \end{bmatrix} = \Phi_A(x_1-x_2)\Phi_A(x_2)\begin{bmatrix} X(0) \\ \lambda(0) \end{bmatrix} + \Gamma_A(x_1-x_2)\Delta_1 - \Phi_A(x_1-x_2)\Gamma_A(x_2)\Delta_2 \tag{24}$$

The following computational scheme is available: (a) Input values for x_1 and x_2; (b) set $\lambda_V(0)=0$ and invoke the boundary conditions (21) to solve (24) for the other 5 initial values; the $\lambda_V(x_1)$ will then be automatically zero; (c) adjust x_2 until $\lambda_V(x_2)$, obtainable from (23), also vanishes. This will yield the optimal solution for the chosen x_1.

The optimal choice of x_1 is most efficiently obtained by appealing to the appropriate

$$\eta_{min} = \frac{1}{4} \;\Rightarrow\; -\frac{3}{5} \le \frac{\delta\eta}{\eta_c} \le \frac{3}{5}$$

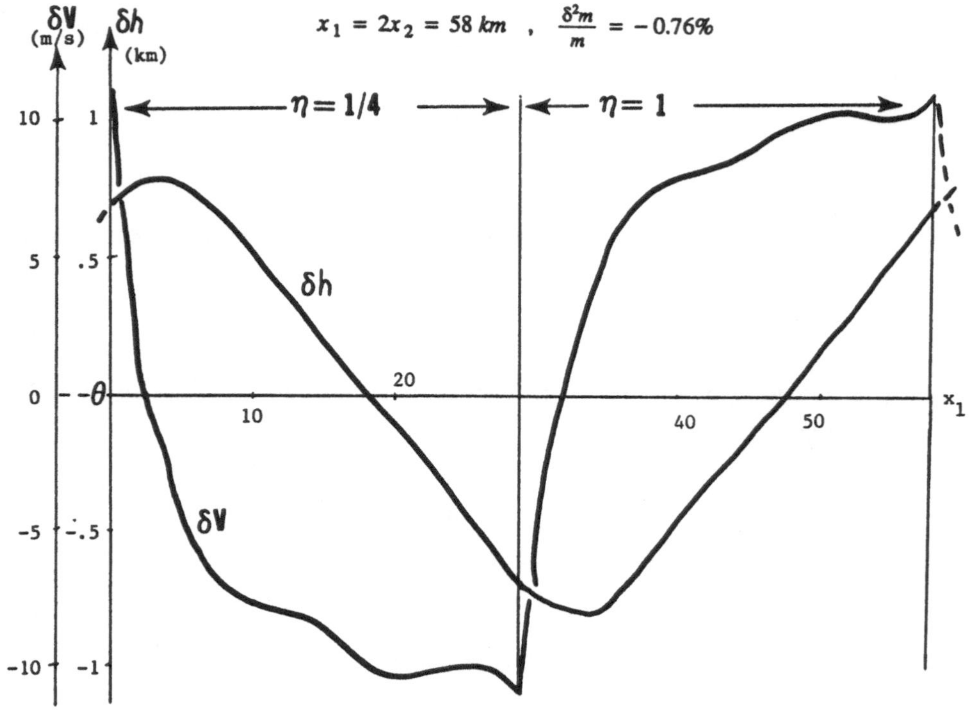

Figure 5.

transversality condition: $H = J$, where H and J are defined by (22) and (17).

The results for the example of reference [8] are now as follows: For $\eta_{min} = 1/4$, so that $\Delta_1 = \Delta_2 = 3/5$:

$$x_1 = 2x_2 = 58\ km \quad , \quad \frac{\delta^2 m}{m} = -0.76\% \quad ,$$

approximately twice the fuel saving obtained with a sinusoidal pattern, as might have been expected. The corresponding δV and δh histories are shown in Fig. 5. For $\eta_{min} = 0$, so that $\Delta_1 = 3/5$, $\Delta_2 = 1$:

$$x_1 = 54\ km \quad , \quad x_2 = 520.3\ km \quad , \quad \frac{\delta^2 m}{m} = -1.26\% \quad ,$$

and the corresponding δV and δh histories are shown in Fig. 6.

Considering the size of the variations involved, the δV and δh histories obtained by our "local" thrust-limited analysis agree surprisingly well with those obtained for various η_{min} from an exact analysis and shown in Fig. 3 of ref. [8].

The optimal wavelength x_1 for $\eta_{min} = 0$ is in close agreement, although the exact fuel saving (1.95%) is somewhat larger than that obtained by the local analysis.

Evidently, the fuel saving obtainable from oscillatory cruise is not substantial! Nevertheless, much larger % savings have been found by Gilbert et al. [9], in a modified problem in which an altitude limit is imposed, far below the altitude which would yield optimal steady cruise. Further oscillations in altitude below the imposed limit can substantially reduce the large penalty incurred by steady cruise at the imposed limit.

$$\eta_{min} = 0 \Rightarrow -1 \leq \frac{\delta\eta}{\eta_c} \leq \frac{3}{5}$$

$$x_1 = 54\ km\ ,\ x_2 = 20.3\ km\ ,\ \frac{\delta^2 m}{m} = -1.26\%$$

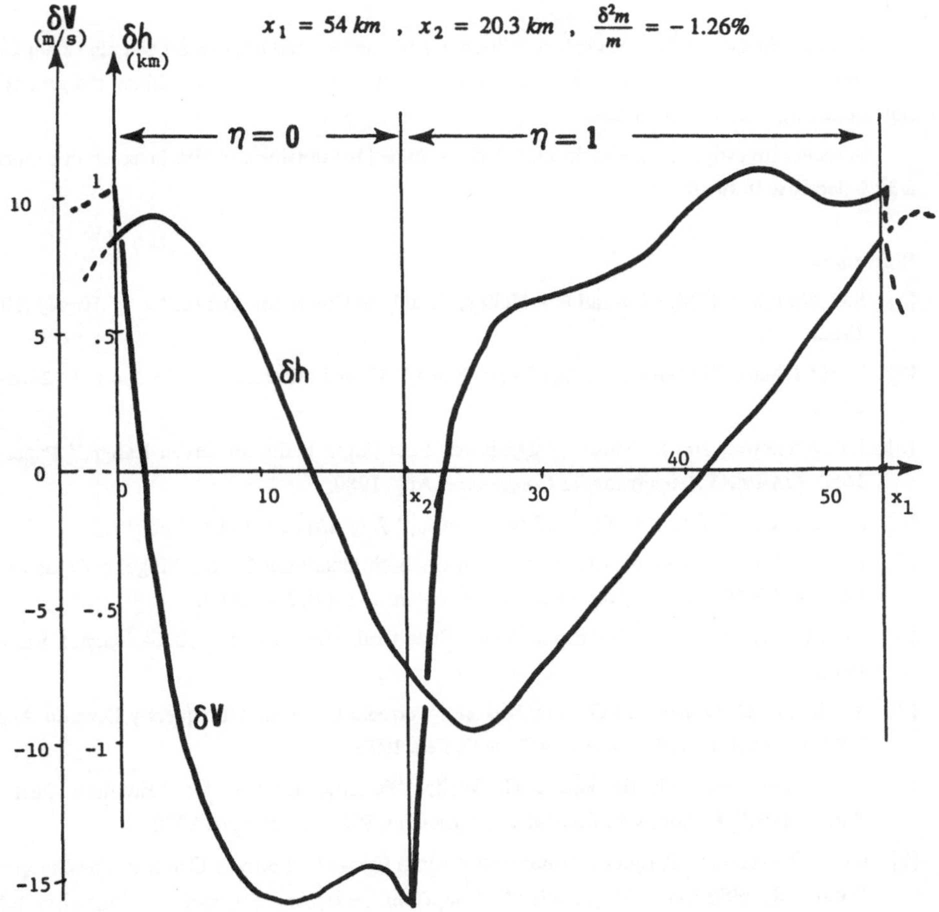

Figure 6.

8. The Endurance Problem

In this last section we shall briefly discuss the possible fuel saving obtainable by oscillatory cruise in a different problem: Minimum fuel/hour instead of Minimum fuel/km.

Here time is the independent variable and optimal steady cruise occurs at $\min_{V,h}(\sigma D)$ rather than $\min_{V,h}(\sigma D/V)$. Naturally this occurs with much lower drag D. Now the fuel-rate is expressible in the form:

$$\dot{m} = \frac{\sigma W}{g}\left[\dot{V} + g\left(\frac{D}{W} + \sin\gamma\right]\right. , \tag{25}$$

and the gravity term $\sigma W \sin\gamma$ again can contribute a 2nd-order saving, since, to 2nd order, $\delta\dot{h} = V\delta\gamma + \delta V\delta\gamma$, so that Δm/cycle includes a term $\sigma W/V\int(\delta\dot{h} - \delta V\delta\gamma)dt$ while $\int\delta\dot{h}dt$ drops out.

Detailed further analysis reveals that there is again an induced drag effect and also a negative $(\delta\gamma)^2$ contribution. However, because W/D is much larger than in the fuel/km problem, the gravity effect is now quite large and predominates.

A recent investigation by Sachs and Christodoulos [10] obtains a 63.5% increase in endurance, i.e. a 39% decrease in fuel/hour.

References

[1] S.C. Houlihan, E.M. Cliff and H.J. Kelley, "Study of Chattering Cruise," *J. of Aircraft,*19(2), Feb. 1982.

[2] T. Edelbaum, "Maximum-Range Flight Paths," United Aircraft Corp. Report R-22465-24, July 1955.

[3] J.V. Breakwell and H. Shoaee, "Minimum Fuel Flight Paths for Given Range," Paper No. 80-1660, *AIAA/AAS Astrodynamics Conference,* Aug. 1980.

[4] J.L. Speyer, "On the Fuel Optimality of Cruise," *J. of Aircraft,* Vol. 9, Feb. 1972.

[5] H.M. Robbins, "A Generalized Legendre-Chebsch Condition for the Singular Cases of Optimal Control," *IBM Journal of Research & Development,* 11(4), July 1967.

[6] J.L. Speyer, "Nonoptimality of the Steady-State Cruise for Aircraft," *AIAA Journal,* Vol. 14, Nov. 1976.

[7] S. Bittanti, G. Fronza and G. Guardabassi, "Periodic Control: A Frequency Domain Approach," *IEEE Trans. Automatic Control,* AC-18(1), Feb. 1973.

[8] W. Grimm, H.J. Oberle and K.H. Well, "Periodic Control for Minimum Fuel Aircraft Trajectories," *J. Guidance, Control and Dynamics,* 9(2):March-April 1986.

[9] E.G. Gilbert and D.T. Lyons, "Improved Aircraft Cruise by Periodic Control: The Computation of Optimal Specific Range Trajectories," *Proc. Conf. on Inform. Sci. and Sys.,* Princeton, NJ, March 1980.

[10] G. Sachs and T. Christodoulos, "Endurance Increase by Cyclic Control," *J. Guidance, Control and Dynamics,* 9(1):Jan-Feb. 1986.

A PLANAR INTERCEPT PROBLEM WITH A CHATTERING JUNCTION OF NON-SINGULAR AND SINGULAR SUBARCS.

Klaus Schnepper
Deutsche Forschungs- und Versuchsanstalt für Luft- und Raumfahrt (DFVLR), Institut für Dynamik der Flugsysteme,
Oberpfaffenhofen, D-8031 Wessling

1. Introduction

A planar pursuit-evasion problem between two opponents is considered in this paper. Kinematical equations are taken to be nonlinear while the pursuer's dynamical equations are modeled as a linear second order system.

Open loop controls are the pursuer's commanded normal acceleration and the evader's normal acceleration. By assuming an evader control time history the pursuit-evasion problem is transformed into an optimal control problem (OPC). It is shown that the solution of this OPC consists of bang-bang type subarcs and possible singular subarcs with a chattering junction [1],[2].

For the numerical example a multiple-shooting algorithm is taken from [3],[4],[5] and an optimal solution with 5 switching points is obtained.

2. Problem Formulation

The problem considered is a two-dimensional constant velocity pursuit-evasion scenario depicted in Figure 1. Here x, y are the evader's E kartesian coordinates relative to the pursuer P. θ_p , θ_e are the pursuer's and evader's heading angles measured from the x-axis, a_p , a_{p2} are the pursuer's lateral acceleration and its time derivative, a_e is the evader's lateral acceleration and v_p , v_e are pursuer's and evader's constant velocities. The pursuer's dynamics are assumed to be of second order; the damping constant is taken to be 1.

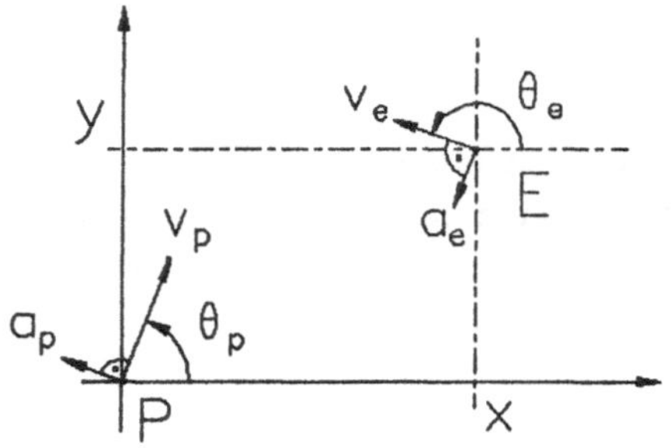

Fig.1 : Pursuit-Evasion Scenario

In this coordinate system the equations of motion are given by:

$$\begin{aligned}
\dot{x} &= v_e\cos\theta_e - v_p\cos\theta_p \\
\dot{y} &= v_e\sin\theta_e - v_p\sin\theta_p \\
\dot{\theta}_e &= \frac{a_e}{v_e} \\
\dot{\theta}_p &= \frac{a_p}{v_p} \\
\dot{a}_p &= a_{p2} \\
\dot{a}_{p2} &= -2\omega a_{p2} - \omega^2 a_p + \omega^2 a_{pc}
\end{aligned} \tag{1}$$

For simplicity an optimal control problem with the evader's lateral acceleration a_e assumed to be a given function of time is considered. The remaining control in this problem is the pursuer's lateral acceleration command a_{pc}. The cost functional is final time t_f.

$$J(a_{pc}) = t_f \tag{2}$$

The following initial- and final conditions are imposed:

$$\begin{aligned}
x(t_0) &= x_0 & x(t_f) &= 0 \\
y(t_0) &= y_0 & y(t_f) &= 0 \\
\theta_e(t_0) &= \theta_{e0} \\
\theta_p(t_0) &= \theta_{p0} \\
a_p(t_0) &= 0 \\
a_{p2}(t_0) &= 0
\end{aligned} \tag{3}$$

The pursuer's lateral acceleration command a_{pc} is limited by:

$$|a_{pc}| \leq a_{pc}^{max} \tag{4}$$

3. Optimal Controls and Junction of Singular- and Nonsingular Subarcs.

3.1. Optimal Controls.

Following [6] the variational Hamiltonian is formed:

$$\begin{aligned} H(t,a_{pc}) = & \lambda_x(v_e\cos\theta_e - v_p\cos\theta_p) + \lambda_y(v_e\sin\theta_e - v_p\sin\theta_p) \\ & + \lambda_{\theta e}\cdot\frac{a_e}{v_e} + \lambda_{\theta p}\cdot\frac{a_p}{v_p} + \lambda_{ap}a_{p2} \\ & + \lambda_{ap2}(-2\omega a_{p2} - \omega^2 a_p + \omega^2 a_{pc}), \end{aligned} \tag{5}$$

where λ_x, λ_y,... are the Lagrange multipliers associated with the differential system (1). The multipliers are obtained from the solution of the adjoint differential equations

$$\begin{aligned} \dot{\lambda}_x &= 0 \\ \dot{\lambda}_y &= 0 \\ \dot{\lambda}_{\theta e} &= v_e(\lambda_x\sin\theta_e - \lambda_y\cos\theta_e) \\ \dot{\lambda}_{\theta p} &= v_p(-\lambda_x\sin\theta_p + \lambda_y\cos\theta_p) \\ \dot{\lambda}_{ap} &= \omega^2\lambda_{ap2} - \frac{1}{v_p}\lambda_{\theta p} \\ \dot{\lambda}_{ap2} &= 2\omega\lambda_{ap2} - \lambda_{ap} \end{aligned} \tag{6}$$

subject to the transversality conditions

$$\begin{aligned} \lambda_{\theta e}(t_f) &= 0 & \lambda_{ap2}(t_f) &= 0 \\ \lambda_{\theta p}(t_f) &= 0 & H(t_f,a_{pc}^*) &= -1 \\ \lambda_{ap}(t_f) &= 0. \end{aligned} \tag{7}$$

Here a_{pc}^* denotes the optimal control. It is calculated from the minimum principle :

$$H(t,a_{pc}^{*}) = \min_{|a_{pc}| \leq a_{pc}^{max}} H(t,a_{pc})$$

For $\lambda_{ap2} \neq 0$ this yields:

$$a_{pc}^{*} = -\text{sign}(\lambda_{ap2}) \cdot a_{pc}^{max} \tag{8}$$

For $\lambda_{ap2} \equiv 0$ on an interval $[t_1,t_2]$, there is a singular control which can be computed by repeated differentiation of $\lambda_{ap2} \equiv 0$ until a_{pc} appears explicitly [6], [7]. This process yields:

$$\begin{aligned} &\lambda_{ap} \equiv 0 && a_p \equiv 0 \\ &\lambda_{\theta p} \equiv 0 && a_{p2} \equiv 0 \\ &-\lambda_x \sin\theta_p + \lambda_y \cos\theta_p \equiv 0 && a_{pc} \equiv 0 \end{aligned} \tag{9}$$

From (9), the order of the singular control is m = 3. As the degree of the singular arc is odd a junction cannot be excluded by the discontinuity of the control at a junction [6],[7].

3.2. Junction of Singular and Nonsingular Arcs.

In this section the junction of singular- and nonsingular arcs is discussed. The result is that for this specific case such a junction is only possible with a chattering control. This result is posed in form of a proposition for wich the outline of a proof is given. A detailed proof can be found in [8].

Proposition :

For $\omega > \dfrac{6a_{pc}^{max}}{\pi\, v_p}$ a singular arc cannot be reached with a finite number of switching points between $\pm a_{pc}^{max}$ from a time t_0 with $a_p(t_0) = a_{p2}(t_0) = 0$

Outline of proof :

The proof is based on a number of lemmas which will be stated but not proven. The proof of these lemmas is a simple but tedius exercise in calculus.

It is assumed that a junction of singular- and nonsingular subarcs exists at some time $t_s > t_0$. Then the behaviour of $a_p, a_{p2}, \lambda_{ap2}$ and $\lambda_{\theta p}$ are studied at bang- bang switching points

$$t_n < t_{n-1} < \dots < t_{i-1} < t_i < \dots < t_1 = t_s$$

It is shown that :

$$\begin{aligned}
&\mathrm{sign}(a_p(t_i)) = -\mathrm{sign}(a_{p2}(t_i)) \quad , \ a_p(t_i) \neq 0\\
&\mathrm{sign}(a_p(t_i)) = \mathrm{sign}(a_{pc}(t_i+\epsilon))\\
&\mathrm{sign}(\dot{\lambda}_{ap2}(t_i)) = \mathrm{sign}(a_p(t_i)\cdot\lambda_x) \qquad (10)\\
&\mathrm{sign}(\dot{\lambda}_{\theta p}(t_i)) = \mathrm{sign}(a_p(t_i)\cdot\lambda_x) \quad , \ \text{for } i = 2,3,\dots\\
&\exists_1 \ t_i^+ \in \,]t_i, t_{i-1}[\ : \lambda_{\theta p}(t_i) = 0 \quad , \ \text{for } i = 3,4,\dots
\end{aligned}$$

where ϵ is some "sufficiently small" positive number. This is a contradiction to the assumption $a_p(t_0) = 0$ and the proposition is proven.

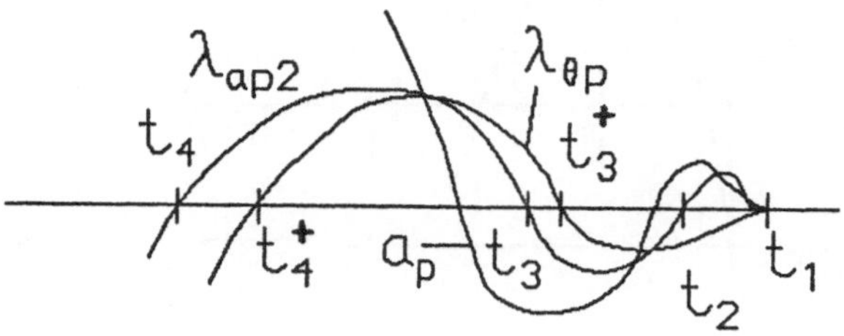

Fig. 2 : Scetch of Behaviour of a_p, a_{p2}, λ_{ap2} and $\lambda_{\theta p}$ According to (10).

Now for the proof of (10).

First of all some properties of a_p are needed.

<u>Lemma 1</u> :

For $a_{pc} \equiv$ const there hold :

$$\lim_{t\to\infty} a_p(t) = a_{pc}$$

$a_p(t)$ has at most one relative extremum and inflection point

For $a_p(t_0) = a_{p2}(t_0) = 0$ and $|a_{pc}| \leq a_{pc}^{max}$:

$$|a_p(t)| \leq a_{pc}^{max} \qquad \text{for all } t \geq t_0.$$

These points can be seen by use of representations of the solution for $a_p(t)$. An immediate consequence of lemma 1 is the existence of at least two switching points t_2 and t_3 before t_s. The next lemma is mainly concerned with an upper bound for the time between two consecutive switching points.

<u>Lemma 2</u> :

The maximum time distance between two consecutive bang- bang switching points is given by

$$t_q = \frac{\pi \cdot v_p}{2\, a_{pc}^{max}} \quad , \quad \text{if } \omega > \frac{6\, a_{pc}^{max}}{\pi\, v_p} .$$

t_q is a lower bound for the time necessary to change θ_p by 90 degrees. This lemma is established by considering t' in Figure 3.

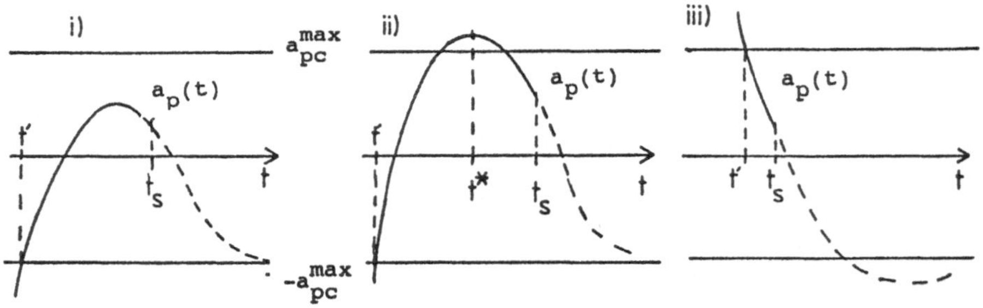

Fig.3 : Possible Behaviour of a_p Before a Switching Point.

Next a coupling between the sign of $\dot{\lambda}_{\theta p}$ and θ_p is established. For it θ_p^o , the constant value of θ_p on a singular arc is introduced. Then the following lemma can be proven.

<u>Lemma 3</u> :

$$\dot{\lambda}_{\theta p} \geq 0 \Leftrightarrow \begin{cases} \theta_p \in \,]-\pi/2, \theta_p^o] \cup [\pi+\theta_p^o, 3/2\cdot\pi] \ , & \text{if } \lambda_x > 0 \\ \theta_p \in [\theta_p^o - \pi, \theta_p^o] & , \text{if } \lambda_x < 0 \end{cases}$$

The proof of lemma 3 uses the generalized Legendre- Clebsch condition [6], [7].

With these lemmas and an explicit solution for $\dot{\lambda}_{ap2}$ in integral form the assertions (10) can be established by induction.

W.l.o.g assume $\lambda_x > 0$ and $a_{pc}(t) = a_{pc}^{max}$ for $t \in]t_2, t_s[$:

<u>Induction start:</u>

The existence of t_2 and t_3 was already shown. By lemma 2 $t_s - t_3 < t_h = 2t_q$ and this yields:

$a_p(t_2) > 0$ $\qquad$ $a_{p2}(t_2) < 0$

$\theta_p(t) < \theta_p^o$, $\qquad$ $\dot{\lambda}_{\theta p}(t) > 0$

$\lambda_{\theta p} < 0$ $\qquad$ for $t \in [t_2, t_s[$.

By the representation for $\dot{\lambda}_{ap2}$ it can be shown that between three successive zeros of λ_{ap2} there is at least one zero t_* of $\lambda_{\theta p}$ and

$$\text{sign}(\dot{\lambda}_{ap2}(t_3)) = \text{sign}(\dot{\lambda}_{\theta p}(t_*)).$$

From this can be deduced the existence of a zero of $\lambda_{\theta p}$ with negative slope in the interval $]t_3, t_2[$. Together with lemma 3 uniqueness of this zero can be shown. It then follows that $\dot{\lambda}_{ap2}(t_3) < 0$ and the sign relations for a_p and a_{p2} hold.

<u>Induction step:</u>

The induction step closely follows the lines of the proof for the start of the induction and will be omitted. □

4. <u>Numerical Example:</u>

The following parameters, initial- and final conditions are used for the example:

$x(t_0)$	$= 2000$ m	$y(t_0)$	$= 0$ m
$\theta_e(t_0)$	$= \pi$	$a_p(t_0)$	$= 0$ m/sec²
$a_{p2}(t_0)$	$= 0$ m/sec³	$x(t_f)$	$= 0$ m
$y(t_f)$	$= 0$ m	v_e	$= 250$ m/sec
v_p	$= 750$ m/sec	ω	$= 10$ 1/sec
a_{pc}^{max}	$= 20 \cdot g$		

a_e is constant with $2 \cdot g \leq |a_e| \leq 8 \cdot g$. The physically meaningful range for initial values of θ_p is given in Figure 4.

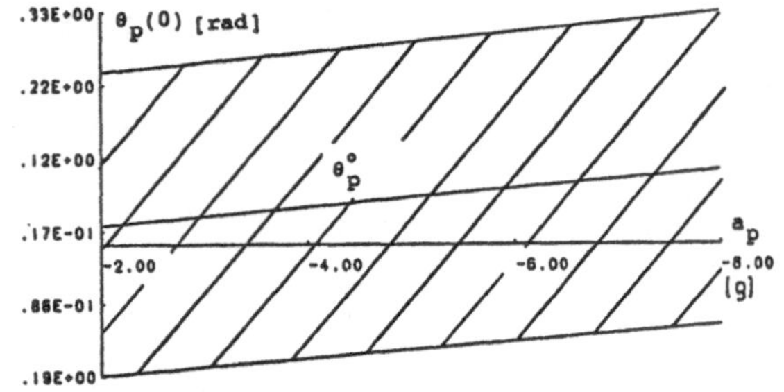

Fig. 4 : Physically Meaningful Range of Initial $\theta_p(t_0)$.

Starting from a boundary of this region optimal trajectories can be generated by continuation. The optimal trajectories are solution of the boundary-value problem (1),(3),(6) and (7), where the control (8) is used. The computation was done with the FORTRAN code BDSCO, developed at TU-MUNICH [3],[4],[5]. During the continuation switching-points are added from the final time. Figures 5 - 8 show some states of a solution with 5 switching points. As can be seen from the plots of the switching function λ_{ap2}, it is hardly possible to proceed any further in this continuation as λ_{ap2} is rapidly decreasing in absolute value. This behavior also gives some indication that in this problem a chattering junction between singular- and nonsingular arcs exists.

For practical purposes one has to resort to approximate solutions prescribing some control structure. Figures 9 - 11 show the results for the same problem as shown in Figures 5 - 8, when a control structure with three switching points and vanishing commanded acceleration at the end is chosen. In this example the cost index for the suboptimal scheme can hardly be discerned from the optimal cost index.

<u>References :</u>

[1] C. Marchal : "Chattering Arcs and Chattering Controls", Journal of Optimization Theory and Applications, Vol. 11, No.5, 1973, p. 441-468.

[2] C. Marchal, P. Contensou : "Singularities in Optimization of Deterministic Dynamic Systems", Journal of Guidance and Control, Vol. 4, No.3, 1981, p 240-252.

[3] J. Stoer, R. Bulirsch : "Introduction to Numerical Analysis", Springer Verlag, New York, Heidelberg, Berlin, 1980.

[4] R. Bulirsch : "Die Mehrzielmethode zur numerischen Lösung von nichtlinearen Randwertproblemen und Aufgaben der Optimalen Steuerung", Report Carl-Cranz-Gesellschaft, Heidelberg, 1971.

[5] W. Grimm, H. J. Oberle, E. Berger : "Benutzeranleitung für das Rechenprogramm BNDSCO zur Lösung beschränkter optimaler Steuerungsprobleme", DFVLR-Mitteilung 85-05, 1985.

[6] A. E. Bryson, Jr., Yu-Chi Ho, "Applied Optimal Control", Hemisphere Publishing Company, Washington, D.C., 1975.

[7] H. J. Kelley, R. E. Kopp, H. G. Moyer : "Singular Extremals" in Topics in Optimization, G. Leitmann, ed., Academic Press, New York, 1967, p. 63-101.

[8] K. Schnepper : "Zeitoptimale Steuerung eines Flugkörpers und Konstruktion suboptimaler Rückkopplungssteuerungen", DFVLR IB 515-83/11, 1983.

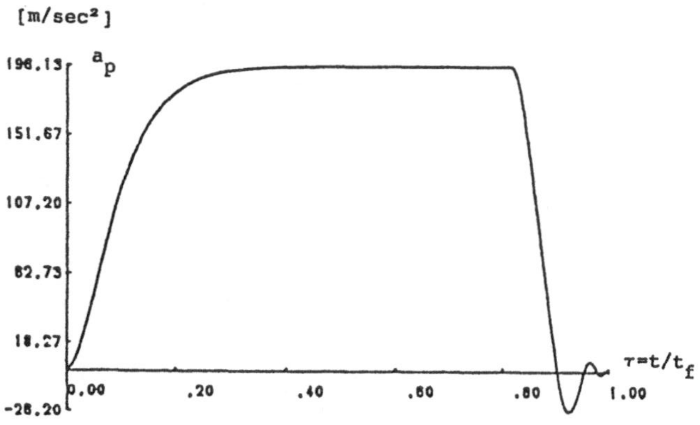

Fig.5 : a_p, Trajectory with 5 Switching Points, a_e=2g

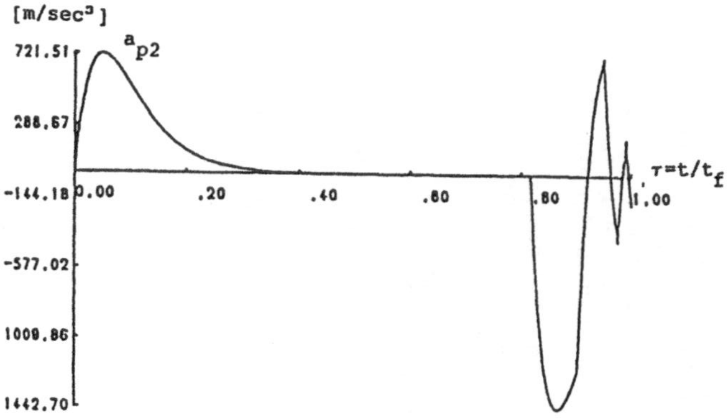

Fig.6 : a_{p2}, Trajectory with 5 Switching Points, a_e=2g

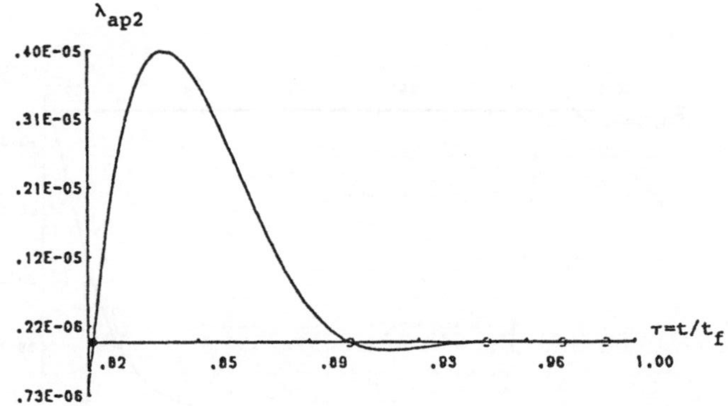

Fig.7 : λ_{ap2}, Trajectory with 5 Switching Points, a_e=2g

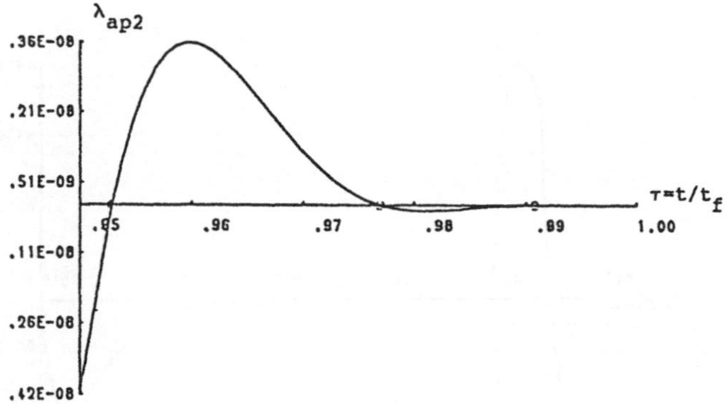

Fig.8 : λ_{ap2}, enlarged, last 3 Switching Points

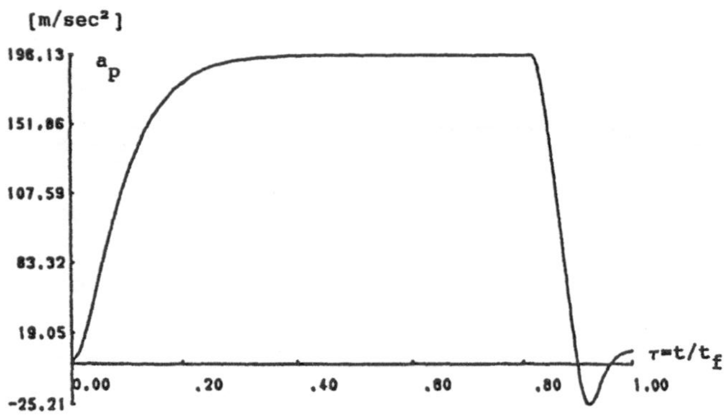

Fig.9 : a_p, Suboptimal Trajectory with 3 Switching Points

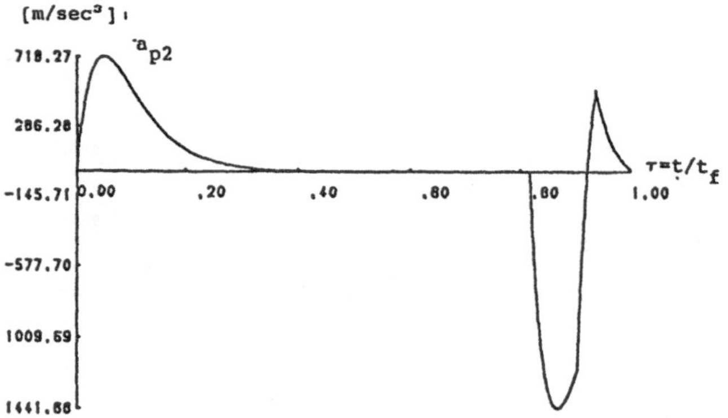

Fig.10: a_{p2}, Suboptimal Trajectory with 3 Switching Points

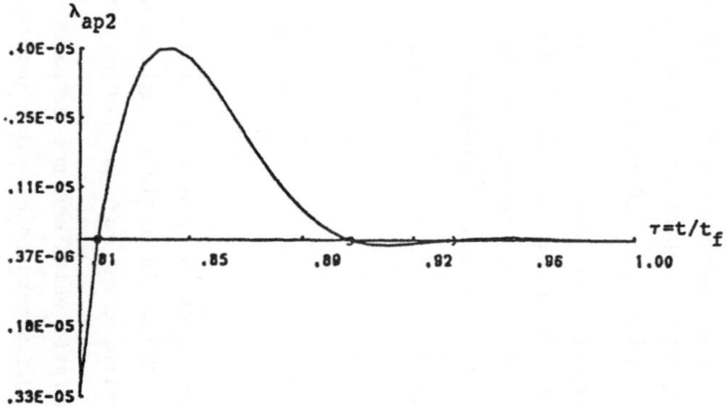

Fig.11: λ_{ap2}, Suboptimal Trajectory with 3 Switching Points

ON THE SYNTHESIS OF OPTIMAL NONLINEAR FEEDBACK LAWS

Houria Bourdache-Siguerdidjane
Laboratoire des Signaux et Systèmes
CNRS/ESE, Plateau du Moulon
91190 Gif-sur-Yvette, France

This paper is concerned with the automatic computation of the optimal nonlinear feedback control law, starting with a short review of the theory. The synthesis of a control system constitutes the main part of an optimization problem.

We have shown in previous papers that the optimal nonlinear feedback control law satisfies a set of partial differential equations. The knowledge of the feedback law can therefore be considered as equivalent to the computation of the hypersurface corresponding to the solution of these equations.

This hypersurface is computed off-line. Attractive features for real time implementation are discussed.

Introduction

The task of running various kinds of physical processes under a performance criterion is an important problem in control engineering. The synthesis of a control system therefore constitutes the main part of an optimization problem.

In nonlinear systems, the analytic solutions are extremely difficult to obtain and numerical solutions involve the two-boundary value problems which may become very cumbersome even with fast computers. However, such solutions would yield the control in open-loop form. In order to avoid the many disadvantages of open-loop controls, feedback controls are desirable.

Up to now, the approaches used have focused on the linearization around a fixed point, or are based on Taylor series expansion of the feedback law [W1], so that they are not useful over a wide region of operation. Furthermore, the difficulty in using power series is the large number of differential equations one needs to solve in order to determine the coefficients of the terms in series, which increases exponentially with the dimension of the process.

Recent results have established a methodology for the computation of optimal nonlinear feedback laws [B1, B2]. It is shown that the feedback control satisfies a set of quasi-linear partial differential equations whose solution is a surface which passes through a given curve.

It is important to realize that the knowledge of the feedback law can therefore be considered as equivalent to the off-line computation of this surface.

The implementation of this surface on a microprocessor promises well for improved performance with respect to present control.

I. Optimal feedback control

Recent results [B1,B2] state the following theorem:

Theorem

Given an optimal problem in Bolza form

$$\begin{cases} \dfrac{dx(t)}{dt} = F(x(t),u(t),t) \\ J = \emptyset\,(x(T)) + \displaystyle\int_0^T F^\circ(x(t),u(t),t)\,dt \end{cases}$$

such that the dimension of the state vector x is equal to the dimension of the control vector u. The state x and the control u are unconstrained. The optimal feedback law u(x,t) then satisfies a system of quasi-linear partial differential equations expressed, in matrix notation, by the following equation:

$$\begin{aligned} &[F^\circ_{u^2} - F^T_{u^2}(I_N \otimes (F_u^{-T} F^\circ_u))]\,[\frac{\partial u}{\partial t} + \frac{\partial u}{\partial x} F] = \\ &F^T_{ut} F_u^{-T} F^\circ_u - F^\circ_{ut} + (F^T_{ux}(I_N \otimes (F_u^{-T} F^\circ_u)))F \\ &-F^\circ_{ux} F + F^T_u F^\circ_x - F^T_u F^T_x F_u^{-T} F^\circ_u \end{aligned} \qquad (1)$$

where $\otimes$ is the Kronecker product. T is the sign transposition. I_N is the N-dimensional identity matrix. We denote by $F^T_{u^2}$ and F^T_{ux} the partitioned matrices $(F^T_{uu_1} | \dots | F^T_{uu_N})$ and $(F_{ux_1}^T | \dots | F^T_{ux_N})$ respectively.

This result can be obtained in two ways. The first uses Lie brackets of vector fields, and the second the Hamilton-Jacobi-Bellman equation. The summary derivation of this equation is given in Appendix 1.

Let $C_{NxN} = [F^\circ_{u^2} - F^T_{u^2} (I_N \otimes F_u^{-T} F^\circ_u)]$.

II. Solution of this PDE

a. General case

-C invertible.

As is well-known, the integration of equation (1) reduces through the method of characteristics (see appendix 2) to a system of first order ordinary differential equations as follows:

$$\begin{cases} \dot{x}_i = F_i \\ \dot{u}_i = P_i \end{cases} \qquad i=1,\dots,N \tag{2}$$

where P_i is the ith components of $P=C^{-1}D$, D is the right-hand side of (1).

The solution of (1) is the hypersurface generated by the integral curves of differential system (2) and which passes through a given curve Γ. The equation for Γ, $\theta(x,u,T)=0$, comes from the equation which minimizes the Hamiltonian and the boundary conditions of the problem. For fixed time and free end-point problem, the boundary condition is $p(T)=\partial\emptyset\ /\partial x$.

Replacing p(T) in $H_u=0$ yields $\theta(x,u,T)=0$.

b. Degenerate case

When the cost function does not depend on the control, the feedback law satisfies the following algebraic equations [B1,B3]:

$$\sum_k F_u^k F^\circ_{x_k} = 0$$

III. Off-line computing procedure

In order to generate automatically the hypersurface solution of the partial differential equation (1), the procedure goes through the following steps:

The user need only make an input of the functions F, F°, θ by calling the appropriate subroutine, and specify the terminal conditions of the states and the terminal time T.

The partial derivative of F° with respect to u is first determined. If $F^\circ_u=0$, go to step 1 otherwise go to step 2.

Step 1

Solve the algebraic equations $\sum_k F_u^k F^\circ_{x_k} = 0$ by calling the programme CALEXP using the symbolic calculus.

Step 2

Determine the expressions of the right-hand side of partial differential equation (1) by calling the programme CALEXP using symbolic calculus as well as the Cauchy condition $\theta(x,u,T)=0$ and solve it as $u(x,T)=\rho(x,T)$ if it is possible.

Step 3

The equations (2) are integrated backward by a suitable method of integration starting with the point $u_i(x,T) = \rho(x_T,T)$, x_T being specified by the user. The values of u, x and t are stored on direct access file in a particular manner and the process begins anew at step 3 with another point $u_{i+1}(x,T) = \rho(x_T+\Delta x,T)$. Δx may be specified by the user as well as a desired stopping point.

This is the numerical solution of the method of characteristics (see Appendix 2 for a short review of this method).

IV. Discussion and implementation

The direct access file with the whole hypersurface is organized in a very special manner in order to restitute the values of the control, for a given state, in as short a time as possible.

This methodology is currently carried out on an IBM Personal Computer with 512K of internal memory. The programmes were written and compiled in FORTRAN. The symbolic manipulation programme is written in REDUCE.

V. Conclusion

The next phase consists of implementing the hypersurface, generated automatically off-line as described above, in a "small machine" which will be called **SCOOP 2000** (Simulateur de **CO**mmande **OP**timal). This control system machine is based on 286A Intel microprocesseur and contains the necessary interfaces. It will permit the real time closed-loop control of the square processes.

This work is supported by the DRET (Direction des Recherches Etudes et Techniques) under Contract 85 34 04000 470 75 01

Appendix 1

A summary of the Lie bracket approach

As usual, the Hamiltonian is defined by

$$H = \langle \tilde{p}, \tilde{F} \rangle = \sum_{k=o}^{N} p_k F^k (x,u,t)$$

where $\tilde{p} = (p_o, p_1, \ldots, p_N)$ is the adjoint vector and $\tilde{F}$ is the vector of components (F^o, F). To the vector fields $\tilde{F}$, associate the new vector field

$$A = \frac{\partial}{\partial t} + \tilde{F}.$$

Recall that a vector field $X:R^N \to R^N$ is defined in a local coordinate chart $x=(x_1,\ldots,x_N)$ as a first order linear differential operator

$$X = \sum_{k=1}^{N} X^k \frac{\partial}{\partial x_k} .$$

Using the Maximum Principle and the Hamiltonian formalism, one obtains

$$\frac{d^\nu H_u}{dt^\nu} = \langle \tilde{p}, ad_A^\nu \tilde{F}_u \rangle = 0. \qquad \nu=0,1,2\ldots$$

where the notation $ad_A^\nu \tilde{F}_u = [A, \tilde{F}_{u_i}]]$ means the Lie bracket $[A, \tilde{F}_{u_i}]$ iterated ν times. Recall that a Lie bracket of two vector fields X,Y is the commutator $[X,Y] = XY-YX$.

Corollary: The optimal feedback control law satisfies the infinite hierarchy of necessary conditions

$$\langle \tilde{p}, ad_A^\nu \tilde{F}_{u_i} \rangle = 0, \qquad \nu=0,1,\ldots \quad i=1,\ldots,m . \tag{3}$$

For m=N, the first order conditions, in other words those for $\nu=0$ and 1 lead to

$$\left\{ \begin{array}{l} \sum_{k=o}^{N} p_k F_{u_i}^k = 0 \\ \sum_{k=o}^{N} p_k [A,\tilde{F}_{u_i}]^k = 0, \qquad i=1,\ldots,N. \end{array} \right. \tag{4}$$

which is a homogeneous system of 2N equations in N+1 unknowns $(p_o, p_1, \ldots, p_N)$. The adjoint vector p is not identically zero, hence (4) has a solution if all the characteristic determinants vanish.

$$\begin{vmatrix} F^\circ_{u_1} & \ldots\ldots\ldots\ldots & F^N_{u_1} \\ & \ldots\ldots\ldots\ldots & \\ & \ldots\ldots\ldots\ldots & \\ F^\circ_{u_N} & \ldots\ldots\ldots\ldots & F^N_{u_N} \\ [A,\tilde{F}_{u_i}]^\circ & \ldots\ldots\ldots\ldots & [A,\tilde{F}_{u_i}]^N \end{vmatrix} = 0; \; i=1,\ldots,N.$$

We obtain a first order system of N quasi-linear partial differential equations expressed by (1).

Remark 1: We treat only the square case here, but we indicate that if m and N are different, the necessary conditions (3) lead to higher order partial differential equations.

Remark 2: We have assumed that the admissible controls and states are not constrained by any boundaries. However, for several problems, one can add a penalty function in the performance index.

Appendix 2

We now briefly discuss how to solve an equation of the form (1) by the method of characteristics.

Method of characteristics (see ref.[C1])

We restrict our discussion to an equation involving two independent variables, although the method is easily extended to handle equations with more independent variables.

The first order quasi-linear partial differential equation

$$a(x,u,t)\frac{\partial u}{\partial t} + b(x,u,t)\frac{\partial u}{\partial x} = c(x,u,t) \tag{A.1}$$

is interpreted geometrically in order to construct a solution.

We assume that a solution $u=u(x,t)$ of (A.1) can be found. $u(x,t)$ is called an integral surface, and we express it in implicit form $\phi(u,x,t)=0$. The gradient vector $\nabla\phi$ $(\frac{\partial u}{\partial t}, \frac{\partial u}{\partial x}, -1)$ is normal to the integral surface ϕ.

The equation (A.1) can be written as the dot product

$$(a,b,c).(\frac{\partial u}{\partial t}, \frac{\partial u}{\partial x}, -1) = 0 \tag{A.2}$$

which implies that the vectors $\vec{\gamma}=(a,b,c)$ and $\nabla\phi=\vec{N}$ are orthogonal. Accordingly, the vector (a,b,c) lies in the tangent plane of the integral surface u=u(x,t) at each point in (x,t,u) space where $\nabla\phi$ is defined and nonzero.

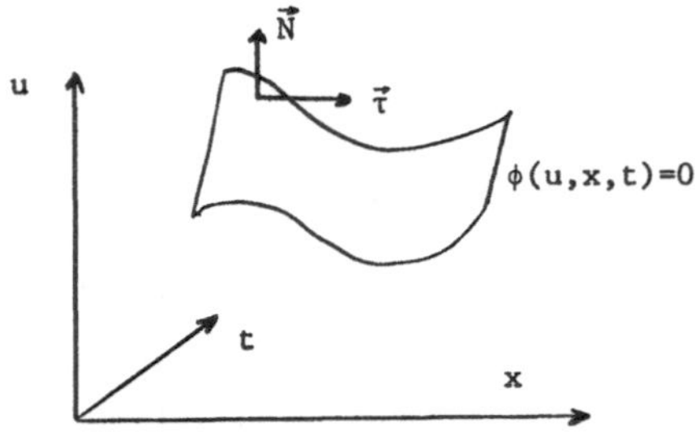

The tangent vector to this surface is given by $\vec{\tau}=(dt,dx,du)$, and from (A.2) then yields

$$\frac{dt}{a} = \frac{dx}{b} = \frac{du}{c} .$$

We get two families of curves

$$\begin{cases} 1) \ \phi_1(x,t,u)=C_1 \\ 2) \ \phi_2(x,t,u)=C_2 \end{cases}$$

which are the characteristic curves.

We are interested in determining a specific solution f(x,t,u)=0 that passes through a given curve Γ. This surface is obtained by eliminating the two arbitrary constants C_1 and C_2 between 1) and 2) using the fact that ϕ_1 and ϕ_2 pass through Γ.

Remark: The problem has a solution if Γ is non characteristic or is not the envelope of characteristic curves.

References

[B1] BOURDACHE-SIGUERDIDJANE, H. "Contribution au calcul des lois de bouclage en commande optimale non linéaire". Thèse de Docteur-Ingénieur, Université Paris XI, 1985.

[B2] BOURDACHE-SIGUERDIDJANE, H., FLIESS, M. "On the computation of optimal nonlinear feedback controls", Proc. Vth IFAC Workshop on Control Application of Nonlinear Programming and Optimization, Capri, Italy, June 11-14, 1985.

[B3] BOURDACHE-SIGUERDIDJANE, H., FLIESS, M. "Optimal feedback control of nonlinear systems", to appear in Automatica.

[C1] COURANT, R., HILBERT, D. Methods of Mathematical Physics, Vol.II. Interscience, New York, 1962.

[W1] WILLEMSTEIN, A.P., "Optimal regulation of nonlinear dynamical systems on a finite interval", SIAM J. Control Optimiz., Vol.15, 1977, pp.1050-1069.

DIRECT AND INDIRECT APPROACH FOR REAL-TIME OPTIMIZATION OF FLIGHT PATHS

Werner Grimm and Peter Hiltmann
Deutsche Forschungs- und Versuchsanstalt für Luft- und Raumfahrt (DFVLR), Institut für Dynamik der Flugsysteme,
Oberpfaffenhofen, D-8031 Wessling

1. Introduction

In commercial aviation flight path selection is normally done prior to take-off. It involves information about destination, aircraft characteristics, winds along the route etc. During the flight itself deviations from the pre-planned trajectory occur and can be corrected using algorithms as f.i. described in [1] , [2] . These references make use of a linearization of the equations of motion about the nominal flight path, thus assuming small deviations from the pre-planned trajectory. Trajectory planning of military aircraft, f.i. fighter aircraft, can normally not be done prior to take-off because destination (target behaviour) changes significantly durig the flight. In mathematical terms: The boundary conditions of the desired optimal flight paths depend on the unknown behaviour of the adversary. A precomputation of nominal trajectories is not possible and the algorithms of [1] and [2] are not applicable. Therefore, the optimization problem has to be solved "on-line", requiring a computational algorithm to generate commands in real time, that is in a fraction of the total flight time, and to generate these commands reliably, that is to ensure the convergence of any iterative scheme used in the algorithm.
This paper presents two different numerical approaches for the adaption of flight path optimization to real-time implementation. Both methods have in common that they begin with a reduced version of the dynamic model, i.e. variables originally belonging to the state vector are assumed to be directly controllable. This kind of simplification is typical for the design of real-time algorithms. In this

work the direction of the velocity vector is regarded as control function. The extremals of this "reduced model" are close to those with complete dynamics, if the flight duration is not too short.
The first algorithm (chapter 4) is a numerical implementation of the "direct approach", i.e. the optimal control is approximated via parameter-optimization. The robustness of the nonlinear program solver is enhanced by a special elimination technique for active constraints.
The second algorithm (chapter 5) is based on the indirect approach, i.e. it employs the necessary conditions of variational calculus. The resulting boundary value problem is solved by collocation.
A primary feature of both methods is a simplified solution of the differential equations. In each case the right-hand sides of the ODE-systems are only evaluated at a few fixed points of the time interval.
Both methods are taken as the core of a feedback control for the complete point mass model of an aircraft. The feedback algorithm is designed to give a near-optimal control for range maximization in fixed time. Simulations driven by the feedback guidance produce highly accurate approximations of the true max-range extremals, as a comparison for some initial conditions shows.

Nomenclature

C_L	lift coefficient
C_D	drag coefficient, function of M and C_L, $D = qS \cdot C_D$
D	drag, dependent on h, V, n,
E	specific energy = total energy per unit weight, $E = h+V^2/(2g)$
g	gravitational acceleration, constant
h	altitude, step length
H	variational Hamiltonian
i	index of a gridpoint in a discretization scheme
M	Mach number
n	load factor
q	dynamic pressure, $q = \rho V^2/2$
R	vector of boundary conditions
S	wing area
T	thrust, $T = \xi T_{max}(h,V)$
t	time
V	velocity
W	aircraft weight

x	down range
y	cross range or state and costate, $=(z,\lambda_z)$
z	state vector, $=(x,h,V)^T$
α	relaxation factor (stepsize)
γ	flight path angle
ϵ	perturbation parameter
λ	costate variable, the corresponding state variable is written as subscript
μ	bank angle
ξ	power setting, see also under T
ρ	air density, $\rho = \rho(h)$
χ	heading angle
$(\ )_o$	initial value
$(\ )_f$	final value
$(\ )_{max}$	upper bound for a variable
$(\dot{\ })$	time derivative
$(\ )_i$	value at the i-th gridpoint

2. Problem Statement

For the motion of an aircraft the following ODE-system holds:

$$\dot{x} = V \cos\gamma \cos\chi \tag{1}$$

$$\dot{y} = V \cos\gamma \sin\chi \tag{2}$$

$$\dot{h} = V \sin\gamma \tag{3}$$

$$\dot{V} = g \left[\frac{\xi T_{max}(h,V) - D(h,V,n)}{W} - \sin\gamma\right] \tag{4}$$

$$\dot{\gamma} = \frac{g}{V} \left[n \cos\mu - \cos\gamma\right] \tag{5}$$

$$\dot{\chi} = \frac{g}{V} \frac{n \sin\mu}{\cos\gamma} \tag{6}$$

Inherent simplifications are:

- constant gravitational acceleration
- flat earth without rotation
- thrust aligned with the velocity vector (assumption of small angles of attack)
- constant weight (neglection of fuel consumption).

The constraints on the load factor are the stall limit, which is not reached in this study, and the structural limit:

$$n \leq n_{max} \quad \text{(set to 9)} \tag{7}$$

The relevant state constraint, which gets active in the numerical experiments, is the dynamic pressure limit:

$$q \leq q_{max} \quad \text{(set to 10000 kgf/m}^2 = 98.0665 \text{ kN/m}^2\text{)} \tag{8}$$

The optimal control problem (P) is to achieve *maximum range in fixed time*:

$$\text{(P)} \quad \min_{n,\mu,\xi} \; - x(t_f)$$

subject to the following constraints:

- equations of motion (1) - (6)
- prescribed initial values for all state variables
- fixed final time
- control and state constraints (7), (8).

3. The Time Scale Separation Concept of the Feedback Algorithms

The ideal feedback control would result from solving (P) starting at the current state. Since this is presently not feasible in real time, (P) is simplified by a "time scale separation" of the state variables. The theoretical background for this approach is provided by the "Singular Perturbation Technique" (SPT; for the application to nonlinear optimal control see [3]). In mathematical terms (1) - (6)

is regarded as a singularly perturbed model where a "small" parameter ϵ (= 1 in reality) multiplies the left-hand sides of eqs. (5) and (6):

$$\dot{x} = \ldots,\ \dot{y} = \ldots,\ \dot{h} = \ldots,\ \dot{V} = \ldots,\ \epsilon\dot{\gamma} = \ldots,\ \epsilon\dot{\chi} = \ldots \ .$$

Theoretically the solution of (P) with the model above can be divided into an "outer solution" valid along the interior part away from the boundaries $t = 0$ and $t = t_f$ and the "boundary layer solutions", which represent the optimal solution locally near $t = 0$ and $t = t_f$. The three solution components are internally patched together by sophisticated "matching conditions". The outer solution as well as the boundary layers can be approximated by asymptotic series obtained by expansion about $\epsilon = 0$. For the feedback control only the terms of zero order are computed. The leading term of the outer expansion is the so-called "reduced solution" obtained from the "reduced problem" (P_r). The state equations for (P_r) are obtained by setting $\epsilon = 0$, which means to regard γ and χ as controls ("pseudo controls"). It is intuitively clear how to choose χ if it was directly controllable; range maximization in x-direction demands

$$\chi \equiv 0. \tag{9}$$

Generally the model reduction about γ and χ leads to a constant optimal value for χ. Hence, (P_r) is always a pure vertical plane problem. Previous examinations of (P) reported in ref. [4] have shown that maximum thrust ($\xi = 1$) is optimal throughout. The following formulation of (P_r) already anticipates the optimal adjustment for ξ and χ:

$$(P_r) \qquad \underset{\gamma}{\text{Min}} \qquad -\,x(t_f)$$

subject to the following constraints:

- reduced ODE-system

$$\dot{x} = V\cos\gamma \tag{10}$$

$$\dot{h} = V\sin\gamma \tag{11}$$

$$\dot{V} = g\left[\frac{T_{max}(h,V) - D(h,V,\cos\gamma)}{W} - \sin\gamma\right] \tag{12}$$

- prescribed initial values for x, h and V
- fixed final time
- dynamic pressure constraint (8).

The only unknown control function is $\gamma(t)$. Note that the substitution $n = \cos\gamma$ is due to setting the right hand sides of (5) and (6) to zero.
The feedback algorithm, which is supposed to provide a near-optimal real time control for (P), consists of two parts:

1) Solution of (P_r) either with the direct approach (chapter 4) or the indirect approach (chapter 5). The initial state for (P_r) is $x = 0$ and the current values for altitude and speed; the fixed final time is equal to the current time-to-go $t_f - t$. The required output of step 1) is the initial value for the pseudo control γ, denoted by $\gamma^0(0)$ and estimates for $\lambda_x(0)$, $\lambda_h(0)$ and $\lambda_V(0)$.

2) Evaluation of closed-form feedback expressions for the actual controls n and μ. The formulae can be derived from the zero order boundary layer term. The input for step 2) consists of the current state including γ and χ (disregarded in step 1) and the output of 1).

For computational efficiency step 1) is only executed in discrete time intervals. In between the results of 1) are held constant or propagated by some type of linear extrapolation. This leads to a somewhat jaggy control history. To demonstrate the near-optimality of the described feedback scheme, some simluations are performed and compared to the true optimal solutions of (P) starting at the same initial conditions.

4. Direct Approach for the Real Time Computation of Optimal Flight Paths

4.1 Parametrization of the control function $\gamma(t)$

This chapter outlines a direct approach for the solution of (P_r). For the sake of a fast on-line computation the easiest parameter-dependent model function for $\gamma(t)$ is chosen: a piecewise constant function relative to a fixed partition $0 = t_0 < t_1 < \ldots < t_m = t_f$. The actual selection of the t_i is $t_i = i \cdot t_f/m$ with $m = 9$. Let (x_i, h_i, V_i) denote the state at $t = t_i$ and γ_i the constant γ-value belonging to the time period $[t_{i-1}, t_i]$.

4.2 Simplified integration of the reduced system

To avoid expensive numerical integration (which causes the greatest deal of computational effort in usual off-line optimization), a special integration formula for the system (10) - (12) is designed. It is based on the observation that a system of the form

$$\dot{x} = V \cos\gamma \tag{13}$$

$$\dot{h} = V \sin\gamma \tag{14}$$

$$\dot{V} = aV^2 + bV + c \tag{15}$$

has a closed-form solution for constant values a,b,c and γ. To exploit this result equ.(12) obviously has to be assimilated to the structure of (15). This is accomplished by setting up the model functions T_{max} and D in a form fitting into the scheme (13) - (15). As an example the structure of the thrust model is presented:

$$T_{max}(M,h) = f_2(h)M^2 + f_1(h)M + f_0(h) \tag{16}$$

An analogous approach is made for the drag polar. Consequently the coefficients a,b,c in (15) are composed of the f_i, the speed of sound, trigonometric functions of γ and further atmospheric and aircraft specific model components. Thus, a,b and c are altitude-dependent functions. The numerical procedure is to apply the solution formula of (13) - (15) to the system (10) - (12) in each subinterval $[t_{i-1}, t_i]$. The functions a(h), b(h), c(h) are replaced by the constants $a(h_{i-1})$, $b(h_{i-1})$, $c(h_{i-1})$.

4.3 Elimination of parameters for active constraints

The dynamic pressure limit (8) serves as an example to demonstrate the simplified treatment of state constraints. Generally the best way to incorporate them in a direct solution approach is to impose them pointwise:

$$\frac{1}{2}\,\rho(h_i)V_i^2 - q_{max} \leq 0\ ,\ 1 \leq i \leq m \qquad (17)$$

Active constraints of this type are satisfied by elimination of control parameters during the integration routine. Thus the optimization algorithm is unburdened from solving a nonlinear equation system, namely the active subset of (17).
According to the solution concept described so far in sections 1, 2 the expression $q(h_i,V_i)-q_{max}$ can be viewed as a function of h_{i-1}, V_{i-1} and γ_i:

$$F(\gamma_i,V_{i-1},h_{i-1}) \leq 0,\ 1 \leq i \leq m \qquad (18)$$

With minor modifications of F an analytic elimination of γ_i becomes possible:

$$\gamma_i \geq G(h_{i-1},V_{i-1}),\ 1 \leq i \leq m \qquad (19)$$

Prior to each integration step condition (19) is checked. If (19) is not satisfied, γ_i is eliminated by $\gamma_i = G(h_{i-1}, V_{i-1})$. Thus an active constraint causes a reduction of the problem dimension. Simultaneously, all trajectories issued by the integration routine are admissible with respect to the constraints.
The approach is essentially transferable to any other state constraint, e.g. Mach number or altitude limits.

4.4 Optimization algorithm

The following scheme is a flow diagram of a single iteration. The index set $I:=(1,...m)$ is divided into two subsets I_f ("free parameters") and I_e ("eliminated parameters") defined by

$$i \in I_e \iff \gamma_i = G(h_{i-1}, V_{i-1}) \tag{20}$$
$$\iff \text{maximum dynamic pressure at } t=t_i.$$

Of course, the partition $I = I_f \cup I_e$ varies during the iterations. The natural initial values are $I_f = I$, $I_e = \phi$. Let Γ_f denote the vector consisting of the elements $\{\gamma_i | i \in I_f\}$ ordered to ascending index number i. For $i \in I_e$ γ_i is determined by (20) and hence not known prior to integration. It can be viewed as a function of Γ_f through the elimination formula (20): $\gamma_i = \gamma_i(\Gamma_f)$ for $i \in I_e$.

a) Integration of the system equations (10)-(12) with a check of the constraints (19) for $i \in I_f$; possible diminution of I_f.

↓

b) Estimates for the Lagrange multipliers of active constraints. If some of them are negative (they must be nonnegative for the optimal solution), I_e is diminished.

↓

c) Choice of a search direction $\Delta\Gamma_f$. Estimate for α_{max}: = $\max\{0<\alpha\leq 1 \mid \text{trajectory for } \Gamma_f + \alpha\cdot\Delta\Gamma_f \text{ satisfies (19)}\}$.

↓

d) Convergence criterion: If $\alpha_{max} = 1$ and $\|\Delta\Gamma_f\|<\delta$, stop.

↓

e) Stepsize selection $0<\alpha\leq\alpha_{max}$. Repeated integration of (10)-(12) with $\Gamma_f+\alpha\cdot\Delta\Gamma_f$ for varying α. I_f and I_e remain unchanged.

↓

go to (a)

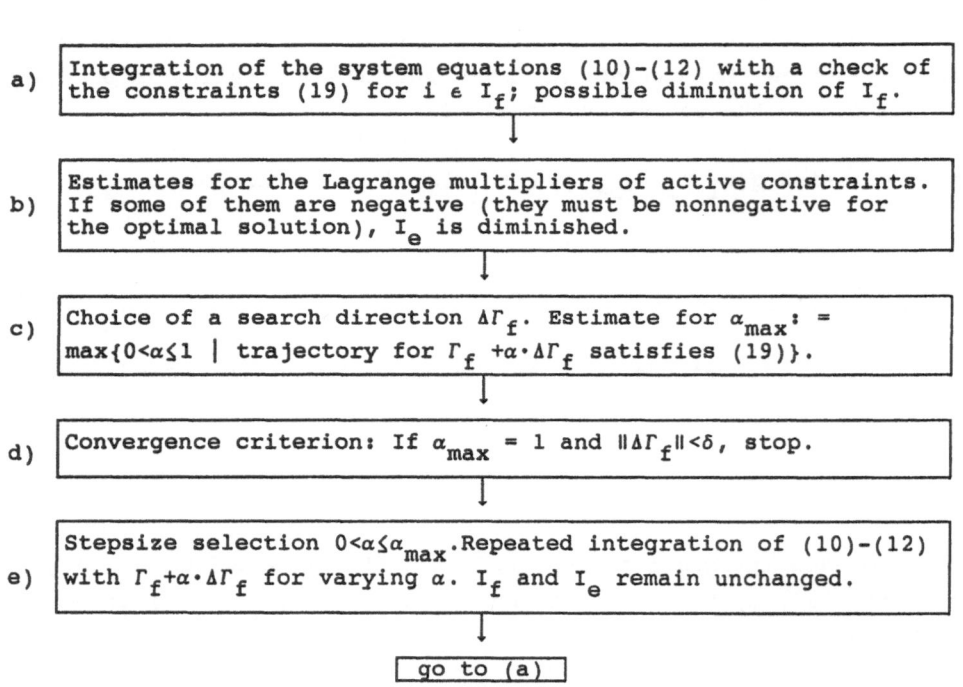

As long as I_f and I_e are held constant, the problem can be considered as unbounded optimization problem in Γ_f. Accordingly the search direction $\Delta\Gamma_f$ is taken from an unbounded optimization method. Possible choices of $\Delta\Gamma_f$ are the negative gradient of the cost function with respect to Γ_f (preferably in the pre-convergence phase with $\alpha_{max} < 1$ and varying index sets I_f and I_e) or the Newton direction (to accelerate convergence after reaching the optimal sets I_f and I_e). In any case partial derivatives of the cost function have to be provided to execute steps a) - e). In the present version of the algorithm all required partials are approximated by finite differences. A more effective way would be an analytical differentiation of the simple integration scheme outlined in 4.2. This would again drastically reduce the computational effort.

4.5 Numerical examples

In sections 4.1 - 4.4 a method is outlined which yields an approximate solution of (P_r). In the numerical examples it is used as an element of the feedback algorithm described in chapter 3. Simulations following the scheme in chapter 3 are performed with a realistic aircraft model.

h_0 and V_0 are selected on three different levels of specific energy: E_0 = 6, 17, 28 km. On each level two different starting points are chosen, one close to the dynamic pressure boundary, the other one far away from it. The boundary conditions t_f = 74 sec, $\gamma_0 = 0^0$, $\chi_0 = 90^0$ hold for all examples. To judge the performance of the feedback algorithm, the extremal of (P) is computed with a standard nonlinear programming approach for all examples, too.

The results are depicted in the altitude/Mach number diagram, fig.1. The general trend obviously is to approach the dynamic pressure boundary. There is a good agreement of optimal and suboptimal trajectories. An exception seems to be the case starting at the highest Mach number. Even in this case there is nearly no difference in the cost function, which indicates a region of "flat minima".

5. Indirect Approach for the Real Time Computation of Optimal Flight Paths

5.1 Necessary conditions

In this approach the necessary optimality conditions for the problem (P_r) are applied and control commands are generated by solving the resulting boundary value problem. For simplicity we only consider problems without constraints. Designating $z = (x,h,V)^T$ as the state vector and $u = (\gamma,n,\xi)^T$ as the control vector and $\lambda_z = (\lambda_x,\lambda_h,\lambda_V)^T$ as the associated Lagrange multiplier vector, the two point boundary value problem (TPBVP) can be stated:

$$\dot{z} = g(z,u) \tag{21}$$

$$\dot{\lambda}_z = -\partial H(z,u,\lambda_z)/\partial z \tag{22}$$

$$0 = \partial H(z,u,\lambda_z)/\partial u \ , \ H_{uu} \text{ pos. def.} \tag{23}$$

$$z(t_0) = z_0, \ \lambda_z(t_f) = (-1,0,0)^T \tag{24}$$

where $H = \lambda_z^T g$ denotes the Hamiltonian and g the right-hand sides of the system (10) - (12). t_0, t_f are prescribed as before.

5.2 Solving the TPBVP

With the notation $y := (z,\lambda_z)^T$ the TPBVP can be stated in the standard form

$$\dot{y} = f(y) \tag{25}$$

$$R\ (y(t_0),\ y(t_f)) = 0 \tag{26}$$

(25) denotes the system (21), (22) with $u = u(y)$ determined by (23). R is defined by (24). The solution $y(t)$ is approximated by piecewise cubic Hermite polynomials $P(t,t_i,t_{i+1})$, $t_i \leq t \leq t_{i+1}$ which are defined over the partitioning grid $t_0, \ldots, t_m = t_f$. This approach was chosen in [5] , [6] in the code PITOHP whereon the code given here is based.

Starting from initial estimates $y^0(t_i)$, $i = 0, \ldots, m$ the solution is iteratively calculated by a modified Newton method (see [7]) such that at the centre θ_i of each interval

$$F_i = \dot{P}(\theta_i) - f(P(\theta_i)) = 0 \tag{27}$$

and such that the boundary conditions $R = 0$ are satisfied.
To obtain an improved estimate

$$Y^{i+1} = Y^i + \alpha\ \Delta Y^i \tag{28}$$

with $Y = (y(t_0)^T, \ldots, y(t_m)^T)^T$ Newton corrections ΔY^i are calculated from

$$DF \cdot \Delta Y^i = - F \tag{29}$$

where

$$F = (F_0, F_1, \ldots, F_{m-1}, R) \tag{30}$$

and the Jacobian DF has the block structure

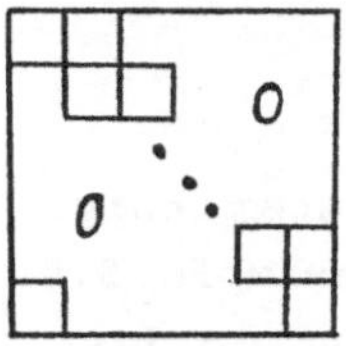

The blocks contain difference approximations of the partial derivatives of F_i and R, w.r.t. the y-values at the nodes. α is a relaxation factor.
To ensure that the discretization errors are small the grid is adapted according to the equidistribution of the local errors at the nodes. The above discretization scheme is equivalent to an implicit RK-formula of order 4. By putting the solution in a suitable 6-th order RK-formula ([8]) one obtains estimates for the local approximation errors at each node and estimates for the new interval lengths

$$h_{new} = h_{old} \, (\sigma \cdot TOL \, / \, (\tau \cdot h_{old}))^{0.2} \tag{32}$$

where TOL is the given tolerance at the nodes, τ the weighted local approximation error and $\sigma = 0.5$ a safety factor.
The solution of (P_r) is obtained by solving the TPBVP getting in particular $\tau(t_0)$, $\lambda_x(t_0)$, $\lambda_h(t_0)$ and $\lambda_V(t_0)$ (see chapter 3).

5.3 Approximation of the solution to slightly disturbed initial values

In a real-time application of the algorithm many solutions to slightly modified initial values $z(t_0)$ for the state must be calculated. Because the difference matrix DF for this neighbouring problem is similar except for higher order terms an approximation may be obtained by a single Newton step

$$\tilde{Y} = Y_{old} - DF^{-1} \cdot F \tag{33}$$

where

$$F = (\Delta y(t_0), 0, \ldots, 0)^T \tag{34}$$

$$\text{with } \Delta y(t_0) = \begin{bmatrix} z(t_0) - z_{old}(t_0) \\ 0 \end{bmatrix} \tag{35}$$

This requires solving a linear system only. The benefit of this technique becomes apparent in the example 5.4. The update time for the reduced solution can be chosen up to five times larger compared to the case with a piecewise constant reduced solution.

5.4 Numerical example

The test example for the indirect approach is the maximum range problem in the vertical plane. Instead of the velocity V the specific energy E is used as state variable and the lift coefficient $CL = n \cdot W/q \cdot S$ instead of n is used as control variable.
The control in the boundary layer can be represented as

$$C_L = (\cos\gamma + \text{sign}\ (\gamma^0 - \gamma)\ \sqrt{A}\ )\ W/(qS) \tag{36}$$

where

$$A = \cos^2\gamma - \cos^2\gamma^0 - ((\cos\gamma^0 - \cos\gamma) + \lambda_h^0(\sin\gamma^0 - \sin\gamma)) \cdot qS/(W \cdot \lambda_E^0 \cdot C_{DCL2}) \tag{37}$$

In this model the drag coefficient has the parabolic structure $C_D(M,C_L) = C_{DO}(M) + C_{DCL2}(M) \cdot C_L^2$. The initial values of the (P_r)-solution are marked by "0". For γ, q and C_{DCL2} the current values are put in. The boundary conditions are $h(t_0) = 5$ m, $V(t_0) = 228$ m/s, $h(t_f)$ = free, $V(t_f)$ = free, $t_f = 1000$ sec.
The update times for the reduced solutions are 10 sec. The feedback solution is compared with the reduced solution in figures 2 and 3.

Except from the initial point the differences between full order and reduced order solution variables E and h are so small that it is not to be seen on a h,M-plot (fig. 4). A significant difference is to be seen between the solution according to energy state model and our reduced solution near V=300m/sec.

6. Conclusions

This paper describes two efforts to adapt numerical trajectory optimization to real-time use. A concrete application would be the automated guidance of future military aircraft.

Two different approaches are taken to approximate the optimal control:

1. the "direct approach" based on parametrization of the control functions and
2. the "indirect approach" leading to a boundary value problem for the state and costate.

One main objective of both methods is to reduce the computational effort compared to usual optimization programs. This is successfully accomplished by a simplified solution of the respective ODE-system. In this connection the collocation technique of method 2. is more general than the integration procedure combined with the "direct approach", which is confined to a certain model of aircraft motion.

While the indirect approach is superior in accuracy, the direct method proves to be more robust (convergence for a primitive initial guess). Furthermore, it allows to incorporate constraints of different types. This still has to be done with the indirect approach.

The test problem in this paper is range maximization in fixed time. Both methods also work for different cost functions and boundary conditions. In this sense they represent generally usable concepts to treat the relevant problems in civil and military aircraft guidance.

References

[1] Pesch, H.J., "Neighboring Optimum Guidance of a Space-Shuttle-Orbiter-Type Vehicle", J. Guidance and Control, Vol. 3, No. 5, 1980, pp. 386-391.

[2] Bock, H.G. and Krämer-Eis, P., "A Multiple Shooting Method for Numerical Computation of Open and Closed Loop Controls in Nonlinear Systems", preprints of the 9th IFAC world congress, Vol. 9, Budapest, Hungary, July 1984, pp. 179-183.

[3] Ardema, M.D., "An Introduction to Singluar Perturbations in Nonlinear Optimal Control", Singular Perturbations in Systems and Control, CISM Courses and Lectures No. 280, M.D. Ardema, ed., International Centre for Mechanical Sciences, Udine, Italy, 1983.

[4] Seywald, J., "Range Optimal Flight Paths for a Supersonic Aircraft", diploma thesis, Technical University of Munich, FRG, 1984 (in German).

[5] Dickmanns, E.D., and Well, K.H., "Approximate Solution of Optimal Control Problems Using Third Order Hermite Polynomial Functions", Lecture Notes in Comp. Science, Vol. 27, 1975, pp. 158-166.

[6] Maier, M., "Die numerische Lösung von Halbleitermodellen mit Hilfe des Kollokationsverfahrens PITOHP unter Verwendung einer automatischen Schrittweitenkontrolle", diploma thesis, Technical University of Munich, FRG, 1979.

[7] Stoer, J. and Bulirsch, R., "Einführung in die numerische Mathematik II", Springer-Verlag, Berlin, 1978.

[8] Gupta, S., "An Adaptive Boundary Value Runge-Kutta Solver For First Order Boundary Value Problems", SIAM J. Numer. Anal., Vol. 22, No.1, 1985, pp. 114-126.

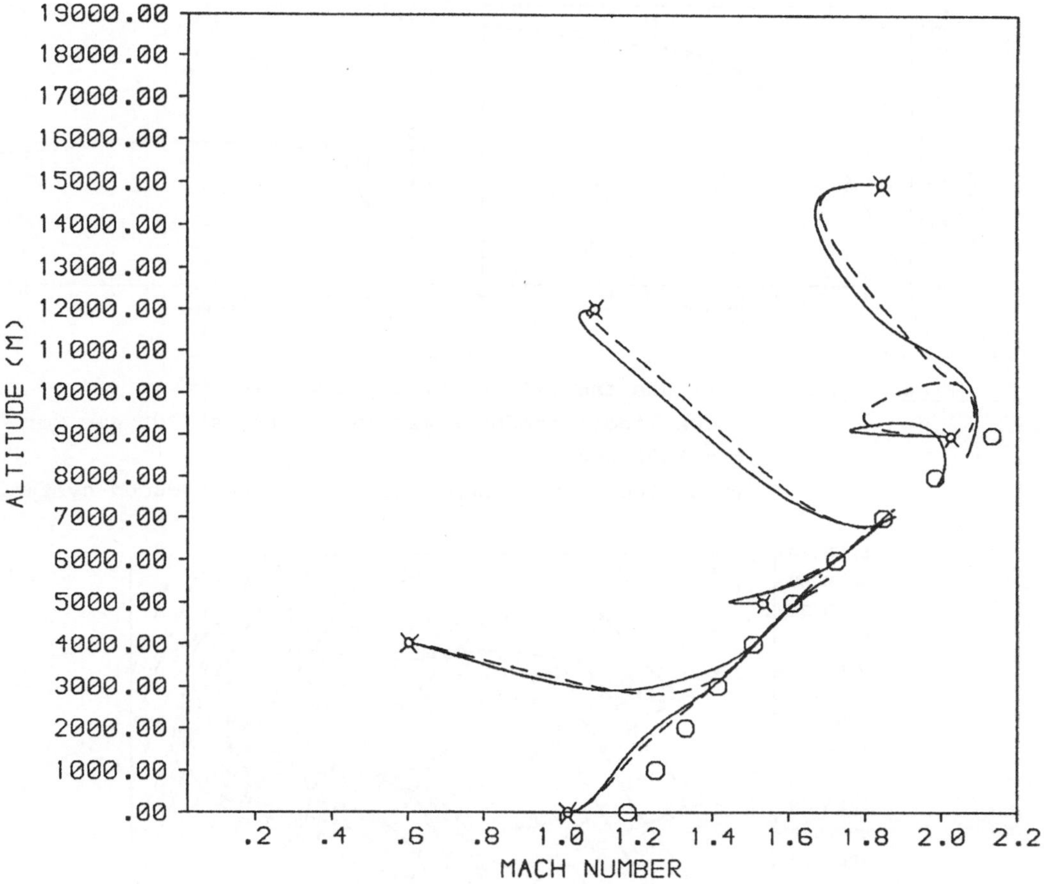

Fig.1: Optimal trajectories and suboptimal feedback simulations for range maximization in fixed time (t_f = 74 sec, χ_0 = 90^0)

——— optimal flight path

------- forward integration with the feedback control

0 0 0 0 q_{max}-limit

⊠ initial state

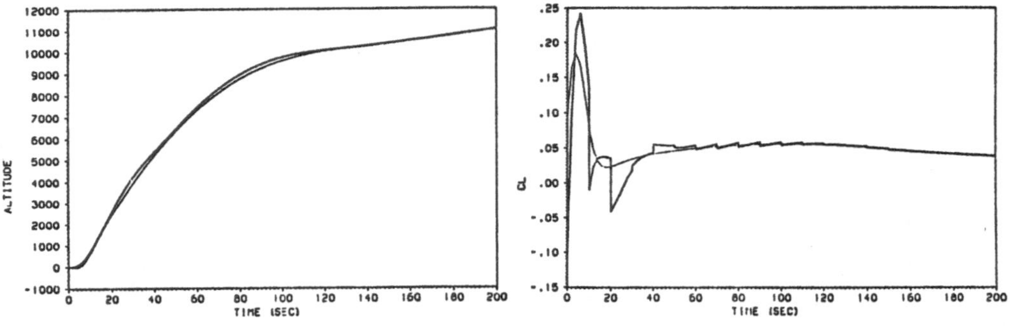

Figs. 2 and 3: Altitude and lift coefficient vs time
Thick lines: feedback simulation (first 200 sec for $t = 1000$ sec).
Thin lines: open loop solution to the reduced system.

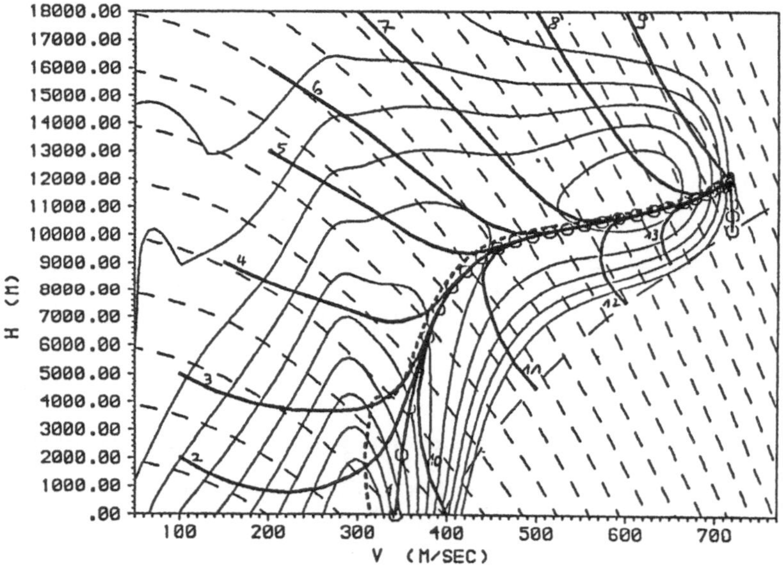

Fig. 4: Flight envelope and reduced solutions to various initial conditions and $t_f = 1000$ sec. The circles mark time steps of 10 sec.
Dashed line: maximum range path for the energy-state-model.

CONTROL SYSTEM DESIGN

POLE PLACEMENT WITH OPTIMIZATION

Richard W. Longman[1] and Gu Zhong-quan[2]
Columbia University
New York, N. Y. 10027 USA

1. Introduction

There are two major difficulties in the use of pole assignment concepts for controller design: 1) Most methods require the use of an observer to reconstruct the state, and hence the use of a high order controller, while very few methods exist for pole placement in prescribed order dynamic feedback control. 2) There is usually freedom in the pole placement process which is used in an arbitrary way in any particular placement method. This usually results in unnecessarily large, and often prohibitively large, feedback gains. This paper addresses these problems by detailing pole placement methods for direct or dynamic output feedback, and by developing algorithms for the optimization of the freedom remaining after pole assignment. The optimization obtains desirable controller properties by minimizing a quadratic cost with respect to this freedom, or by minimizing a quadratic penalty function of the feedback gains. This latter penalty function is found to be particularly promising as a design approach. Use of the remaining freedom to accomplish spillover suppression or decoupling is also discussed.

2. A possible controller design philosophy

By way of introduction, the motivation or underlying thinking is discussed here that led to the pole placement and optimization results presented. Bear in mind that these results are useful by themselves, and need not be thought of in the context of the controller design philosophy suggested in this section. Consider the properties of the two most basic modern control design methods, the Linear-Quadratic method, and the pole placement method.

1. A major portion of the work reported here was done while this author was at the Lockheed Missiles and Space Company, Sunnyvale, California, 1984, and their support is greatfully acknowledged.

2. Visiting Scholar, on leave from Nanjing Aeronautical Institute, Nanjing, Peoples Republic of China.

The Linear-Quadratic design method is a reasonably convenient design tool. It has parameters to adjust to get the desired behavior, and the relationship between the parameters and their effect on behavior is rather simple. However, use of LQ results usually involves a high order controller in the form of a Kalman filter or Luenberger observer, in order to reconstruct the state from measurements. Such high order controllers are often impractical. There exists evidence indicating they are unnecessary -- that controllers of lower dimension can give essentially equivalent performance in most cases. The authors, as well as others, have studied the design of prescribed order controllers that are optimal relative to a quadratic cost (e.g. [1-3]). The resulting steepest descent algorithm can be prohibitively slow, making the design of LQ fixed order controllers much more difficult than the standard LQ full order controllers, and they can exhibit a profusion of troublesome local minima that don't exist in the full order version.

The other major design method of modern control, pole placement [4-8], also has difficulties: 1. There are many design methods if state feedback is available, or if one reconstructs the state with an observer, but when direct output feedback or a fixed order controller is to be used, there is a dearth of design approaches. 2. There is no obvious way to pick the plant poles, in contrast to a Kalman filter which sometimes succeeds in placing the poles based on physical considerations. 3. It is not obvious what would be reasonable to request for plant poles. In classical control theory the root locus plot graphically demonstrates the relationship between root locations and physically reasonable gains, but this connection becomes obscured in modern control pole assignment because of the proliferation of gains. 4. Once the pole assignment has been completed, there is freedom in the design process that has been chosen arbitrarily by the assignment method, and hence the resulting design can ask for unnecessarily large gains, and be impractical. Very little has been published on the optimization of the freedom remaining after pole assignment(e.g. [7,9,10]), which is the main topic of this paper.

The above discussion suggests the following design sequence to try to address some of the problems raised:

1. First apply LQ theory and tune for reasonable performance, as if the state were available for feedback. This takes advantage of the relative ease of LQ design, and the ability to adjust the control penalty terms to obtain resonable magnitude control actions. The LQ design process is used here to discover eigenvalue locations that are reasonable to assign in light of physical controller saturation limits.

2. Fix the order of the controller, and then perform the pole placement for the direct or dynamic output feedback system. The plant poles are those obtained above, and the extra controller poles (if any) are either chosen or left free -- to be chosen

in the optimization below.

3. Optimize the freedom remaining after pole placement is accomplished. This can be done with respect to the original quadratic cost functional (which now reduces to a cost function), keeping the placed eigenvalues fixed. Or one can choose a different optimization criterion, such as minimizing a quadratic function of the controller gains. In terms of the methods discussed in this paper one can optimize with respect to L, or to J, or to L followed by J, see equations (17) and (21) below. One can also optimize the condition number of the feedback system matrix, or the eigenvalue sensitivity to parameter variations, etc., see [7,10].

4. Check the design to see that the required control magnitudes are reasonable. If too large control actions are required for certain control signals (because the full state feedback assumed in 1 is not available), one can first adjust the parameters in the optimization in 3, and then if necessary return to step 1, adjust the control penalty in the quadratic cost so that these control signals are smaller (relaxing the eigenvalue locations being chosen), and repeat the remaining steps.

To use this design approach one needs algorithms to accomplish the following:

1. A pole placement method for dynamic output feedback, placing some or all of the poles.

2. Algorithms for optimizing the freedom remaining after the pole assigment is accomplished.

This paper is devoted to developing algorithms for both of these tasks.

3. A pole placement algorithm for direct or dynamic output feedback

There are relatively few method available for pole placement with direct or dynamic output feedback. One method is presented here and shown to be effective in examples (see also [7,10]). The associated mathematics is needed later in the optimization step. The method is theoretically equivalent to that of [7], but avoids taking powers of matrices. Another method, the unity rank method [4], is more direct when it applies, but it has fewer free parameters and hence there are cases when pole placement can be accomplished but the unity rank method fails. In the sequel, another pole placement method will be discussed which can extend the usefulness of the unity rank method.

The system to be controlled is

$$\dot{x} = Ax + Bu$$
$$y = Cx \tag{1}$$

where $x \in R^{n_x}$, $u \in R^{n_u}$, $y \in R^{n_y}$. We consider, as the basic problem, one involving the deterministic system (1) with Gausian random initial conditions

$$E\,x(0) = 0 \quad ; \quad E\,x(0)x^T(0) = X_0 \, , \tag{2}$$

but many other versions of the problem can be addressed as discussed in [1]. The dynamic feedback controller is given by

$$\begin{aligned} \dot{s} &= Ds + Fy \\ u &= Gs + Hy \\ s_0 &= s(0) \end{aligned} \quad ; \quad P = \begin{bmatrix} H & G \\ F & D \end{bmatrix} \tag{3}$$

where $s \in R^{n_s}$ is the controller state vector, and note that this includes direct output feedback as a special case when $P = H$. It may be desirable to use a canonical form in representing the controller in order to decrease the number of parameters to be chosen in D.

The feedback system combining (1) and (3) can be written in terms of the joint state vector $z = [\,x^T \; s^T]^T$ as

$$\dot{z} = (A_a + B_a P C_a) z = A_c z \tag{4}$$

where the matices involved are defined as partitioned block diagonal matrices of appropriate dimension

$$\begin{aligned} A_a &= \mathrm{diag}(A, 0) \quad ; \quad & B_a &= \mathrm{diag}(B, I) \\ C_a &= \mathrm{diag}(C, I) \quad ; \quad & A_c &= A_a + B_a P C_a \end{aligned} \tag{5}$$

and I and 0 are identity and zero matrices.

The pole placement problem is to pick matrix P to obtain some or all of the eigenvalues of the closed loop system matrix A_c in desired locations. The relationship between the eigenvalues and the elements p_{ij} of P is a nonlinear one which complicates the pole placement problem, but it is possible to obtain local linear relationships between these quantities. We choose to use such linear relationships and a continuation method to transfer from any present pole locations to the desired locations. Before detailing this method some notation is needed.

For any given P, let X_i be the right eigenvector of A_c, and Y_i the left, associated with eigenvalue λ_i, let X and Y be associated square matrices of these eigenvectors, and it greatly facilitates the computations to let the normalization on Y be such that $Y = X^{-1}$, i.e. $X_i Y_j = \delta_{ij}$ where δ_{ij} is the Kronecker delta. Form column matrix p from rectangular matrix P by entering the elements row by row, so that element p_{ij} of P appears as p_k, the kth element of p, where $k = (i-1)(n_s + n_y) + j$.

(In the event that a canonical form is used for D, there will be certain elements of P which are fixed, and these should not be included in forming p.) Define the column matrix of eigenvalues $\lambda = [\lambda_1, \lambda_2, \ldots, \lambda_{n_x+n_c}]^T$ which we assume are all distinct, and form the matrix $\partial\lambda/\partial p$ whose i,j^{th} element is $\partial\lambda_i/\partial p_j$. Then the linearized relationship for λ as a function of p can be written as

$$\lambda(p+\delta p) = \lambda(p) + [\partial\lambda/\partial p]\,\delta p \tag{6}$$

where δp is a change in p. The elements of $\partial\lambda/\partial p$ can be computed using the eigenvalue derivative formula to obtain

$$\frac{\partial\lambda_i}{\partial p_{jk}} = Y_i^T \frac{\partial(A_a+B_aPC_a)}{\partial p_{jk}} X_i = [Y_i^T B_a(j)][C_a(k)X_i] \tag{7}$$

where $B_a(j)$ is the jth column of B_a, and $C_a(k)$ is the kth row of C_a.

The elements of $\partial\lambda/\partial p$ are in general complex valued. Define $\bar{\lambda}$ as λ with the first entry of a complex conjugate pair of eigenvalues replaced by its real part, and the entry for its complex conjugate replaced by the imaginary part. If the corresponding replacements are madefor the corresponding rows in $\partial\lambda/\partial p$ to produce $\partial\bar{\lambda}/\partial p$, then $\delta\bar{\lambda} = [\partial\bar{\lambda}/\partial p]\,\delta p$.

In the event that not all eigenvalues are prescribed, let $\bar{\lambda}'$ be $\bar{\lambda}$ with all eigenvalues that are not prescribed deleted, and let $\partial\bar{\lambda}'/\partial p$ be $\partial\bar{\lambda}/\partial p$ with the corresponding rows deleted so that the linearized relationship between changes δp of the parameters, and the resulting changes in the real and imaginary parts of the eigenvalues to be placed is given by

$$\delta\bar{\lambda}' = [\partial\bar{\lambda}'/\partial p]\,\delta p \tag{8}$$

Given any initial controller P with its set of associated closed loop eigenvalues, equation (8) tells locally how to change P to move the present eigenvalues toward the desired eigenvalue positions. In order for the eigenvalues $\bar{\lambda}'$ to be locally assignable [7], the dimension of δp should be greater than or equal to that of $\delta\bar{\lambda}'$. If the dimension is greater, then there is usually no unique solution for δp associated with a prescribed change $\delta\bar{\lambda}'$ and this is the freedom used for optimization in the sequel. One solution, in particular that solution requiring a minimum change δp in a Euclidian norm sense, is given in terms of the pseudoinverse

$$\delta p = [\partial\bar{\lambda}'/\partial p]^{\dagger}\,\delta\bar{\lambda}' \tag{9}$$

The proposed algorithm for pole assignment is based on using a sequence of these linearized changes. A specific path is chosen for each eigenvalue, connecting the eigenvalues for any initially chosen controller to the desired eigenvalue positions. Each path is initially divided into a fixed number of steps with the aim that the transition from one step-to the next should remain within the range of validity of the linearized relationship (9) for each step. At the kth step, denote the feedback parameter matrix

p by $p^{(k)}$, and the associated closed loop eigenvalues $\bar{\lambda}'^{(k)}$. According to the chosen path, denote the desired eigenvalue location at step $k+1$ by $\bar{\lambda}_d'^{(k+1)}$. One computes the pseudoinverse of (9) representing a linearization about $p^{(k)}$, and obtains

$$p^{(k+1)} = p^{(k)} + [\partial\bar{\lambda}'/\partial p]_k^{+} (\bar{\lambda}_d'^{(k+1)} - \bar{\lambda}'^{(k)}) \tag{10}$$

To monitor whether the step size chosen is too large, compute $\| \bar{\lambda}'^{(k+1)} - \bar{\lambda}_d'^{(k+1)} \|$, and adjust the step size accordingly.

Experience with the above algorithm indicates that it often works quite well. However, cases have been encountered where the step size adjustment asked for very small steps, and a modified approach was then found useful. The path chosen for each eigenvalue to follow in the homotopy, and the distance each eigenvalue is to progress on its path for a given step as compared to the other eigenvalues at the same step, are rather arbitrary choices made be the user. Rather than force the eigenvalues to adhere to these choices, one can let a troublesome eigenvalue deviate from the chosen path by picking δp at stage k to minimize

$$\bar{J} = (\delta\bar{\lambda}' - [\partial\bar{\lambda}'/\partial p]\delta p)^T W (\delta\bar{\lambda}' - [\partial\bar{\lambda}'/\partial p]\delta p) \tag{11}$$

that is, choose

$$\delta p = ([\partial\bar{\lambda}'/\partial p]^T W [\partial\bar{\lambda}'/\partial p])^{-1} [\partial\bar{\lambda}'/\partial p]^T W \delta\bar{\lambda}' \tag{12}$$

where W is a weighting matrix. Picking W to be diagonal, then one sets the weight to a small number for any eigenvalue causing difficulty, allowing it to move whatever direction it chooses. One can repeatedly establish new paths from the resulting poles to the desired pole locations.

4. Characterization of the freedom remaining after pole placement

Before the freedom remaining after pole assignment has been accomplished, can be used to optimize system performance, one must have a mathematical characterization of it. This is difficult, because the freedom is a nonlinear relationship among the parameters, but it is possible to characterize the freedom locally about any set of parameters For such a p, compute $\partial\bar{\lambda}'/\partial p$ as in the previous section, and obtain its singular value decomposition, so that

$$\delta\bar{\lambda}' = U \Sigma V^T \delta p \tag{13}$$

$$\Sigma = \begin{bmatrix} S & 0 \\ 0 & 0 \end{bmatrix} \tag{14}$$

where U and V are square unitary matrices of the dimension of $\bar{\lambda}'$ and p respectively, and S is the diagonal $r \times r$ square matrix of nonzero singular values. Usually there will be no zero rows in Σ below the S submatrix, because such a row indicates that not all eigenvalues in $\bar{\lambda}'$ can be adjusted arbitrarily (locally) by adjusting p.

Any possible parameter change δp is a vector which can be expressed in any chosen coordinate system. The columns of V represent an orthonormal basis of a Euclidian space of the dimension of δp, and the product $V^T\delta p$ is then the vector δp expressed as a column matrix of components of δp on this orthonormal basis. Let $V = [V_1\ V_2]$ where V_1 represents the first r columns, and V_2 the remaining columns. Because of the form of Σ, any $V^T\delta p$ of the form

$$V^T\delta p = \begin{bmatrix} O_r \\ \gamma \end{bmatrix} \tag{15}$$

maps into zero (O_r is an r dimensional zero vector). The set of corresponding $\delta p = V[O_r^T\ \gamma^T]$ forms the null space of the transformation $\partial\bar{\lambda}'/\partial p$. Equivalently, any δp of the form

$$\delta p = V_2\gamma \tag{16}$$

will not disturb the eigenvalue placement, to within linear terms, for any choice of vector γ, and the set of scalar elements of γ are the free parameters for local optimization about the given parameter set p.

5. Pole placement with optimization of a quadratic function of the control gains: $L = \frac{1}{2}\,\mathrm{tr}(P^TMP)$

The somewhat nonstandard objective function $L = \frac{1}{2}\,\mathrm{tr}(P^TMP)$ introduced in [1] has certain computational advantages over more typical quadratic cost functionals. (The actual cost function used can be slightly more general than the form given above for L, as indicated in (18).) Consider the weighting matrix M to be diagonal with elements m_i and let $P^{(i)}$ be the ith row of the control parameter matrix P. Then L can be written as

$$\begin{aligned} L &= \tfrac{1}{2}\,\mathrm{tr}(P^TMP) \\ &= \tfrac{1}{2}\sum_i m_i \|P^{(i)}\|^2 \end{aligned} \tag{17}$$

Hence, the first n_u scalars m_i can be adjusted to directly affect the magnitudes of the control efforts u by penalizing the associated feedback gains. These penalty factors are analogous to the diagonal elements of the control penalty matrix in the standard quadratic cost functional, and can be used in the same manner in the design process. The remaining m_i penalize use of large elements in the system matrices for the controller dynamics, and hence have a relationship to limiting the bandwidth or the time constants of the feedback compensation.

Once the pole placement has been accomplished, by whatever method, one has a matrix of feedback gains p. We wish to make changes in p to minimize L while maintaining the eigenvalues $\bar{\lambda}'$ of the feedback system. Using the local characterization of the free parameters γ, we can establish a steepest descent method, re-evaluating $\partial\bar{\lambda}'/\partial p$ as p is adjusted.

Rewrite the cost function in terms of p instead of P

$$L = \tfrac{1}{2} p^T \bar{M} p$$
$$\bar{M} = \mathrm{diag}\left(\mathrm{diag}(m_1), \mathrm{diag}(m_2), \ldots\right) \tag{18}$$

where $\bar{M}$ is a block diagonal matrix composed of diagonal blocks in which the same element m_i is repeated. Note that this new form of L is what is actually used in the computations, and that one can, if desired, make the cost more general by picking each of the diagonal elements of $\bar{M}$ separately. The gradient $\partial L/\partial\gamma$ is given by

$$(\partial L/\partial \gamma) = [\partial p/\partial \gamma](\partial L/\partial p)$$
$$= V_2^T \bar{M} p \tag{19}$$

where $[\partial p/\partial\gamma]$ is the matrix whose ij element is $\partial p_j/\partial\gamma_i$.

Then, at the kth step of the gradient search algorithm, one has a set of parameters in column vector form $p^{(k)}$ and in matrix form $P^{(k)}$. One computes $\partial\bar{\lambda}'/\partial p$ making use of equation (7), and then forms its singular value decomposition. The improved parameter values $p^{(k+1)}$ are determined using gradient (19) according to

$$p^{(k+1)} = p^{(k)} - \varepsilon V_2 V_2^T \bar{M} p^{(k)} \tag{20}$$

where ε is a step size parameter that can be adjusted to keep the changes within the range of validity of the linearized relations. Periodic checks should be made to determine that the linearized steps taken have not accumulated significant error in maintaining the desired eigenvalues $\bar{\lambda}'$; and if error has accumulated, the pole placement algorithm given previously can conveniently return the eigenvalues as needed, since most of the computations required are already being performed in the algorithm.

Some comments are needed concerning the choice of the initial condition s_0 of the controller. The cost function L in this section is not a function of these parameters, and hence they must be chosen from other considerations. For the problem as stated with zero mean initial conditions, equations (1-3), $s_0 = 0$ is a logical choice (as will be seen in section 7).

6. The advantages and uses of optimization function L -- including an alternative pole placement algorithm

6.1 Advantages

The standard cost functional of modern linear control theory is the quadratic cost functional J. The next section is devoted to the optimization of this standard cost with respect to the free parameters available after accomplishing the desired pole placement. Numerical experience indicates that the steepest descent algorithm to minimize L is somewhat easier and better behaved than for J. In particular, minimizing J starting from a pole placement controller containing some large elements, can easily result in a local minimum solution which is very far from the global minimum. On these same problems, minimization with respect to L yielded much better results.

Since the optimality function L is numerically better behaved than J, and since it has many of the advantages associated with cost function J, such as ease of adjustment of parameters in the cost to desaturate a controller, it can be desirable to design controllers based on L optimality. The price one pays for this comparative ease of computation is that the part of L penalizing the coefficients of the controller dynamics, is not as clearly connected with desirable performance characteristics as one might like.

6.2 Use of L optimization in minimizing J

Numerical experience indicates that when one wishes to use the remaining freedom after pole placement to minimize J, it is often advisable to minimize with respect to L first, and then use the resulting controller P as a starting point in the steepest descent algorithm on J. As mentioned above, without doing this, the steepest descent algorithm on J can easily end in a local minimum which still requires very large control signals. The optimization with respect to L first, makes use of the better behavior of this optimization problem, in order to get to a reasonable region of control gains accomplishing the pole placement, and the minimization of J starting from this solution often gives much smaller values of J.

6.3 Three uses of controller order adjustment available using L

Consider what happens when the last element of the diagonal matrix M tends to infinity. Then the last row of P (see equation (17)) will go to zero if it is possible to do so while maintaining the pole placement. Hence, the last rows of D and F will go to zero (if possible) in the controller (3). Therefore, the associated controller state, call it s_j, will satisfy $\dot{s}_j \approx 0$, while its initial condition is $s_j(0) = 0$. The conclusion is that, letting this element of M tend to infinity produces a controller of lower dimension (when possible). This procedure of decreasing the dimension of the controller has been found to have several important uses.

6.3.1 An alternative pole placement algorithm

The unity rank method of pole placement [4] is a constructive design method which can make it easier to use than the continuation method discussed in section 3 -- but only for those problems for which the unity rank method applies. When the method does not apply, one can of course use section 3, but an alternative is to increase the order of the controller temporarily, in order to allow the unity rank method to apply, and then use the L optimization procedure above to reduce the order again once the pole placement has been accomplished.

6.3.2 Evaluation of the penalty paid for reducing controller order

As mentioned previously, the most common modern control design methods usually want full order controllers, but that it is often possible to get essentially the same performance from lower order controllers which would be easier to implement. It is therefore of interest to study the penalty in performance that one must pay for each reduction in order, and to determine the minimum possible order controller that can accomplish the pole placement task. The controller order reduction technique described above forms a homotopy method to connect controllers of different dimensions, and hence can be used for studying these questions of performance versus controller order and minimum controller order. Caution must be exercized in interpreting the results because the homotopy is a homotopy on a local optimal solution and cannot claim to give global information, but the results nevertheless can be of very practical importance.

6.3.3 Easing the numerical computation process

Numerical experience suggests that when there are relatively few free parameters, the pole placement and optimization process can be difficult, and that increasing the order of the controller temporarily, without prescribing extra pole locations for the controller, can make the pole placement and optimization process go more smoothly. Then one reduces the order of the controller to the desired order using the above homotopy.

7. Pole placement with optimization based on quadratic cost

With the aim of alleviating the controller dimension problem, the authors [1-3] as well as others have considered the problem of designing fixed order controllers that are optimal with respect to the standard quadratic cost (in this case we consider the deterministic system with random initial conditions, (1) and (2))

$$J = E\int_0^\infty (x^TQx + u^TRu)\,dt \tag{21}$$

i.e. J is minimized with respect to the controller parameter matrix P of (3). As the controller dimension is increased, the number of parameters in P increases rapidly, and one is then looking for a minimum in a large dimensional parameter space. Numerical experience shows that one can easily have difficulty obtaining good results using the gradient algorithm either because of slow convergence in a large dimensional gradient search, or because the algorithm converges to a local minimum which is far from the global minimum in terms of the value of J obtained.

In this section we modify the problem statement so that pole placement is first performed, and then the remaining freedom is used to minimize J, i.e. J is minimized over γ instead of P. The aim is to decrease the dimension of the parameter space in which the minimization is being performed, and in doing so, increase the likelihood of obtaining good results.

When pole placement is performed before minimizing J, one might consider the first term in the cost to be superfluous, and numerical examples in fact often indicate that the contribution of this term changed very little during the optimization. However, since one does not necessarily place all the poles, and because even if one does it is possible to modify performance by changing the eigenvectors using γ while keeping the eigenvalues fixed, we develop the steepest descent algorithm with both terms present.

It will be convenient to convert the expression for the free parameter, equation (16), from column vector form to the rectangular form of P. By picking one element of γ to be unity, and the remaining elements zero, the δp of (16) becomes one of the orthonormal columns of V_2. This δp corresponds to a δP in rectanguar matrix form. Denote the δP obtained from the ith column of V_2 as P_i. Then the set of all δP's in the null space of $[\partial\lambda'/\partial p]$ is obtained from

$$\delta P = \sum_i P_i \gamma_i \tag{22}$$

by considering all possible γ_i.

The development of the steepest descent algorithm for minimization of J with respect to γ is very similar to minimizing with respect to P. The quadratic cost J can be rewritten, making use of (4), as

$$\begin{aligned} J &= E\int_0^\infty z^TQ_c z\,dt = E\left\{ z_0 \int_0^\infty e^{A_c^T t} Q_c e^{A_c t}\,dt\; z_0\right\} \\ &= E(z_0^T K z_0) = E\,\mathrm{tr}(K z_0 z_0^T) \\ &= \mathrm{tr}(K Z_0) \end{aligned} \tag{23}$$

where

$$Z_o = \begin{bmatrix} X_o & 0 \\ 0 & 0 \end{bmatrix}, \quad Q_a = \begin{bmatrix} Q & 0 \\ 0 & 0 \end{bmatrix}, \quad R_a = \begin{bmatrix} R & 0 \\ 0 & 0 \end{bmatrix}$$

$$Q_c = Q_a + C_a^T P^T R_a P C_a \tag{24}$$

The integral K satisfies

$$A_c^T K + K A_c = e^{A_c^T t} Q_c e^{A_c t} \Big|_0^\infty = -Q_c \tag{25}$$

as can be seen be substitution, provided that the value of P used is stabilizing so that $\exp(A_c t)$ is zero in the limit. The problem can now be restated: we wish to minimize $J = tr(KZ_o)$ with respect to γ, subject to the constraint of satisfying the Liapunov equation (25). A Lagrange multiplier can be introduced for each scalar constraint equation in (25) according to

$$J' = tr[KZ_o + \Lambda(A_c^T K + K A_c + Q_c)] \tag{26}$$

and one then wishes to satisfy $\partial J'/\partial K = 0$, $\partial J'/\partial \Lambda = 0$, $\partial J'/\partial s_o = 0$, and $\partial J'/\partial \gamma = 0$. Using the fact that $\partial[tr(ABC)]/\partial B = (CA)^T$ for any matrices A, B, C one obtains the necessary conditions

$$(A_a + B_a P C_a)^T K + K(A_a + B_a P C_a) + C_a^T P^T R_a P C_a + Q_a = 0 \tag{27}$$

$$(A_a + B_a P C_a)\Lambda + \Lambda(A_a + B_a P C_a)^T + Z_o = 0 \tag{28}$$

$$s_o = 0 \tag{29}$$

and

$$\frac{\partial J'}{\partial \gamma_i} = tr\left[\left(\frac{\partial J'}{\partial P}\right)^T \frac{\partial P}{\partial \gamma_i}\right] = tr\left[\left(\frac{\partial J'}{\partial P}\right)^T P_i\right]$$

$$= 2\, tr[C_a \Lambda (K B_a + C_a^T P^T R_a) P_i] \tag{30}$$

The steepest descent algorithm is then to start with a stabilizing initial value of P [3], solve the linear Liapunov equations (27) and (28) for K and Λ, and use these results to evaluate the gradient (30). Choose a change in γ, $\delta\gamma$, along the steepest descent direction. Monitor the change in the cost $J = tr(KZ_o)$ at each step to insure that the change in γ chosen is not so large that the cost has increased, and adjust the size of the change if needed.

As has already been mentioned, numerical experience with the algorithm indicates the need for good starting values for P in order to avoid converging to a poor local minimum, and that use of optimization with respect to L first can be very useful for this purpose.

8. Optimization for spillover reduction or decoupling

The problem of shape control of large flexible spacecraft raised the issue of control and observation spillover of residual vibration modes that were not considered in the controller design. The extra freedom remaining in pole placement design can be used to minimize spillover of modelled residual modes. The same technique can also have application to designing systems with decoupled controllers.

Consider the system

$$\begin{aligned} \dot{x} &= Ax + Bu \\ \dot{x}_r &= A_r x_r + B_r u \\ y &= Cx + C_r x_r \end{aligned} \tag{31}$$

where x is the state vector for the controlled modes, and x_r is the state vector for the residual modes. Suppose that a pole placement controller design has been performed for the controlled system

$$\begin{aligned} \dot{x} &= Ax + Bu \\ y &= Cx \\ \dot{s} &= Ds + Fy \\ u &= Gs + Hy \end{aligned} \tag{32}$$

When the resulting controller is applied to the original system (31) which includes the residual modes, observation spillover $C_r x_r$ will disturb the control computation, andcontrol spillover $B_r u$ will disturb the residual modes. The combination of these effects can destabilize the system. The behavior of the feedback system is governed by the differential equations

$$\begin{bmatrix} \dot{x} \\ \dot{s} \\ \dot{x}_r \end{bmatrix} = \begin{bmatrix} A+BHC & BG & BHC_r \\ FC & D & FC_r \\ B_r HC & B_r G & A_r + B_r HC_r \end{bmatrix} \begin{bmatrix} x \\ s \\ x_r \end{bmatrix} \tag{33}$$

If HC_r and FC_r can be made zero matrices, then no destabilization can occur, and the controlled modes (as well as the uncontrolled modes) maintain the intended eigenvalue locations. This requirement is often difficult to satisfy, so it is reasonable to pose a problem where the freedom γ associated with the pole placement problem (32) is used to minimize a quadratic function of the elements of these two matrices

$$\bar{J} = tr[(P\bar{C}_r)^T(P\bar{C}_r)] \tag{34}$$

where

$$P\bar{C}_r = \begin{bmatrix} H & G \\ F & D \end{bmatrix} \begin{bmatrix} C_r & 0 \\ 0 & 0 \end{bmatrix} = \begin{bmatrix} HC_r & 0 \\ FC_r & 0 \end{bmatrix} \tag{35}$$

The gradient needed for a steepest descent algorithm to minimize $\bar{J}$ with respect to γ is

$$\frac{\partial \bar{J}}{\partial \gamma} = \left[\frac{\partial p}{\partial \gamma}\right] \frac{\partial \bar{J}}{\partial p} \tag{36}$$

where the elements of the column matrix $\partial \bar{J}/\partial p$ come from those of

$$\frac{\partial \bar{J}}{\partial P} = 2P\bar{C}_r\bar{C}_r^T \tag{37}$$

Numerical experience demonstrates the effectiveness of this approach.

9. Conclusions

The two most basic modern control design approaches, the Linear-Quadratic or Linear-Quadratic-Gausian (LQG) method, and the pole placement method, have certain undesirable properties which the design philosophy and algorithms developed here are aimed at alleviating. The LQG method suffers from what can be called the "curse of modern control", i.e. the desire to use a full order controller to reconstruct the state. Such high order controllers are often unnecessary. Here we design fixed order controllers, and develop methods of reducing the order and evaluating the penalty paid for each order reduction. Robustness of LQG controllers is also an issue, and the companion paper [10] addresses such issues by approaches related to those in this work.

Pole placement design has several difficulties: 1) There are very few methods to perform pole placement when the full state is not available either by measurement or by use of state reconstruction with the resulting high controller order. Two methods to solve the controller dimension difficulty are presented here -- i.e. two methods are presented to perform pole placement for direct output or fixed order dynamic output feedback. 2) There is usually no unique solution to the pole placement problem, but most methods only find you one solution, which can ask for unreasonable control gains. Here the freedom remaining after accomplishing pole placement has been characterized mathematically, and then used to optimize control system performance according to several optimality criteria. 3) There is no obvious and compelling way to choose the controller poles introduced in dynamic output feedback. The methods presented here allow one to let these poles be picked by the optimization process. 4) Unlike the root locus plots of classical control theory, the generalization to pole placement in a multivariable setting obscures the relationship between requested pole locations and the magnitudes of the signals that will have to be generated by the controller. Hence, the designer often needs guidance in picking appropriate pole locations to request. The design philosophy suggested here picks the plant poles from LQ theory, making use of the ease of adjustment of cost function parameters to obtain reasonable control signal magnitudes. 5) Robustness issues associated with pole placement are discussed in the companion paper [10].

References

[1] Kabamba, P.T. and Longman, R.W., "An Integrated Approach to Reduced Order Control Theory," Proceedings of the 3d VPI&SU/AIAA Symposium on Dynamics and Control of Large Flexible Spacecraft, Blacksburg, VA, June 1981.

[2] Kabamba, P.T. and Longman, R.W., "An Integrated Approach to Reduced Order Control Theory," Optimal Control Applications and Methods, Vol. 4, 1983, pp. 405-415.

[3] Kabamba, P.T., Longman R.W., and Sun J-G, "On Feedback Stabilization of Linear Systems," Journal of Guidance, Navigation and Control, to appear. Short version appears in the Proceedings of the American Control Conference, Seattle, Washington, 1986.

[4] Ahmari, R. and Vacroux, A.G., "Approximate Pole Placement in Linear Multivariable Systems Using Dynamic Compensators," International Journal of Control, Vol. 18, No. 6, 1973, pp. 1329-1336.

[5] Munro, N. and Novin-hirbod, S., "Pole Assignment Using Full-Rank Output-Feedback Compensators," International Journal of System Science, Vol. 10, No. 3, 1979, pp. 285-306.

[6] Sirisena, H.R. and Choi, S.S., "Optimal Pole Placement in Linear Multivariable Systems Using Dynamic Output Feedback," International Journal of Control, Vol. 21, No. 4, 1975, pp. 661-671.

[7] Kabamba, P.T. and Longman, R.W., "Exact Pole Assignment Using Direct or Dynamic Output Feedback," IEEE Transactions on Automatic Control, Vol. AC-27, 1982, pp. 1244-1246.

[8] Lee, G.K.F. and Godbout Jr., L.F., "Output Feedback Pole Placement under System Variation," Optimal Control Applications and Methods, Vol. 4, 1983, pp. 253-264.

[9] Mahi, M.C. and Van de Vegte, J., "Optimization of Multiple-Input Systems with Assigned Poles," IEEE Transactions on Automatic Control, Vol. AC-19, No. 2, 1974, pp. 130-133.

[10] Gu Z.-q. and Longman, R.W., "Design of Dynamic Output Feedback Controllers with Parameter Variation Insensitivity," Proceedings of the AIAA Guidance, Navigation and Control Conference, Williamsburg, VA, August 1986, pp. 303-313.

NONLINEAR SYSTEM ANALYSIS BY DIRECT COLLOCATION

D. Kraft
Institute for Flight Systems Dynamics
German Aerospace Research Establishment,

Oberpfaffenhofen, D-8031 Wessling, Federal Republic of Germany

Abstract. A heuristic analysis tool for nonlinear dynamical systems is described, which is based on the solution of a sequence of optimal control problems. These are solved by a direct collocation method, which uses information of system simulation to update the control parameters into a desired direction. It is possible to consider vector optimal control problems.

Keywords. Optimal control, nonlinear programming, vector optimization, nonlinear systems, system analysis.

Introduction

A necessary prerequisite for the design of control systems for nonlinear dynamical processes is a careful analysis of the time-response of the process due to the control inputs. Choice and dimensioning of the controls as well as the corresponding investment costs are based on these analysis results. Such investigations also give informations on the reachable operation region of the process and on deviations of certain output values from their required values resulting from systems disturbances.

A widely used design methodology consists in system linearization and eigenvalue/eigenvector assignment by pole-shifting or optimization techniques. The results obtained for the linear model are then tested on the nonlinear model by system simulation. To this end special simulation software tools have been developed in the past (ACSL, CSMP, ECSSL, etc.). Common to all of them is that the user has to specifiy certain input functions and initial values of the system state variables; and within the simulation the initial value problem with differential equations is solved yielding the time history of the state variables.

This procedure is somewhat unsystematic as the analyzing engineer is not interested primarily in a state trajectory resulting from some predetermined input. He would rather like to specify the output and determine the (eventually constrained) input that is necessary to generate the desired output. This means that a solution for the inverse simulation problem is searched. An overview of this idea is given in Fig. 1.

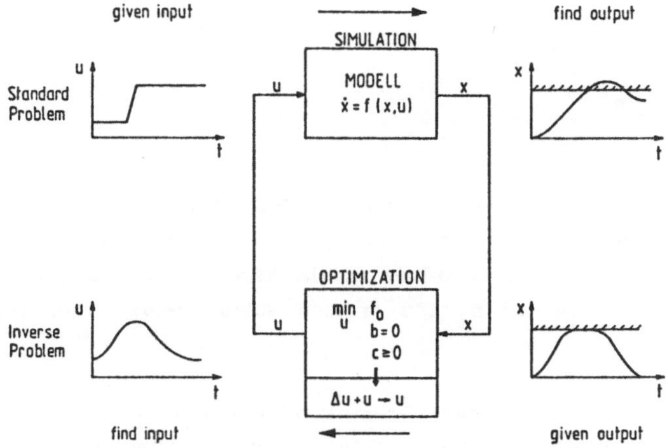

Fig. 1. The inverse simulation problem.

The adequate mathematical tool for this is the solution of constrained boundary value problems, together with the minimization of some cost functional to give a (locally) unique solution. This optimal control problem will be solved sequentially for systematically varied constraints to gain insight into the system sensitivity, as is shown in Fig. 2.

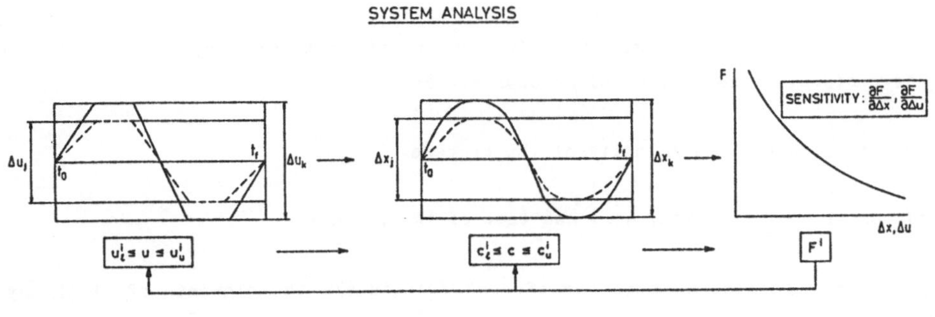

Fig. 2. System sensitivity analysis.

Generally the boundary values for the problem are given by the operation phases of the process while the constraints are defined by the operation region. It should be clear that state constraints as well as control constraints are permitted.

The analysis and design procedure outlined above will be exemplified on the problem of controlling the cryogenic windtunnel of the DFVLR at Cologne (Kraft, 1986).

Analysis Problem

In analyzing dynamic systems we have the following primary problem: determine the set of controls within a given admissible control domain which carry the system states from certain initial values at the beginning of the process to prescribed final values at the end of the process while keeping them within a feasible region.

The secondary problem consists in minimizing a cost functional which depends on the states and/or the controls within the entire time interval of the process. While optimizing the process this secondary task guarantees local uniqueness of the primary problem. The third problem is to determine the sensitivity of the cost with respect to changes in the admissible control domain and to find a best compromise between the cost associated with the process and that in realizing the control.

Examples of criteria in optimum engineering design are for control and state region specification and payoff, respectively:

- values and rates of mass flows in valves, of rotational speed of fans and servomotors, of deflection of aerodynamic devices,

- accuracy, safety and velocity of the process,

- consumption of material, cost and time of the process.

The payoff may consist of a vector of cost functionals the elements of which may behave in a conflicting manner.

The statements above can be formalized as follows: determine piecewise continuous control functions $u(t) \in \mathbb{R}^m$, $t \in [t_o,t_f]$ and parameters $p \in \mathbb{R}^p$ to solve the constrained boundary value problem

$$\begin{aligned} \dot{x} - f(x,u,p) &= 0 \in \mathbb{R}^n, \; t \in [t_o,t_f], \\ b(x(t_o),x(t_f)) &= 0 \in \mathbb{R}^q, \\ c_l \leq c(x,u,p) &\leq c_u \in \mathbb{R}^r, \; t \in [t_o,t_f], \\ u(t) \in U &\subset \mathbb{R}^m. \end{aligned}$$

and to simultaneously minimize the cost vector

$$J(u,p) = M(x(t_f),p).$$

Then by systematically variing $c_l(t)$ and $c_u(t)$ determine the sensitivity of the cost vector with respect to these variations. Note that for every variation a complete optimal control problem is to be solved. It should be clear that an efficient and robust numerical solver of this problem has to be the core of this analysis methodology.

Iterative Numerical Solution

A direct collocation method is applied to the approximate solution of the above problem (Kraft, 1985). For this reason the interval of the independent variable is discretized by the knot sequences

$$\Delta_k := \{t_o < t_1^k < t_2^k \; \dots \; t_{f-1}^k < t_f\},$$

one for each component of the involved functions. This leads to parameters y_i represented by the values of the control functions, their derivatives, and the state functions, respectively, associated with the knots t_j^k, these knots themselves and the parameters p of the given problem:

$$\{y_i\} = \{x_1(t_o),x_2(t_o),\dots,x_n(t_o),u_1(t_o),\dots,u_m(t_o),\dot{u}_1(t_o),\dots,\dot{u}_m(t_o),$$
$$x_1(t_1^1),x_2(t_1^2),\dots,\dot{u}_m(t_f),t_o,t_1^1,\dots,t_{f-1}^m,t_f,p_1,\dots,p_p\}.$$

The quadruples $(x_j^k,u_j^k,\dot{u}_j^k,t_j^k)$ approximate the problem functions by piecewise cubic Hermite polynominals P_j on each subinterval $[t_j,t_{j+1}]$, $j=0,\dots,f-1$

$$u(t) \approx P_j(t), \qquad t \in [t_o, t_f], \quad j=0,\dots,f-1,$$
$$x(t) \approx P_j(t),$$

with $P_j(t) := a_j+b_j(t-t_j)+c_j(t-t_j)^2+d_j(t-t_j)^3$.

We require <u>continuous</u>, <u>interpolating</u> approximations, characterized by

$$u(t_j) = P_j(t_j), \quad x(t_j) = P_j(t_j),$$

and

$$P_{j-1}(t_j) = P_j(t_j),$$

which also have <u>continuous first derivatives</u>

$$\dot P_j(t_j) = \dot u_j(t_j), \quad \dot P_j(t_{j+1}) = \dot u_j(t_{j+1}),$$

and

$$\dot P_j(t_j) = \dot x_j(t_j), \quad \dot P_j(t_{j+1}) = \dot x_j(t_{j+1}),$$

These conditions determine the coefficients a,...,d of P uniquely, e.g.,

$$a_j = u_j, \quad b_j = \dot u_j,$$
$$c_j = 3(u_{j+1}-u_j)/h_j^2 - (\dot u_{j+1}+2\dot u_j)/h_j,$$
$$d_j = 2(u_{j+1}u_j)/h_j^3 + (\dot u_{j+1}+\dot u_j)/h_j^2,$$

with u_j abbreviating $u(t_j)$ and $h_j = t_{j+1}-t_j$. The analog formulae hold for x, where

$$\dot x_j = f(x_j, u_j, p),$$

and thus the differential equations are satisfied at the discretizing knots.

To check the approximations within the subintervals the <u>defect</u> of the differential equations is evaluated at the center of each subinterval $\tau_j = (t_j+t_{j+1})/2$ for $j=0,\dots,f-1$:

$r_j = r_j(x_j,u_j,\dot{u}_j,x_{j+1},u_{j+1},\dot{u}_{j+1},p) := \dot{\zeta}(\tau_j)-f(\zeta_j,\nu_j p)$, with $\zeta_j := \zeta(\tau_j)$ and $\nu_j := \nu(\tau_j)$ representing the values of the Hermite polynominal at τ_j:

$$\zeta_j = (x_j+x_{j+1})/2 + h_j(f_j-f_{j+1})/8,$$

$$\dot{\zeta}_j = 3(x_{j+1}-x_j)/(2h_j) - (f_j+f_{j+1})/4,$$

$$\nu_j = (u_j+u_{j+1})/2 + h_j(\dot{u}_j-\dot{u}_{j+1})/8,$$

$$\dot{\nu}_j = 3(u_{j+1}-u_j)/(2h_j) - (\dot{u}_j+\dot{u}_{j+1})/4,$$

where $f_j = f(x_j,u_j,p)$.

The algorithm tries to drive the defect to zero:

$$r_j = 0, \qquad j = 0,\dots,f-1,$$

thus collocating the right hand side of the differential equation with the derivative of the approximating polynomial. The idea of this procedure is graphically interpreted in Fig. 3.

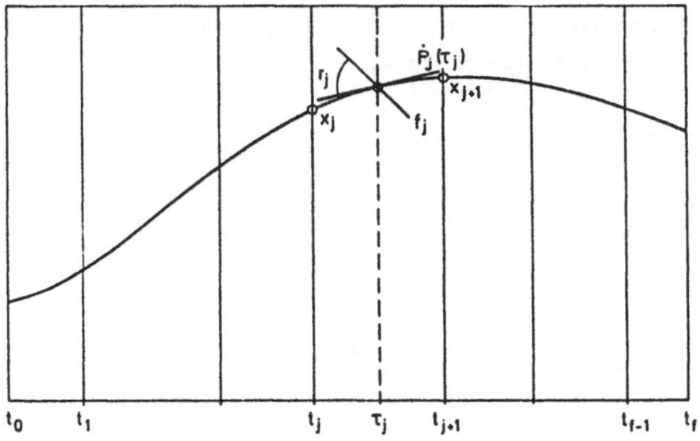

Fig. 3. Schematic collocation procedure.

In addition to the above nxf equality constraints the boundary conditions

$$b_i(x_1,x_f) = 0, \qquad i=1,\dots,q,$$

have to be satisfied together with the control and/or state constraints

$$c_{ij}^{l} \le c_{ij}(x_j,\dot{x}_j,u_j,\dot{u}_j,p) \le c_{ij}^{u}, \qquad \begin{array}{l} i=1,\ldots,r, \\ j=2,\ldots,f-1. \end{array}$$

Thus with

$$\min J_i(y), \quad i \in I := \{i : i=1,\ldots,l\},$$

the vector optimal control problem is reduced to a finite-dimensional vector nonlinear program.

This problem is solved by a constrained variant of the method of Kreißelmeier and Steinhauser (1979), which finds the efficient or Pareto-optimal points (Jahn, 1985) with the help of the following interactive algorithm:

a) choose a suitable starting approximation y^o, together with an upper bound u^o on J such that

$$J(y^o) < u^o,$$

with the notations $a < b := a_i \le b_i, i=1,\ldots,l$, and $a_i < b_i$ for at least one i,

for k=1 step 1 until (convergence) do

b) $$\min_{y^k \varepsilon Y} \max_{i \varepsilon I} J_i(y^k)/u_i^{k-1},$$

c) choose u^k such that $J(y^k) < u^k < u^{k-1}$.

Remarks: Step b) is solved by the minimax algorithm of Murray and Overton (1980). In step c) the decision-maker interactively reduces the sequence of upper bounds $\{u^k\}$ in such a way that the best possible compromise between the conflicting cost elements is eventually found.

The main analysis step now consists in the systematic variation of the upper and lower bounds specifying the set Y and in the determination of the sensitivity matrix $\partial J_i/\partial c_j$. From this information the decision-maker chooses specifications for the performance of control actuators (Kraft, 1986).

Analysis Example

As an example for the application of the analysis procedure we choose the optimal control of a cryogenic wind tunnel (Kraft, 1986).

Figure 4 shows a cross-section of the cryogenic wind tunnel of the DFVLR at Cologne (KKK = Kryo-Kanal Köln) together with the four control elements: fan, liquid and gaseous nitrogen injection and gaseous nitrogen ejection. These control the following state variables of the tunnel gas: velocity w, temperature T, and pressure p, where the values are those in the tunnel test section. The mathematical model describing the connection of the control inputs with the state variables which are to be controlled within their constraints is established through thermo-fluidmechanical balance principles.

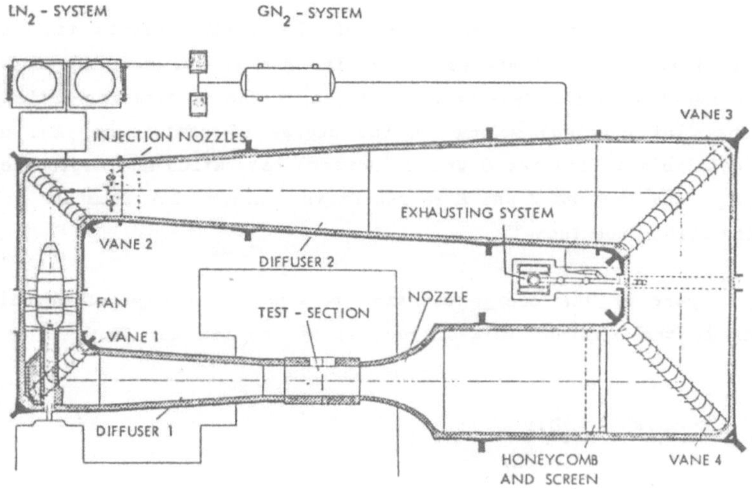

Fig. 4. Cross section of the cryogenic wind tunnel.

The balance of mass flows yields the continuity equation

$$\dot{m} = \dot{m}_1 + \dot{m}_2 - \dot{m}_3,$$

where $\dot{m}$ is the total mass flow and indices 1 to 3 indicate the mass flows of liquid nitrogen injection, gaseous nitrogen injection, and gaseous nitrogen ejection, respectively.

The caloric part of the energy equation gives an expression for the change of temperature

$$
\begin{aligned}
m\dot{T} \; = \; & \dot{m}_1(t-\tau_1)[T-\gamma T_S+r/c_V] \\
+ \; & \dot{m}_2(t-\tau_2)[T-\gamma T_U] \\
+ \; & \dot{m}_3(t-\tau_3)[\gamma-1]T \\
+ \; & k\rho w^3 A/_2 \\
- \; & \alpha_W A_W(T-T_W) \\
- \; & \alpha_M A_M(T-T_M),
\end{aligned}
$$

where $\dot{m}_i(t-\tau_i)$ indicates the influence of the transportation time lag in the controls with respect to their action on the gas in the test section. A is the test cross section, ρ the density of nitrogen, r the evaporation enthalpy, c_V the specific heat at constant volume, η the degree of efficiency, and α the heat transfer coefficient. Indices S and U indicate saturation and ambient conditions, respectively, and indices W and M represent the energy flow from the wall and the metal parts within the tunnel.

The kinetic part of the energy equation relates the change of velocity of the tunnel gas to the external work introduced by the fan and the head loss of the gas

$$\rho l\dot{w} \; = \; g(T)n^2 - k\rho w^2/2$$

with fan rotational velocity n and g(T) as fan characteristic.

The gas pressure can be derived from the introduced quantities and with R as specific gas constant via the state equation for a perfect gas

$$p \; = \; \rho RT,$$

or in differential form

$$\dot{p}/p = \dot{m}/m + \dot{T}(1/T+\rho_T/\rho) + \dot{w}\rho_w/\rho + \dot{\alpha}\rho_\alpha/\rho,$$

where ρ_x denotes $\partial\rho/\partial x$, and $\dot{\alpha}$ is the time derivative of the angle-of-attack of the model to be tested in the tunnel. Kraft (1986) gives further information on the development of the above model equations.

The nonlinear differential equations are strongly coupled, and the control and state variables entering them have to satisfy accuracy and safety constraints (termed as operating conditions) in the entire time interval as can be seen from Tables 1 and 2. It is therefore evident that careful control analysis must precede operation of the tunnel and that the tunnel gas has to be controlled by an automatic control system in the entire operating region

$$0 \leq w \leq 100 \text{ m/s}$$

$$100 \leq T \leq 300 \text{ K}$$

and in all operating phases (cool down, stand-by, accelerate, test, decelerate, warm-up).

Operating Phase	Δw(%)	ΔT(K)	Δp(%)
Cool-down	L	F	±2.0
Stand-by	L	±5.0	±2.0
Run-up	F	±2.0	±2.0
Testing	±1.0 (±0.5)	±1.5 (±0.5)	±1.0 (±0.05)
Run-down	F	±2.0	±2.0
Warm-up	L	F	±2.0
F = Optimized Input, L = Idle.			

Tab. 1. State constraints

Control	Control Constraint min	max	Control Derivative Constraint min	max
Fan Rpm	0 U/min	500 U/min	- 5 U/min/s	+ 5 U/min/s
Nitrogen Injection	0 kg/s	12 kg/s	- 0.375 kg/s²	+ 0.375 kg/s²
Blow-in	0 kg/s	4 kg/s	- 0.400 kg/s²	+ 0.400 kg/s²
Ejection	0 kg/s	12 kg/s	- 2.40 kg/s²	+ 2.4 kg/s²

Tab. 2. Control and control rate constraints

The control design involves a trade-off between two conflicting demands, namely the above mentioned satisfaction of the operating conditions and the requirement for high cost effectiveness and large tunnel productivity. For tighter specifications the operation time required to achieve them is longer and therefore the tunnel productivity in terms of possible test runs will be lower, and vice versa.

This control trade-off will be illustrated for cryogenic temperatures (T = 100 K) and for the operating phase "acceleration" of the tunnel working gas from idle to maximum test velocity at this temperature (0 m/s < w < 70 m/s).

We will consider minimum time problems only, as other cost effectiveness factors like injected mass and consumed electric energy are of minor importance w.r.t. optimization because they are nearly completely prescribed by the operating conditions (boundary values). This is shown in Kraft (1986).

The analysis is organized in the following steps. Initially the state tolerances are varied at nominal control constraints and the influence of this variation on the operation time is checked. In the second step the control tolerances will be varied at nominal state constraints. Again operation time is the criterion of interest. Finally the influence of valve location is investigated for nominal operating conditions.

The results of the first step are summarized in Fig. 5, where the operation time t_f is given for increasing tolerances ΔT and Δp of temperature and pressure, respectively.

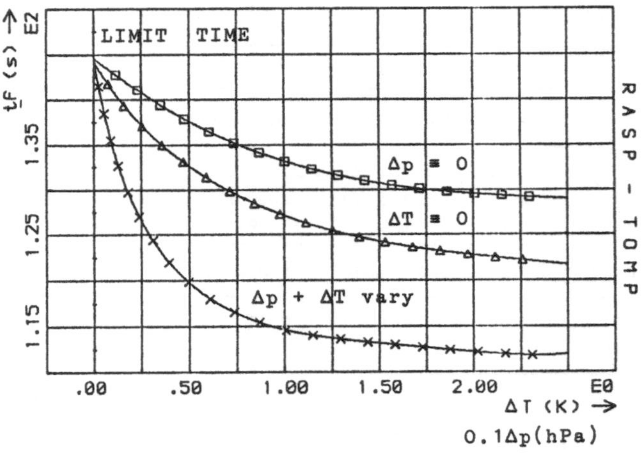

Fig. 5. Influence of state constraints on operation time.

In the lower graph (x) ΔT and Δp are linearly related with the indicated values. In both the other graphs the state variables ΔT (Δ) and Δp ( ) are not allowed to deviate from their nominal values ($T = 100$ K, $p = 1000$ hPa). In these cases as well as in the limiting case $\Delta T = 0$ and $\Delta p = 0$ the corresponding differential equations change to nonlinear algebraic equations from which the corresponding control functions can be determined. Thus a considerable reduction of computer time can be achieved. A quantitative derivation of these results is given in Kraft (1986).

If the constraints on the state variables are tightened by decreasing the tolerances more dynamic freedom is taken from the process and thus in all three cases the operation time increases up to its limiting value where ΔT and Δp vanish simultaneously.

In step two of the control analysis we concentrate on the variation of the control tolerances of $\dot{m}_2$ because it has a striking influence on the operation duration. The remaining control tolerances are kept at their nominal values. As in step one, restricting the control takes freedom from the process and prolongs its duration. This is indicated in Fig. 6 for stationary ($\dot{x}_b = 0$) and nonstationary ($\dot{x}_b \neq 0$) boundary conditions. For the former conditions the time gained when injecting gaseous nitrogen with ambient temperature at nominal values ($\dot{m}_2 = 4$ kg/s) as opposed to noninjecting is 250 percent.

The last step in the control analysis demonstrates an alternative possiblility of this analysis approach. Instead of variing the boundaries of the controls and states certain design parameters that define the process are changed and the influence of this on the cost is observed and evaluated for the design specifications.

This possibility is exemplified by the variation of the valve location for liquid and gaseous nitrogen. The influence of the distance l_1 of the liquid nitrogen valve from the test section on the process duration is shown in Fig. 7 for nominal and optimal location l_2 of the gaseous nitrogen valve.

For extended physical interpretations of all observed phenomena to Kraft (1986) is referred.

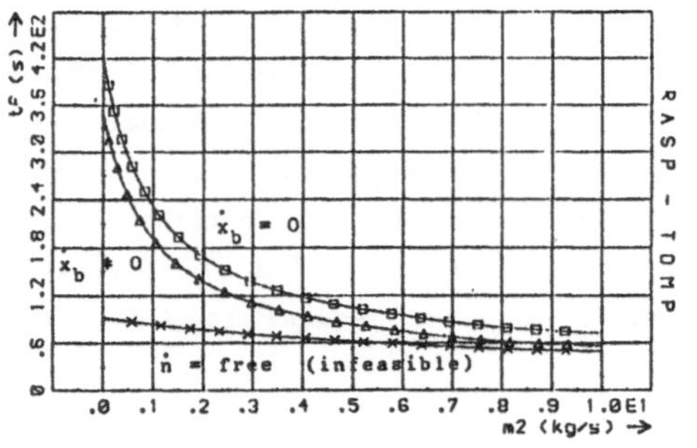

Fig. 6. Influence of constraints on gaseous nitrogen injection on operation time.

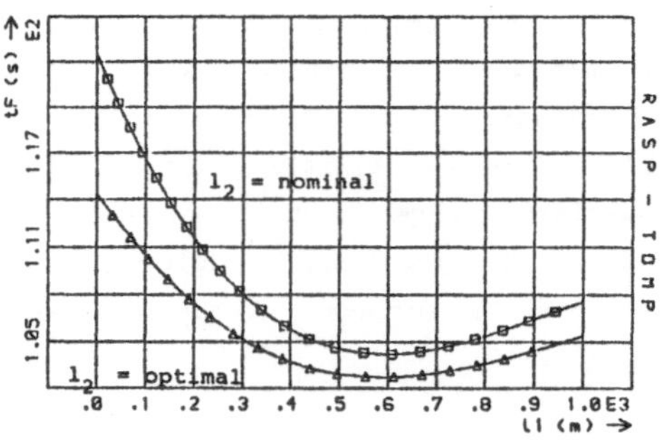

Fig. 7. Influence of valve location on operation time.

Conclusions

The analysis results of the last section often show surprising results in the behaviour of nonlinear systems. These effects can rarely be obtained by trial and error simulation. Instead of this a systematic analysis tool which combines simulation and optimization and which has been described in this contribution is needed in practical system analysis and design.

References

Jahn, J. (1985). Some characterizations of the optimal solution of a vector optimization problem. OR Spektrum, 7, 7-17.

Kraft, D. (1986). Optimalsteuerungen - ein systematisches Hilfsmittel zur rechnergestützten Erforschung der dynamischen Möglichkeiten eines Tieftemperaturwindkanals. DFVLR-FB 86-23, 1986.

Kraft, D. (1985). On converting optimal control problems into nonlinear programming problems. In K. Schittkowski (Ed.), Computational Mathematical Programming. Springer, Berlin.

Kreißelmeier, G. and Steinhauser, R. (1979). Systematische Auslegung von Reglern durch Optimierung eines vektoriellen Gütekriteriums. Regelungstechnik, 27, 76-79.

Murray, W., and Overton, M.L. (1980). A projected Lagrangian algorithm for nonlinear minimax optimization. SIAM J. Sci. Stat. Comp., 1, 345-370.

ROBOT CONTROL

CONTROL OF A ROBOTIC MANIPULATOR ON A PRESCRIBED PATH SUBJECT TO OPTIMIZATION CONDITIONS AND ADDITIONAL CONSTRAINTS

Ulrich Leiner

Mathematisches Institut der TU München
D-8000 München

The most common way to derive dynamic equations for a robotic manipulator model is to use the Lagrange equations of motion, e.g. [2], [7]:

$$(1) \qquad \frac{\partial}{\partial t}\frac{\partial L}{\partial \dot{\alpha}_i} - \frac{\partial L}{\partial \alpha_i} = Q_i + M_i \qquad i = 1,\ldots,N$$

- L : difference between potential and kinetic energy
- Q_i : gravity torque
- M_i : torque due to actuators
- N : number of actuators.

For a standard model of a rigid robot with three rotatory degrees of freedom (dof) as described in [8] three differential equations of the following kind are obtained:

$$(2) \qquad \begin{pmatrix} M_1 \\ M_2 \\ M_3 \end{pmatrix} = I(\alpha)\begin{pmatrix} \ddot{\alpha}_1 \\ \ddot{\alpha}_2 \\ \ddot{\alpha}_3 \end{pmatrix} + R\begin{pmatrix} \dot{\alpha}_1 \\ \dot{\alpha}_2 \\ \dot{\alpha}_3 \end{pmatrix} + C(\alpha)\begin{pmatrix} \dot{\alpha}_1^2 \\ \dot{\alpha}_2^2 \\ \dot{\alpha}_3^2 \end{pmatrix} + D(\alpha)\begin{pmatrix} \dot{\alpha}_1\,\dot{\alpha}_2 \\ \dot{\alpha}_1\,\dot{\alpha}_3 \\ \dot{\alpha}_2\,\dot{\alpha}_3 \end{pmatrix} + G(\alpha)$$

$I(\alpha)$	:	Matrix of inertia	$D(\alpha)$	:	Coriolis forces
R	:	Friction forces	$G(\alpha)$	:	Gravity forces
$C(\alpha)$	:	Centrifugal forces	α	:	joint angles .

The torques $M_i(t)$ are considered to be control variables for the motion of the manipulator. Their computation under certain constraints will be our aim.

The first restricting condition is to prescribe the path along which the top of the manipulator shall be moved.
That is done according to [5] by :

$$(3) \qquad f(s) \in \mathbb{R}^3 \quad s(t) \in [0,1] \quad \text{path parameter depending on time.}$$

The angles α_i and their derivatives can be expressed in terms of s to obtain differential equations depending only on s and its first and second derivative. There are additional simplifications, for example neglecting coriolis forces which have also to be made, in order to obtain the equations

$$(4) \qquad u_i(t) = M_i(t) = a_i(s)\ddot{s} + b_i(s)\dot{s}^2 + c_i(s) \qquad i = 1,2,3,$$

if the prescribed path is substituted.

After that substitution there is only one dof left in (4), which is $s(t)$. As a next step one $u_i(t)$ can be computed using one out of the three equations in (4) and an additional optimization criterion to determine $s(t)$. Minimum time or minimum energy consumption are possible criterions for that. Necessary boundary conditions are added to these equations. The other two $u_i(t)$ will be calculated from the remaining two equations in (4).

This method was published by [4] and [6]. They use either standard integration methods,which treat the problem as an initial value problem with time minimization after an additional transformation or use Bellman's dynamic programming method with time-energy mixed optimization criterion.

It was now the idea to use the multiple shooting method [1] in order to solve the time or energy optimal path problem.

To look for a time-optimal solution is only reasonable if the range of the control variables $u_i(t) = M_i(t)$ is limited, e.g.:

$$(5) \qquad M_{i\,min} \le M_i(t) \le M_{i\,max} \quad i = 1,2,3 \quad t \in [0,t_f].$$

Because of the linearity of the control it is expected to gain an optimal solution with bang-bang control. The results show indeed, that there is always one of the three control-variables which is on its bound, whereas the other two controls are depending variables and vary within their extremals.

The actual extremal control can change throughout the prescribed path. Considering these points as switching points, the theory of optimal control leads to a system of differential equations with boundary conditions:

$$\dot{s} = v \qquad\qquad s(0) = s_o \quad ; \quad s(t_f) = s_f$$

$$\dot{v} = (u_j - b_j v^2 - c_j)/a_j \qquad\qquad v(0) = v_o \quad ; \quad v(t_f) = v_f$$

$$\dot{\lambda}_s = \frac{\lambda_v}{a_j}\left(\frac{db_j}{ds} v^2 + \frac{dc_j}{ds} - \frac{da_j}{ds}\left(v^2 b_j + c_j - u_j\right)/\, a_j\right)$$

$$\dot{\lambda}_v = -\lambda_s + 2\lambda_v v b_j/a_j\,, \qquad j = 1 \text{ or } j = 2 \text{ or } j = 3.$$

With the Hamiltonian H the switching function will be

(7) $$H_{u_i} = \frac{\lambda_v}{\alpha_i} .$$

Solving this system of differential equations for certain robot data using the FORTRAN-routine BOUNDSCO [3] similar results as in [4] and [6] were obtained.

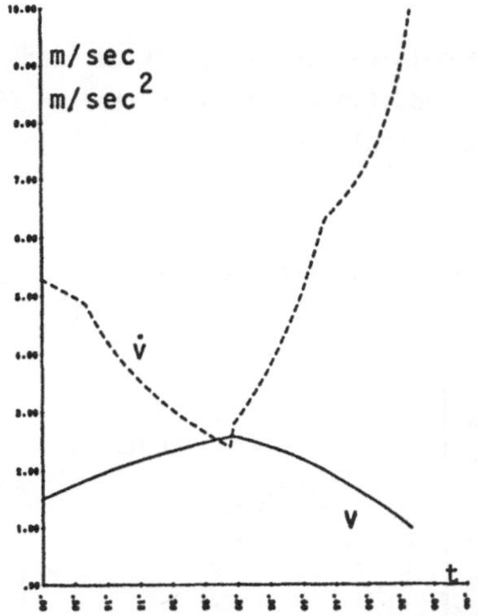

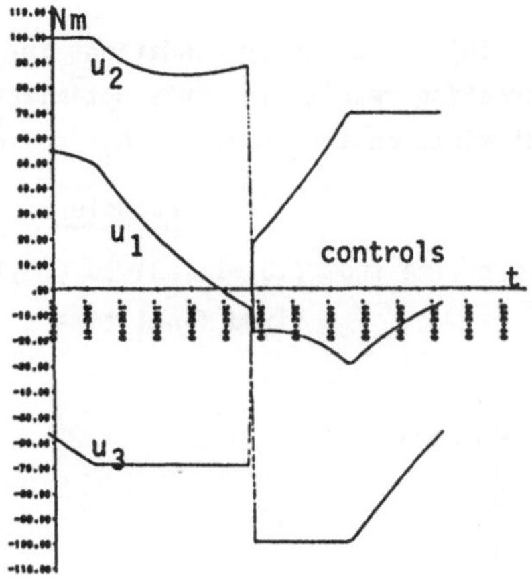

Example 1

straight line from (.3,.3,.5) to (.7,.7,.5) $v_o = 1.5$, $v_e = 1.0$

$|u_1| \le 100, \ |u_2| \le 100, \ |u_3| \le 70$

Instead of time optimal solutions, minimizing an energy consumption criterion may also be of some interest. With

(8) $$\int_0^{t_f} \sum_{i=1}^{3} u_i^2(t)\,dt$$

as optimization index and u_1 as the independent control variable (u_2 and u_3 are then the depending variables) the Hamiltonian holds:

(9) $$H = \lambda_s v + \lambda_v(u_1 - b_1 v^2 - c_1)/a_1 + u_1^2 +$$
$$+ \left[\frac{a_2}{a_1}\left(u_1 - b_1 v^2 - c_1\right) + b_2 v^2 + c_2\right]^2 + \left[\frac{a_3}{a_1}\left(u_1 - b_1 v^2 - c_1\right) + b_3 v^2 + c_3\right]^2 .$$

The adjoint equations $\dot{\lambda}_s = -H_s$ and $\dot{\lambda}_v = -H_v$ become a bit more complicated and numerical differentiation is necessary for the evaluation.

The condition $H_u = 0$ yields the control:

$$(10)\quad -u_1 = \left[\frac{\lambda_v}{2a_1} + \left(\left(\frac{a_2}{a_1}\right)^2 + \left(\frac{a_3}{a_1}\right)^2\right)\left(-b_1v^2 - c_1\right) + \frac{a_2}{a_1}\left(b_2v^2 + c_2\right) + \frac{a_3}{a_1}\left(b_3v^2 + c_3\right)\right]$$

$$* \quad \left(1 + \left(\frac{a_2}{a_1}\right)^2 + \left(\frac{a_3}{a_1}\right)^2\right)^{-1} .$$

With (10) and boundary conditions for s and v the multiple shooting method leads to interesting results for this optimization problem. Additional to the criterion (8) constraints on the controls u_i may or may not be postulated.

<u>Example 2</u>

straight line from (.3,-1.,.3) to (.4,1.,.4); optimization criterion (8); fixed final time $t_f = 2.0$ sec.

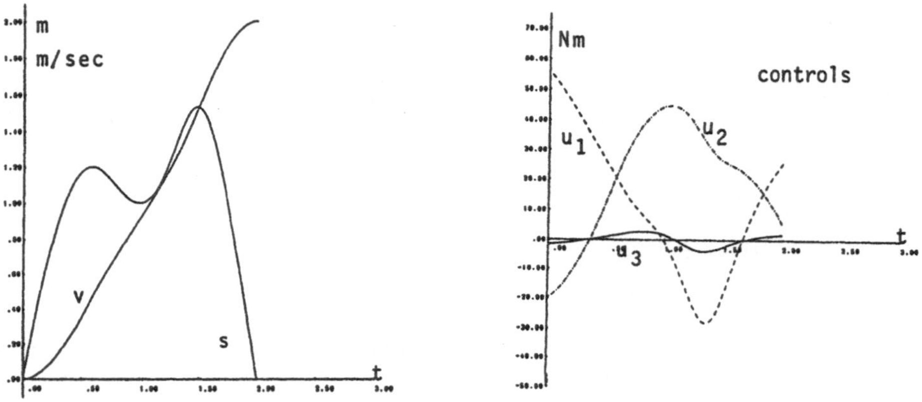

a) no additional constraints

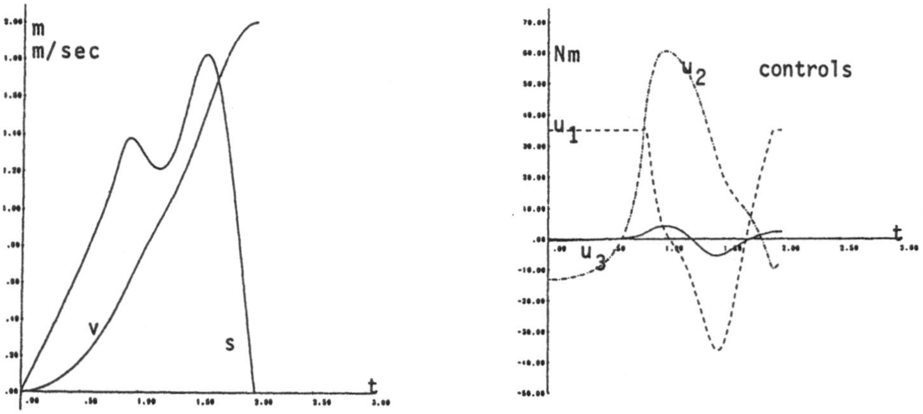

b) additional constraints: $|u_1| \leq 35$

LITERATURE

[1] Bulirsch, R.:
Die Mehrzielmethode zur numerischen Lösung von nichtlinearen Randwertproblemen und Aufgaben der optimalen Steuerung.
Heidelberg: Carl-Cranz-Lehrgang über Flugbahnoptimierung, 1971
Nachdruck: München, Technische Universität, Math. Institut, 1985

[2] Coiffet, F.; Johanni, R.:
Robot Technology, Volume 1: Modelling and Control;
Cogan Page, London 1983

[3] Oberle, H.J.; u.a.:
BNDSCO: Rechenprogramm zur Lösung beschränkter optimaler Steuerungsprobleme;
Technische Universität München, TUM-M8509, 1985

[4] Pfeiffer, F.; Johanni, R.:
A Concept for Manipulator Trajectory Planning
regular paper for IEEE, Journal of Robotics and Automation, 1985

[5] Pfeiffer, F.:
Beitrag zu optimalen Roboterbahnen
GAMM-Tagung 1984, ZAMM, Band 65, Heft 4/5, T 321

[6] Shin, K.G.; McKay, N.D.:
Minimum Time Control of Robotic Manipulators with Geometric Path Constraints.
IEEE Trans. on Automatic Control, Vol AC-30, No.6
June 1985, S.531 - 541

[7] Snyder, W.E.:
Industrial Robots: Computer Interfacing and Control,
Prentice-Hall Inc., Englewood Cliffs, New Jersey 07632, 1985

[8] Vukobratović, M.; Potkonjak, V.:
Scientific Fundamentals of Robotics 1; Dynamics of Manipulation Robots;
Theory and Application;
Springer Verlag Berlin, Heidelberg, New York, 1982

NUMERICAL COMPUTATION OF SINGULAR CONTROL FUNCTIONS FOR A TWO-LINK ROBOT ARM

Hans Joachim Oberle
Universität Hamburg, Institut für Angewandte Mathematik
Bundesstraße 55, D-2000 Hamburg 13

1. Introduction

In recent years there has been growing interest in modelling and optimization of robot movements which is caused by monifold employments of robots in industry and technique. Special attention is payed to the computation of energy- of time-optimal movements of robot arms.

In [1] Bryson and Weinreb published a model of a horizontal articulated two-link robot arm. They used this model for the computation of minimum-time movements. Here, the distance of the movement is prescribed and the robot arm has to be in rest at the begin and at the end of the motion. The initial- and the final position of the arm, however, are kept free, and they are determined such that the transfer time is minimized. The solutions presented in [1] are characterized by control functions of the bang-bang type and it remained an open question, whether there exist singular time-optimal control functions.

In this paper the problem is treated by means of a special multiple shooting algorithm which is suitable for the numerical solution of optimal control problems with bang-bang and singular subarcs. For the data-set taken from [1] optimal solutions are computed depending on the distance parameter x_L. It turns out that, for certain values of x_L, the solution trajectory contains a singular subarc. In this case, the bang-bang solution, which can be computed as well, violets an optimality condition.
The numerical computations were performed on the Siemens computer 7.882 of the Computer Center of the University of Hamburg in double precision arithmetic.

2. The Model

The following model describes a two-link robot arm which moves frictionless in a horizontal plane. The model, the nomenclature and the formulation of the problem were taken from [1]. The same model but a different formulation of the problem was considered by Geering et al. [2].

The geometry of the robot arm is shown in Fig.1. The arm consists of two links (with mass m_i and length l_i, i=1,2) and a payload M which is situated at the tip of link 2.

The motion is described by four state varibles:

α : elbow angle

θ : angular rotation of link1 relative to its initial position

$\omega_{1/2}$: angular velocity of link 1/2 .

The control variables are the actuator torques U_1 and U_2 for the two links related to shoulder and elbow axis, respectively.

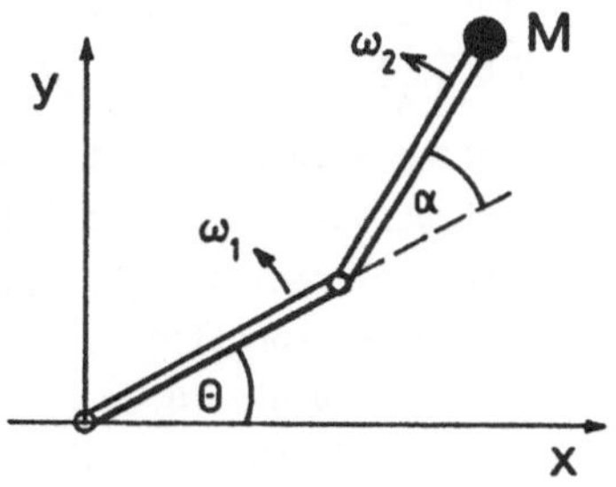

Fig.1: Geometry of the arm

The equations of motion are given as follows:

$$\begin{aligned}\dot{\alpha} &= \omega_2 - \omega_1\\ \dot{\theta} &= \omega_1\end{aligned}$$

$$\begin{bmatrix} J_{11} & J_{12}\cos\alpha \\ J_{12}\cos\alpha & J_{22}\end{bmatrix}\begin{bmatrix}\dot{\omega}_1\\ \dot{\omega}_2\end{bmatrix} = \begin{bmatrix} J_{12}\sin\alpha\ \omega_2^2 \\ -J_{12}\sin\alpha\ \omega_1^2\end{bmatrix} + \begin{bmatrix} U_1 - U_2 \\ U_2\end{bmatrix} \tag{1}$$

Here, the J_{ik} denote mass moments of inertia:

$$J_{11} = J_1 + (m_2 + M)\, l_1^2 , \qquad J_{12} = m_2\, a_2\, l_1 + M\, l_1\, l_2 ,$$

$$J_{22} = J_2 + J_3 + M\, l_2^2 ,$$

$J_{1/2}$: mass moment of inertia of the link 1/2 with respect to shoulder- and elbow axis, respectively,

J_3 : mass moment of inertia of the payload with respect to the tip,

a_2 : distance between the center of gravity of the second link and the elbow.

The control variables are bounded: $|U_i(t)| \leq U_{i\,max}$, $i=1,2$. The data are (cf. [1]):

$$m_1 = m_2 = M = 1,\quad l_1 = l_2 = 1,\quad a_2 = 0.5,\quad J_1 = J_2 = 1/3,\quad J_3 = 0,$$
$$U_{1\,max} = U_{2\,max} = 1. \tag{2}$$

For these data the coefficient matrix in (1) turns out to be nonsingular with the determinant:

$$\Delta = J_{11}\, J_{22} - J_{12}^2 \cos^2\alpha > 0 . \tag{3}$$

We want to determine control functions $U_1(t)$, $U_2(t)$, $0 \leq t \leq t_f$, such that the tip of the robot arm moves along a certain prescribed distance within minimal time. With the coordinates of the tip (cf. Fig.1):

$$\begin{aligned} x(t) &= l_1 \cos\theta + l_2 \cos(\theta+\alpha)\\ y(t) &= l_1 \sin\theta + l_2 \sin(\theta+\alpha) ,\end{aligned} \tag{4}$$

we have the following *optimal control problem* :

Determine $U_i(t)$, $0 \leq t \leq t_f$, i=1,2 such that the functional:

$$J[U_1,U_2] = t_f$$

is minimized with respect to the constraints:

- equations of motion: (1)
- control region: $|U_i(t)| \leq U_{i\,max}$, i=1,2 (5)
- boundary conditions:

$$\theta(0) = 0 \tag{6}$$

$$\omega_1(0) = \omega_2(0) = 0 \tag{7}$$

$$\omega_1(t_f) = \omega_2(t_f) = 0 \tag{8}$$

$$\sqrt{(x(t_f)-x(0))^2 + (y(t_f)-y(0))^2} = x_L . \tag{9}$$

Here, x_L denotes the prescribed distance. This parameter ranges within $0 \leq x_L \leq 2(l_1+l_2)$. The boundary conditions (7-8) state that the robot arm is in rest at the begin and the end of the motion. Notice, that the initial and the final position of the arm (given by the data $\alpha(0)$, $\alpha(t_f)$, and $\theta(t_f)$) are not prescribed; i.e. these data are determined as well as the control variables such that the transfer time is minimized. Within this feature our problem differs from the problem considered in [2]. Here, the authors prescribe $\alpha(0) = \alpha(t_f) = 0$ and $\theta(t_f)$ (parameter of the problem), which reduces the degree of freedom.

3. The Necessary Conditions

In this section we apply the necessary conditions of optimal control theory in order to establish a boundary-value problem (BVP) with switching conditions. Special attention is payed to the case of a singular subarc with respect to the control variable U_1 (cf. [3],[4]). Supposed that the switching structure of the solution can be estimated, the BVP can be solved numerically by multiple shooting technique ([5],[6]).

Due to Eq. (3) the differential equations (1) can be transformed into an explicit system. Let λ_α, λ_θ, λ_1 and λ_2 denote the adjoint variables with respect to α, θ, ω_1 and ω_2, respectively. Then, the Hamiltonian can be expressed by

$$H = \lambda_\alpha(\omega_1 - \omega_2) + \lambda_\theta\,\omega_1 + \frac{1}{\Delta}\{ S_1\,(J_{12}\sin\alpha\,\omega_2^2 + U_1 - U_2) - S_\Delta\,(J_{12}\sin\alpha\,\omega_1^2 - U_2)\}. \tag{10}$$

Δ is given by equation (3), and the auxiliary functions S_1 and S_Δ are defined by

$$S_1 = J_{22}\,\lambda_1 - J_{12}\cos\alpha\,\lambda_2 \;, \quad S_\Delta = J_{11}\,\lambda_2 - J_{12}\cos\alpha\,\lambda_1 . \tag{11}$$

Furtheron, we define:

$$S_2 = S_\Delta - S_1. \tag{12}$$

Then, due to $\Delta > 0$, the functions S_1 and S_2 can be used as switching functions with respect to the control variables U_1 and U_2.

The adjoint differential equations are derived from Eq. (10). One obtains:

$$\dot{\lambda}_\alpha = \frac{J_{12}}{\Delta} \{ S_1 (-\cos\alpha\, \omega_2^2 + \frac{J_{12}\sin 2\alpha}{\Delta} (J_{12}\sin\alpha\, \omega_2^2 + U_1 - U_2)) + S_\Delta (\cos\alpha\, \omega_1^2 - \frac{J_{12}\sin 2\alpha}{\Delta} (J_{12}\sin\alpha\, \omega_1^2 - U_2)) \} + \frac{J_{12}\sin\alpha}{\Delta} \{ \lambda_1 (J_{12}\sin\alpha\, \omega_1^2 - U_2) - \lambda_2 (J_{12}\sin\alpha\, \omega_2^2 + U_1 - U_2) \} \quad (13)$$

$$\dot{\lambda}_\theta = 0$$

$$\dot{\lambda}_1 = \lambda_\alpha - \lambda_\theta + 2\frac{J_{12}\sin\alpha}{\Delta} \omega_1 S_\Delta$$

$$\dot{\lambda}_2 = -\lambda_\alpha - 2\frac{J_{12}\sin\alpha}{\Delta} \omega_2 S_1 .$$

This system of first order differential equations together with the Eqs. (1) establishes the basic system for the numerical integration. By the standard transformation of the independent variable

$$t = t_f \cdot \xi , \quad \frac{d}{d\xi} = t_f \cdot \frac{d}{dt} , \quad (14)$$

these equations are transformed into a system with the independent variable ξ ranging in the interval $0 \leq \xi \leq 1$, and the additional trivial differential equation:

$$\frac{d}{d\xi} t_f = 0. \quad (15)$$

The natural boundary conditions are determined in the main by the boundary condition (9). This equation can by expressed in the form:

$$\rho(\alpha(0),\alpha(t_f),\theta(t_f)) := (x(t_f)-x(0))^2 + (y(t_f)-y(0))^2 - x_L{}^2 = 0 . \quad (16)$$

In this form the constraint can be adjoined to the augmented performance index (cf. [5]). By elimination of the corresponding Lagrange parameters one obtains:

$$\lambda_\theta(t_f) \frac{\partial \rho}{\partial\alpha(0)} + \lambda_\alpha(0) \frac{\partial \rho}{\partial\theta(t_f)} = 0$$

$$\lambda_\theta(t_f) \frac{\partial \rho}{\partial\alpha(t_f)} - \lambda_\alpha(t_f) \frac{\partial \rho}{\partial\theta(t_f)} = 0, \quad (17)$$

where the partial derivatives are given by:

$$\frac{1}{2}\frac{\partial \rho}{\partial\alpha(0)} = (x(t_f)-x(0))\, l_2 \sin\alpha(0) - (y(t_f)-y(0))\, l_2 \cos\alpha(0) \quad (18)$$

$$\frac{1}{2}\frac{\partial \rho}{\partial\alpha(t_f)} = -(x(t_f)-x(0))\, l_2 \sin(\alpha(t_f)+\theta(t_f)) + (y(t_f)-y(0))\, l_2 \cos(\alpha(t_f)+\theta(t_f))$$

$$\frac{1}{2}\frac{\partial \rho}{\partial\theta(t_f)} = x(0)\, y(t_f) - y(0)\, x(t_f) .$$

A third natural boundary condition is obtained with respect to the free final time:

$$H|_{t=t_f} = -1 \ . \qquad (19)$$

Making use of the boundary conditions (6-8) this relation can be rewritten:

$$\{ S_1 U_1 + S_2 U_2 + \Delta \} |_{t=t_f} = 0 \ . \qquad (20)$$

With these conditions the BVP is formally completed. We have nine first order differential equations (1), (13) and (15) and corresponding to them the two-point boundary conditions (6-9), (17) and (20). Now, it remains to find expressions for the control variables. This is already done in the case of a bang-bang control:

$$U_j = \begin{cases} U_{j\,max} \ , & \text{if } S_j < 0 \\ -U_{j\,max} \ , & \text{if } S_j > 0 \end{cases} \qquad (j=1,2) \qquad (21)$$

For the numerical solution one uses an estimated switching structure, i.e. the number and the relative position of the bang-bang subarcs are fixed. The switching points τ_k are parameters of the BVP which are determined by the switching conditions

$$S_i(\tau_k) = 0. \qquad (22)$$

Now, the case of a singular subarc with respect to U_1 is considered. We assume, that

$$S_1(t) = 0, \quad \tau_1 \leq t \leq \tau_2 \qquad (23)$$

holds for a certain subintervall $[\tau_1,\tau_2]$. The underlying philosophy for the treatment of singular subarcs is to differentiate the switching function up to a certain order $q = 2p$, such that the total time-derivative $S_1^{(q)}$ contains the control variable U_1 explicitly; cf. [4]. For the problem considered one finds $p = 1$, and with the assumption (23) one obtains the following relations for the derivatives:

$$S_1^{(1)} = (J_{22}+J_{12}\cos\alpha)\ \lambda_\alpha - J_{22}\ \lambda_\theta + J_{12}\sin\alpha\ (\omega_1+\omega_2)\ \lambda_2 \qquad (24)$$

$$S_1^{(2)} = A_1 + B_1\ U_1 \qquad (25)$$

$$\begin{aligned} A_1 = {} & -2J_{12}\sin\alpha\ \omega_2\ \lambda_\alpha + J_{22}(\omega_2^2-\omega_1^2)\ \lambda_1 + \frac{J_{12}^2\sin^2\alpha}{\Delta}\ \{ \omega_1^2(J_{22}\lambda_1-J_{11}\lambda_2) + \\ & + \omega_2^2\ J_{22}(\lambda_2-\lambda_1) \} + J_{12}\omega_1^2\ (\cos\alpha - \tfrac{1}{\Delta}\ J_{12}^2\ \sin 2\alpha\ \sin\alpha)(\lambda_1+\lambda_2) + \\ & + \frac{J_{12}\ \sin^2\alpha}{S_\Delta}\ (\lambda_1\omega_1^2 - \lambda_2\omega_2^2)\ (\lambda_1+\lambda_2) + \frac{J_{12}\ \sin\alpha}{\Delta}\ \{ J_{11}\lambda_2 + \\ & + (2J_{22} + J_{12}\cos\alpha)\ \lambda_1 \} \cdot U_2 \end{aligned} \qquad (26)$$

$$B_1 = -\frac{2J_{12}\ \sin\alpha}{\Delta}\ J_{22}\ \lambda_1 \ . \qquad (27)$$

Notice, that A_1 contains the control variable U_2, which has been assumed to be of the bang-bang type. More precisely, we assume that $S_\Delta \neq 0$, $t \in [\tau_1,\tau_2]$.

If $B_1 \neq 0$ holds, one can use (25) for the computation of the singular control:

$$U_{1\,sing} = - A_1/B_1 \ . \qquad (28)$$

The switching points τ_1 and τ_2 are fixed by the switching conditions:

$$S_1(\tau_1) = S_1^{(1)}(\tau_1) = 0 . \qquad (29)$$

With these conditions the BVP with switching conditions is completed.

It should be remarked, that additional sign-conditions have to be checked after the solution of the BVP has been computed. These sign-conditions are:

(S1) Test for the correct sign of the switching functions S_i, i=1,2 in the interior of the bang-bang subarcs (corresponding to Eq.(21))

(S2) Generalized Legendre-Clebsch condition: $B_1 < 0$ on singular subarcs.

These sign-conditions are satisfied for the numerical solutions presented in Section 5.

4. Multiple Shooting Technique for BVPs with Switching Conditions

In the previous section we saw that the necessary conditions lead to a BVP with switching conditions. Such problems are of the general structure:

$$\dot{y}(t) = f(t,y,u) , \quad 0 \leq t \leq 1 \qquad (30)$$

$$u(t) = u_k(t,y(t)) , \quad \tau_k \leq t \leq \tau_{k+1} , \quad k=0,1,\ldots,s \qquad (31)$$

$$r_k(y(0),y(1)) = 0 , \quad k=1,..,n_1 \qquad (32)$$

$$r_k(\tau_{s_k},y(\tau_{s_k})) = 0 , \quad k=n_1+1,..,n+s . \qquad (33)$$

Here, y(t) is a n-vector function which is sufficiently smooth in the subintervalls $[\tau_k,\tau_{k+1}]$ (one-side limits at τ_k and τ_{k+1}). The switching points τ_k, k=1,..,s are parameters of the problem and satisfy:

$$0 = \tau_0 < \tau_1 < \ldots . < \tau_s < \tau_{s+1} = 1. \qquad (34)$$

According to the degree of freedom, the boundary conditions (32) and the switching conditions (33) establish a (n+s)-dimensional system of possibly nonlinear equations.

For the numerical treatment of the problem we use multiple shooting technique. The special code, called BOUNDSCO, is based on the earlier developed multiple shooting codes BOUNDSOL and OPTSOL due to Bulirsch (cf. [6],[7]). We briefly sketch the main steps of the algorithm; for the details the reader is refered to [8].

We consider a fixed mesh of multiple shooting nodes t_j which satisfy

$$0 = t_1 < t_2 < \ldots . < t_{m-1} < t_m = 1. \qquad (35)$$

Let Y_j denote an estimate of the solution $y(t_j)$ (j=1,..,m-1) and let $\tau=(\tau_1,..,\tau_s)$ denote an estimate of the switching points. Then, these values are corrected by the following iteration scheme.

Step 1 Numerical solution of (m-1) initial-value problems (IVPs):

$$\dot{y} = f(t,y,u) , \quad y(t_j) = Y_j , \quad t_j \leq t \leq t_{j+1}, \quad j=1,\ldots,m-1 \qquad (36)$$

This step can be performed by the use of any conventional IVP solver with automatic stepsize control. In this paper the Bulirsch-Gragg-Stoer extrapolation code DIFSYS (cf. [9]) was used.

Step 2 Let $y(t;t_j,Z_j)$ denote the solution of the IVP (36), where $Z_j := (Y_j,\tau)$. Then, the BVP is equivalent to the following system of nonlinear equations:

$$F_j(Z_j,Z_{j+1}) \;:=\; z(t_{j+1};t_j,Z_j) - Z_{j+1} \;=\; 0\;,\quad j=1,2,\ldots,m-2 \tag{37}$$

$$F_{m-1}(Z_1,\ldots,Z_{m-1}) \;:=\; \begin{bmatrix} r_k(Y_1,y(t_m;t_{m-1},Z_{m-1})) & 1\le k\le n_1 \\ r_k(\tau_{s_k},y(\tau_{s_k};t_{j_k},Z_{j_k})) & n_1+1\le k\le n+s \end{bmatrix} = 0 \tag{38}$$

The indices j_k are fixed by:

$$t_{j_k} \;<\; \tau_{s_k} \;<\; t_{j_k+1}\;,\quad k = n_1+1,\ldots,n+s\;. \tag{39}$$

Notice, that the switching conditions in (38) can be considered as multipoint boundary conditions which have to be satisfied at the multiple shooting node t_j.

The system (37-38) is solved numerically by means of a modified underrelaxed Newton-method (cf. [10],[11]) with a special Broyden-update of the Jacobian matrix DF(Z).

For the numerical treatment of the robot arm problem the choice of a rather coarse multiple shooting grid was sufficient. We choose m = 4 with the nodes:

$$t_1 = 0\;,\quad t_2 = 0.3\;,\quad t_3 = 0.6\;,\quad t_4 = 1\;.$$

The prescribed relative tolerance for the multiple shooting method was TOL = 1.E-7 .

5. Numerical Results

The numerical computation of the time-optimal motions of a two-link robot arm has been performed for the data-set (2). A standard homotopy method has been applied using the distance parameter x_L as the homotopy parameter for the BVP.

Roughly spoken, the solutions obtained can be subdivided into two branches. One homotopy branch, which is characterized by asymmetric control functions, covers "small" distance parameters. The other branch is characterized by symmetric control functions (symmetry with respect to the center of the time interval). Both branches intersect on a certain region of x_L, i.e. for certain parameters x_L there exist two trajectories which both satisfy all the necessary conditions. The "optimal" one is indicated by a lower value of the tranfer time t_f. In Figure 2 the value of the performance index t_f is plotted in dependance on the homotopy parameter x_L.

Within these features our results coincide with those published by Bryson and Weinreb. They differ, however, with respect to the switching structure of the asymmetric solution branch. We found that for certain values of the distance parameter x_L the optimal control function U_1 contains a singular subarc. In Table 1 the

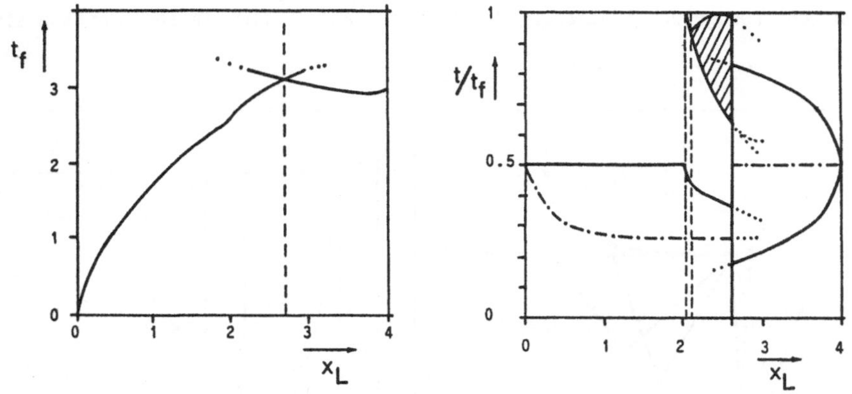

Fig. 2/3: Performance index and switching structure in dependance on x_L
—— : switching points w.r.t. U_1 —·—·: switching points w.r.t. U_2
▨ : singular subarcs

switching structure is listed in dependance on the parameter x_L. Notice, that each link of the robot arm has the length one, such that x_L ranges within $0 \leq x_L \leq 4$.

x_L	switching structure	
$0 < x_L \leq 2.02$	U_1: (-1,1);	U_2: (1,-1)
$2.04 \leq x_L \leq 2.119$	U_1: (-1,1,-1);	U_2: (1,-1)
$2.119 \leq x_L \leq 2.473$	U_1: $(-1,1,U_{1\ sing},-1)$;	U_2: (1,-1)
$2.473 \leq x_L \leq 2.614$	U_1: $(-1,1,U_{1\ sing},1)$;	U_2: (1,-1)
$2.614 \leq x_L < 3.99$	U_1: (-1,1,-1,1);	U_2: (1,-1)
$3.99 \leq x_L < 4.0$	U_1: (-1,1);	U_2: (1,-1)
$x_L = 4.0$	U_1: (-1,1);	U_2: (1,-1,1,-1)

Table 1: Switching structure in dependance on x_L

The appearance of a singular subarc is detected by consideration of the homotopy path $2.04 \leq x_L \leq 2.119$. For the distance parameter $x_L^* = 2.1195\,050$ (numerically determined) the derivative $S_1^{(1)}(t)$ changes the sign just at the switching point τ with $U_1(\tau^-) = 1$ and $U_1(\tau^+) = -1$. Thus, the situation at τ corresponds exactly to the situation on a singular subarc (cf. the switching conditions (29)) and it indicates that the solutions for $x_L > x_L^*$ may contain a singular subarc.

If one neclects this indication one can compute bang-bang solutions even for distances $x_L > x_L^*$. However, all these trajectories violate the sign-condition (S1) and, thus, they are not solutions of the optimal control problem.

In the Figures 4-9 the solution trajectory for the distance parameter $x_L = 2.6$ is shown. One recognizes that about 30% of the transfer time is spent on the singular subarc.

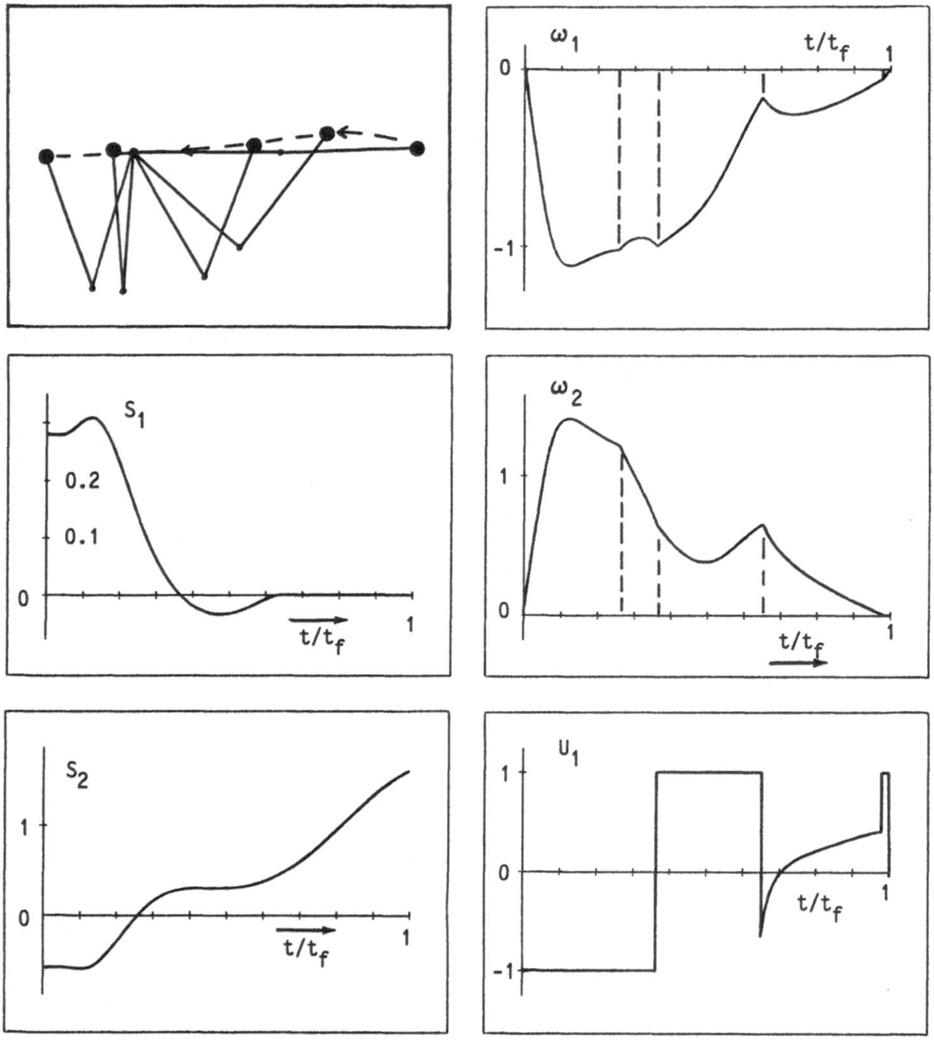

Figs. 4-9: Solution trajectories for the distance $x_L = 2.6$;
$\omega_{1,2}$: angular velocities of link 1,2;
$S_{1,2}$: switching functions;
U_1 : control of link 1

References

[1] Weinreb, A., Bryson, A.E.: Minimum-Time Control of a Two-Link Robot Arm; presented at the 5-th IFAC Workshop on Nonlinear Programming and Optimization, Capri, 1985.

[2] Geering, H.P., Guzzella, L., Hepner, S.A.R., Onder, C.H.: Time-optimal Motions of Robots in Assembly Tasks. IEEE Transactions on Automatic Control, Vol. AC-31, No.6, pp. 512-518, 1986.

[3] Kelley, H.J., Kopp, R.E., Moyer, H.G.: Singular Extremals; in: Leitmann, G. (ed.): Topics in Optimization, Academic Press, New York, 1967.

[4] Maurer, H.: Numerical Solution of Singular Control Problems using Multiple Shooting Technique. Jour. of Optim. Theory and Appl., Vol. 18, No. 2, pp. 235-257, 1976.

[5] Neustadt, L.W.: Optimization. Princeton University Press, 1976.

[6] Bulirsch, R.: Die Mehrzielmethode zur numerischen Lösung von nichtlinearen Randwertproblemen und Aufgaben der optimalen Steuerung, Report of the Carl-Cranz Gesellschaft, Oberpfaffenhofen, West Germany, 1971.

[7] Stoer, J, Bulirsch, R.: Introduction to Numerical Analysis, Springer Verlag, Berlin, Germany, 1979.

[8] Oberle, H.J.: Numerical Computation of Singular Control problems with Application to Optimal Heating and Cooling by Solar Energy; Applied Mathematics and Optimization, Vol. 5, pp. 297-314, 1979.

[9] Bulirsch, R., Stoer, J.: Numerical Treatment of Ordinary Differential Equations by Extrapolation Methods; Numer. Math., Vol. 8, pp. 1-13, 1966.

[10] Deuflhard, P.: A Modified Newton Method for the Solution of Ill-conditioned Systems of Nonlinear Equations with Application in Multiple Shooting; Numer. Math., Vol. 22, No. 4, pp. 289-315, 1974.

[11] Deuflhard, P.: A Relaxation Strategy for the Modified Newton Method; in: Bulirsch, R., Oettli, W., Stoer, J. (eds.): Optimization and Optimal Control, Springer Verlag, Berlin, Germany, 1975.

WATER RESOURCES MANAGMENT

DECENTRALIZED CONTROL FOR AN UNCERTAIN MULTI-REACH RIVER SYSTEM[1]

G. Leitmann
Department of Mechanical Engineering
University of California
Berkeley, CA 94720 USA

C.S. Lee
Department of Mathematics
University of Malaya
59100, Kuala Lumpur, Malaysia

Y.H. Chen
Department of Mechanical and Aerospace Engineering
Syracuse Univesity
Syracuse, New York 13244 USA

Abstract

We employ a deterministic approach to study a multi-reach water quality system subject to uncertain but bounded disturbances at both the input and output levels as well as to uncertain parameter variations. Decentralized controls in the form of rate variation of B.O.D. concentration from an "optimal" steady state value in the effluent and in-stream aeration rate are used to drive the river system to and maintain it at the desired state. If the full state of the river system can be measured, possibly with measurement error, a decentralized state feedback saturation control based on measured state is employed; if only partial state can be measured, an appropriate observer is designed to estimate the state on which the decentralized state feedback saturation control is based.

Keywords. Stability, large scale uncertain water quality system, decentralized control.

1. Introduction

Many real-life problems dealing with systems in water resources, management, the environment, economy, society and industry are highly complex, "large" in dimension and uncertain. To manage these systems, the use of conventional (centralized) techniques of modeling, analysis, control design and optimization becomes inappropriate since it is not only difficult to optimize numerically owing to Bellman's "curse of dimensionality", but also difficult to implement policies when various compartments

---

[1]Based on research supported by NSF and AFOSR.

(subsystems) of the systems are separated geographically as in the case of a river system. Thus, the lack of centralized computing capability, the time required for on-line dynamic optimization, the cost of communication of information and the reliability of communication links can make the implementation of a centralized control very difficult, if not impossible [1].

Recently, many authors have modeled water resources systems as large scale systems, and then utilized hierarchical and decentralized controls in their management [2] - [11]. In [12] and [13] a deterministic approach was employed to illustrate the control of a one-reach water quality system subject to uncertainties (by these we mean disturbances a both the input and output levels, and sometimes, uncertain parameter variations as well). In this paper we endeavor to extend the results obtained in [12] and [13] to a multi-reach river system subject to uncertainties. In each reach, we introduce decentralized controls in the form of effluent waste discharge rate and in-stream aeration rate to manage the river system whose state consists of two important indices of water quality, namely, biochemical oxygen demand (B.O.D.) and dissolved oxygen (D.O.), the dynamics of which are subject to uncertainties.

The theory employed here is a generalization of the deterministic control schemes for uncertain dynamical systems; for instance, see [14] - [16].

2. Problem Formulation

We consider an N-reach river system. By a reach we mean a stretch of a river, of some convenient length, which has a waste treatment facility located at the top of the reach. As has been mentioned earlier, biochemical oxygen demand and dissolved oxygen are two important indices of water quality which constitute the state in each reach of the river system.

Let $z_i(t)$ and $q_i(t)$ denote the concentrations per unit volume of B.O.D. and D.O., respectively, in the i-th reach at time t. If we assume that the flow rate[2] is constant and water is well mixed in each reach, then by mass balance considerations, the following equations describe the dynamics of B.O.D. and D.O. in the i-th reach:

$$\dot{z}_i(t) = - k_1^i(t)z_i(t) + \frac{Q_{i-1}z_{i-1}(t) + m_i Q_{iE} - Q_i z_i(t)}{V_i} + \frac{\Delta m_i(t) Q_{iE}}{V_i} + v_1^i(t)$$

[2]We assume that the flow rate in each reach is measured at the end of the reach.

$$\dot{q}_i(t) = - k_3^i(t)z_i(t) + k_2^i(t)(q_i^s - q_i(t)) + \frac{Q_{i-1}q_{i-1}(t) - Q_iq_i(t)}{V_i} + r_i(t) + v_2^i(t),$$

$$i = 1, \ldots, N \qquad (2.1)$$

where $k_1^i(\cdot)$, $k_2^i(\cdot)$ and $k_3^i(\cdot)$ are Lebesgue measurable positive functions of time; they denote, respectively, the B.O.D. decay rate, D.O. reaeration rate and the B.O.D. deoxygenation rate in the i-th reach. q_i^s denotes the i-th reach D.O. saturation concentration, while Q_i and Q_{iE} denote the stream flow rate in the i-th reach and the effluent flow rate into the i-th reach, respectively. V_i denotes the volume of water in the i-th reach, $v_1^i(t)$ and $v_2^i(t)$ denote uncertain disturbances that affect the rates of change of B.O.D. and D.O., respectively, in the i-th reach. m_i is a constant whose "optimal" value is the effluent B.O.D. concentration into the i-th reach corresponding to steady state conditions at desired levels of D.O. concentration. $\Delta m_i(t)$ represents the additional controlled variation of B.O.D. concentration from its "optimal" value, m_i^*, while $r_i(t)$ represents the in-stream aeration rate in the i-th reach; these are the controls introduced to drive the river system to and maintain it at the desired state in the presence of uncertainties.

We assume that the rate coefficients $k_1^i(t)$, $k_2^i(t)$ and $k_3^i(t)$, $i = 1, \ldots, N$, are of the form

$$k_j^i(t) = h_j^i + \Delta h_j^i(t), \; i = 1, \ldots, N, \qquad j = 1, 2, 3,$$

where h_j^i, $i = 1, \ldots, N$, $j = 1, 2, 3$, are known positive constants, while $\Delta h_j^i(\cdot)$, $i = 1, \ldots, N$, $j = 1, 2, 3$, are unknown Lebesgue measurable functions of time with known bounds. In the absence of uncertainty and control, (2.1) reduces to

$$\dot{z}_i(t) = - h_1^i z_i(t) + \frac{Q_{i-1}z_{i-1}(t) + m_iQ_{iE} - Q_iz_i(t)}{V_i}$$

$$\dot{q}_i(t) = - h_3^i z_i(t) + h_2^i(q_i^s - q_i(t)) + \frac{Q_{i-1}q_{i-1}(t) - Q_iq_i(t)}{V_i} \, ,$$

$$i = 1, \ldots, N. \qquad (2.2)$$

If the desired steady state values of D.O. concentration in the N reaches have been specified, say q_1^*, q_2^*. ... q_N^*, and if[3] the values of Q_o, $z_o(t) = z_o$, $q_o(t) = q_o$, as well as V_i, Q_i Q_{iE} and q_i^S, $i = 1, \ldots, N$, are known, then the corresponding steady state values of B.O.D. concentration, z_i^*, $i = 1, \ldots, N$, and the effluent B.O.D. concentration in the N reaches, m_i^*, $i = 1, \ldots, N$, can be obtained by setting $\dot{z}_i(t) = \dot{q}_i(t) = 0$, in (2.2); that is,

---

[3]Subscript o denotes the reach preceding the first reach.

$$- h_1^i z_i(t) + \frac{Q_{i-1} z_{i-1}(t) + m_i Q_{iE} - Q_i z_i(t)}{V_i} = 0$$

$$- h_3^i z_i(t) + h_2^i (q_i^s - q_i^*) + \frac{Q_{i-1} q_{i-1}^* - Q_i q_i^*}{V_i} = 0,$$

$$i = 1, \ldots, N. \qquad (2.3)$$

Solving (2.3), we obtain

$$z_i^* = \frac{1}{h_3^i} [h_2^i (q_i^s - q_i^*) + \frac{Q_{i-1} q_{i-1}^* - Q_i q_i^*}{V_i}]$$

$$m_i^* = \frac{1}{Q_{iE}} [h_1^i z_i^* V_i - Q_{i-1} z_{i-1}^* + Q_i z_i^*] ,$$

$$i = 1, \ldots, N. \qquad (2.4)$$

Thus, if in each reach, the effluent B.O.D. concentration is monitored with respect to the "optimal" value m_i^*, $i = 1, \ldots, N$, then the uncertain river system (2.1) becomes

$$\dot{z}_i(t) = - k_1^i(t)\, z_i(t) + \frac{Q_{i-1} z_{i-1}(t) + m_i^* Q_{iE} - Q_i z_i(t)}{V_i} + \frac{\Delta m_i(t) Q_{iE}}{V_i} + v_1^i(t)$$

$$\dot{q}_i(t) = - k_3^i(t) z_i(t) + k_2^i(t)(q_i^s - q_i(t)) + \frac{Q_{i-1} q_{i-1}(t) - Q_i q_i(t)}{V_i} + r_i(t) + v_2^i(t),$$

$$i = 1, \ldots, N. \qquad (2.5)$$

Consider the transformation

$$\Delta z_i(t) = z_i(t) - z_i^*$$

$$\Delta q_i(t) = q_i(t) - q_i^* \quad , \qquad i = 1, \ldots, N \qquad (2.6)$$

and let[4] $\Delta z_0(t) \equiv 0$ and $\Delta q_0(t) \equiv 0$.

For the sake of simplicity, we also let $\delta m_i(t) \triangleq \frac{\Delta m_i(t) Q_{iE}}{V_i}$, $i = 1, \ldots, N$.

In view of (2.4) and (2.6), (2.5) reduces to

---

[4]We assume that the water flowing into reach 1 is "clean."

$$\Delta\dot{z}_i(t) = -\Delta h_1^i(t)(\Delta z_i(t) + z_i^*) - h_1^i\Delta z_i(t) + \frac{Q_{i-1}\Delta z_{i-1}(t) - Q_i\Delta z_i(t)}{V_i} + \delta m_i(t) + v_1^i(t)$$

$$\Delta\dot{q}_i(t) = -\Delta h_3^i(t)(\Delta z_i(t) + z_i^*) - h_3^i\Delta z_i(t) + \Delta h_2^i(t)(q_i^s - \Delta q_i(t) - q_i^*) - h_2^i\Delta q_i(t)$$
$$+ \frac{Q_{i-1}\Delta q_{i-1}(t) - Q_i\Delta q_i(t)}{V_i} + r_i(t) + v_2^i(t), \quad i = 1, \ldots, N; \tag{2.7}$$

that is,

$$\begin{pmatrix} \Delta\dot{z}_i(t) \\ \Delta\dot{q}_i(t) \end{pmatrix} = \begin{pmatrix} -(h_1^i + \frac{Q_i}{V_i}) & 0 \\ -h_3^i & -(h_2^i + \frac{Q_i}{V_i}) \end{pmatrix} \begin{pmatrix} \Delta z_i(t) \\ \Delta q_i(t) \end{pmatrix} + \begin{pmatrix} -\Delta h_1^i(t) & 0 \\ -\Delta h_3^i(t) & -\Delta h_2^i(t) \end{pmatrix} \begin{pmatrix} \Delta z_i(t) \\ \Delta q_i(t) \end{pmatrix}$$
$$+ \begin{pmatrix} \frac{Q_{i-1}}{V_i} & 0 \\ 0 & \frac{Q_{i-1}}{V_i} \end{pmatrix} \begin{pmatrix} \Delta z_{i-1}(t) \\ \Delta q_{i-1}(t) \end{pmatrix} + \begin{pmatrix} 1 & 0 \\ 0 & 1 \end{pmatrix} \begin{pmatrix} \delta m_i(t) \\ r_i(t) \end{pmatrix}$$
$$+ \begin{pmatrix} -z_i^*\Delta h_1^i(t) + v_1^i(t) \\ -z_i^*\Delta h_3^i(t) + \Delta h_2^i(t)(q_i^s - q_i^*) + v_2^i(t) \end{pmatrix},$$
$$i = 1, \ldots, N. \tag{2.8}$$

3. Decentralized State Feedback Saturation Control

In this section we assume that the full state of the river system can be measured, possibly with measurement error, and that the state feedback control is based on the measured state. In each reach, let $(x_1^i(t), x_2^i(t)) \triangleq (\Delta z_i(t), \Delta q_i(t))$. Suppose that the measurement of the full state $x^i(t)$ is possible with a measurement error $w^i(t)$. Here, $x^i(t) \triangleq (x_1^i(t), x_2^i(t))^T$ and $w^i(t) \triangleq (w_1^i(t), w_2^i(t))^T$. Let $u^i(t) \triangleq (u_1^i(t), u_2^i(t))^T$, $v^i(t) \triangleq (v_1^i(t), v_2^i(t))^T$, $\sigma^i(t) \triangleq (\Delta h_1(t), \Delta h_2(t), \Delta h_3^i(t))^T$ and $y^i(t) \triangleq (y_1(t), y_2(t))^T$ denote, respectively, the control, the uncertain input disturbance, the model parameter uncertainty and the measured state vectors in the i-th reach; that is, $u_1^i(t) \triangleq \delta m_i(t)$, $u_2^i(t) \triangleq r_i(t)$ and $y^i(t) \triangleq x^i(t) + w^i(t)$. Then (2.8) may be written as

$$\begin{pmatrix} \dot{x}_1^i(t) \\ \dot{x}_2^i(t) \end{pmatrix} = \begin{pmatrix} -(h_1^i + \frac{Q_i}{V_i}) & 0 \\ -h_3^i & -(h_2^i + \frac{Q_i}{V_i}) \end{pmatrix} \begin{pmatrix} x_1^i(t) \\ x_2^i(t) \end{pmatrix} + \begin{pmatrix} -\Delta h_1^i(t) & 0 \\ -\Delta h_3^i(t) & -\Delta h_2^i(t) \end{pmatrix} \begin{pmatrix} x_1^i(t) \\ x_2^i(t) \end{pmatrix}$$

$$+ \begin{pmatrix} \frac{Q_{i-1}}{V_i} & 0 \\ 0 & \frac{Q_{i-1}}{V_i} \end{pmatrix} \begin{pmatrix} x_1^{i-1}(t) \\ x_2^{i-1}(t) \end{pmatrix} + \begin{pmatrix} 1 & 0 \\ 0 & 1 \end{pmatrix} \begin{pmatrix} u_1^i(t) \\ u_2^i(t) \end{pmatrix}$$

$$+ \begin{pmatrix} 1 & 0 \\ 0 & 1 \end{pmatrix} \begin{pmatrix} -z_i^* \, \Delta h_1^i(t) + v_1^i(t) \\ -z_i^* \, \Delta h_3^i(t) + \Delta h_2^i(t)(q_i^s - q_i^*) + v_2^i(t) \end{pmatrix}, \quad i = 1, \ldots, N \tag{3.1}$$

or, in vector notation,

$$\dot{x}^i(t) = A_i x^i(t) + \Delta A_i(\sigma^i(t))\, x^i(t) + B_i u^i(t) + C_i x^{i-1}(t) + D_i e^i(\sigma^i(t), v^i(t)), \quad i = 1, \ldots, N \tag{3.2}$$

where

$$A_i \triangleq \begin{pmatrix} -(h_1^i + \frac{Q_i}{V_i}) & 0 \\ -h_3^i & -(h_2^i + \frac{Q_i}{V_i}) \end{pmatrix}, \quad \Delta A_i(\sigma^i(t)) \triangleq \begin{pmatrix} -\Delta h_1^i(t) & 0 \\ -\Delta h_3^i(t) & -\Delta h_2^i(t) \end{pmatrix},$$

$$B_i \triangleq \begin{pmatrix} 1 & 0 \\ 0 & 1 \end{pmatrix} \quad C_i \triangleq \begin{pmatrix} \frac{Q_{i-1}}{V_i} & 0 \\ 0 & \frac{Q_{i-1}}{V_i} \end{pmatrix}, \quad D_i \triangleq \begin{pmatrix} 1 & 0 \\ 0 & 1 \end{pmatrix},$$

$$e^i(\sigma^i(t), v^i(t)) \triangleq \begin{pmatrix} -z_i^* \Delta h_1^i(t) + v_1^i(t) \\ -z_i^* \Delta h_3^i(t) + \Delta h_2^i(t)(q_i^s - q_i^*) + v_2^i(t) \end{pmatrix}.$$

The following decentralized control scheme renders system (3.2) globally practically stable (e.g., see Appendix A): Given $\varepsilon_i > 0$,

$$u^i = p_{\varepsilon_i}(y^i) = \begin{cases} -\dfrac{\mu_i(y^i)}{\|\mu_i(y^i)\|}\,\rho_i(y^i) & \text{if } \|\mu_i(y^i)\| > \varepsilon_i \\ & \qquad i = 1, \ldots, N \\ -\dfrac{\mu_i(y^i)}{\varepsilon_i}\,\rho_i(y^i) & \text{if } \|\mu_i(y^i)\| < \varepsilon_i \end{cases} \tag{3.3}$$

where $\mu_i(y^i) \triangleq B_i^T P_i y^i \rho_i(y^i)$, and P_i is the solution of the Lyapunov equation

$$P_i A_i + A_i^T P_i + M_i = 0$$

for a given constant positive definite symmetric 2×2 matrix M_i. Here,

$$\rho_i(y^i) \triangleq \max_{\sigma^i \in \Sigma_i} \|\Delta A_i(\sigma^i) y^i\| + \max_{\substack{\sigma^i \in \Sigma_i \\ w^i \in \Omega_i}} \|\Delta A_i(\sigma^i) w^i\| + \max_{\substack{\sigma^i \in \Sigma_i \\ v^i \in \Lambda_i}} \| D_i e^i(\sigma^i, v^i)\|$$

where Σ_i, Ω_i and Λ_i are the bounding sets of σ^i, w^i and v^i, respectively. Simulation results for a five reach river system may be found in [19].

4. Decentralized Estimated State Feedback Saturation Control

In the previous section, we assumed that the full state of the river system in each reach could be measured, possibly with measurement error, and that the decentralized state feedback saturation control is based on the measured state. However, as pointed out by Hassan and Younis [20], the measurement of B.O.D. concentration is difficult and time consuming. In fact, the current state of measurement technology, the cost involved and the time required for measuring the full state of the system are but some of the factors that make the use of observers for full state estimation a more practical approach. Thus, in this section we introduce an appropriate observer to estimate the full state in each reach (subsystem) of the river system, and decentralized state feedback saturation control is then based on the estimated state in each reach.

Consider again the uncertain river system which consists of N interconnected subsystems described by

$$\dot{x}^i(t) = A_i x^i(t) + \Delta A_i(\sigma(t))x^i(t) + B_i u^i(t) + C_i x^{i-1}(t) + D_i e^i(\sigma^i(t), v^i(t)), \qquad (4.1)$$

$$x^i(t_o) = x^{io}$$

with output $\hat{y}^i(t) = H_i x^i(t) + \delta^i(t)$, $i = 1, \ldots, N$. (4.2)

Here, A_i, ΔA_i, B_i, C_i and e^i are the same as those defined in Section 3. $\hat{y}^i(t)$ and $\delta^i(t)$ denote, respectively, the output vector and the output measurement error in the i-th reach. H_i is a constant matrix of appropriate dimension. We let $\zeta^i(t)$ and $x_e^i(t)$ denote, respectively, the observer vector and the estimated state for the i-th subsystem, where $\zeta^i(t)$ is the solution of

$$\dot{\zeta}^i(t) = \overline{D}_i \zeta^i(t) + T_i B_i u^i(t) + L_i \hat{y}^i(t)$$
$$\zeta^i(t_o) = \zeta_o^i \qquad (4.3)$$

Here, $\overline{D}_i$ is the observer matrix, T_i and L_i are gain matrices, all of appropriate dimensions.

The estimated state $x_e^i(t)$ is a linear combination of the observer vector and the output vector, that is,

$$x_e^i(t) = W_1^i \zeta^i(t) + W_2^i \hat{y}^i(t) \qquad (4.4)$$

where W_1^i and W_2^i are constant matrices.

The following decentralized control scheme renders system (4.1) and (4.3) globally practically stable (see Appendix B): Given an $\varepsilon_i > 0$,

$$u^i = \hat{P}_{\varepsilon_i}(x_e^i) = \begin{cases} -\dfrac{\hat{\mu}_i(x_e^i)}{\|\hat{\mu}_i(x_e^i)\|}\hat{\rho}_i(x_e^i) & \text{if } \|\hat{\mu}_i(x_e^i)\| > \varepsilon_i \\ & i = 1, \ldots, N \\ -\dfrac{\hat{\mu}_i(x_e^i)}{\varepsilon_i}\hat{\rho}_i(x_e^i) & \text{if } \|\hat{\mu}_i(x_e^i)\| \leq \varepsilon_i \end{cases} \qquad (4.5)$$

where $\hat{\mu}_i(x_e^i) \triangleq B_i^T P_i x_e^i \hat{\rho}_i(x_e^i)$, and P_i is the solution of the Lyapunov equation

$$P_i A_i + A_i^T P_i + M_i = 0$$

for a given constant positive definite symmetric 2×2 matrix M_i. Here,

$$\hat{\rho}_i(x_e^i) \triangleq \max_{\sigma^i \in \Sigma_i} \|\Delta A_i(\sigma^i)x_e^i\| + \max_{\substack{\sigma^i \in \Sigma_i \\ \delta^i \in \Delta_i}} \|\Delta A_i(\sigma^i)\delta^i\| + \max_{\substack{\sigma^i \in \Sigma_i \\ v^i \in \Lambda_i}} \|D_i(\sigma^i, v^i)\|$$

where Σ_i, Δ_i and Λ_i are the bounding sets of σ^i, δ^i and v^i, respectively.

In view of the fact that the measurement of B.O.D. concentration is difficult and time consuming, we take $H_i = (0,1)$ for $i = 1, \ldots, N$. Furthermore, in order to satisfy the observer constraints (see [21]), we choose $T_i = (t_i, 0)$, $t_i \neq 0$,

$$L_i = 0,\ \overline{D}_i = -\left(h_1^i + \frac{Q_i}{V_i}\right),\ W_1^i = (t_i^{-1}, 0)^T \text{ and } W_2^i = (0,1)^T .$$

Simulation results for a five reach river system may be found in [21].

5. Appendix A

In this Appendix we consider an uncertain dynamical system, S, composed of N subsystems, S_i, as follows:

$$S_i:\ \dot{x}^i(t) = [A_i + \Delta A_i(\sigma^i(t))]x^i(t) + B_i u^i(t) + D_i e^i(\sigma^i(t),\ v^i(t)) + \sum_{\substack{j=1 \\ j\neq i}}^{N} g_{ij}\,(x^j(t),\ \sigma^i(t), t)\ ,\quad x^i(t_0) = x^{io} \qquad (5.1)$$

with

$$y^i(t) = x^i(t) + w^i(t)$$

for all $i \in I \triangleq \{i \mid i = 1, \ldots, N\}$, where $t \in R$ is time, $x^i(t) \in R^{ni}$ is the state, $u^i(t) \in R^{mi}$ is the control, $\sigma^i(t) \in R^{pi}$ is the model parameter uncertainty, $v^i(t) \in R^{qi}$ is the input disturbance, $y^i(t) \in R^{ni}$ is the measured state and $w^i(t) \in R^{ni}$ is the state measurement error. A_i, ΔA_i, B_i, D_i, e^i and g_{ij} are matrices of appropriate dimensions.

In compact form, the system S may be expressed as

$$S:\ \dot{x}(t) = [A + \Delta A(\sigma(t))]x(t) + B\,u(t) + D\,e(\sigma(t),\ v(t)) + G(x(t),\ \sigma(t), t),\quad x(t_0) = x^0 \qquad (5.2)$$

with

$$y(t) = x(t) + w(t),$$

where $x = (x^{1T}, x^{2T}, \ldots\ldots, x^{NT})^T \in R^n, \quad n = \sum_{i=1}^{N} n_i$

$y = (y^{1T}, y^{2T}, \ldots\ldots, y^{NT})^T \in R^n,$

$e = (e^{1T}, e^{2T}, \ldots\ldots, e^{NT})^T \in R^n,$

$w = (w^{1T}, w^{2T}, \ldots\ldots, w^{NT})^T \in R^n,$

$u = (u^{1T}, u^{2T}, \ldots\ldots, u^{NT})^T \in R^m, \quad m = \sum_{i=1}^{N} m_i$

$\sigma = (\sigma^{1T}, \sigma^{2T}, \ldots\ldots, \sigma^{NT})^T \in R^p, \quad p = \sum_{i=1}^{N} p_i$

$v = (v^{1T}, v^{2T}, \ldots\ldots, v^{NT})^T \in R^q, \quad q = \sum_{i=1}^{N} q_i$

$A = \text{diag}\,\{A_1, A_2, \ldots\ldots, A_N\}$

$\Delta A = \text{diag}\,\{\Delta A_1, \Delta A_2, \ldots\ldots, \Delta A_N\}$

$B = \text{diag}\,\{B_1, B_2, \ldots\ldots, B_N\}$

$D = \text{diag}\,\{D_1, D_2, \ldots\ldots, D_N\}$

$G(\cdot) = [g_{ij}(\cdot)]_{N\times N}$, with $g_{ii}(\cdot) = 0$

$x^o = (x^{1o^T}, x^{2o^T}, \ldots\ldots, x^{No^T})^T$

Before introducing a class of feedback controls, we state some definitions and assumptions.

Definition 5.1 (Caratheodory function) A function $f(\cdot): \mathcal{D}\times R \to R^s$, $\mathcal{D} \subset R^\ell$, is Caratheodory iff for each $t \in R$, $f(\cdot, t)$ is continuous; for each $x \in \mathcal{D}$, $f(x, \cdot)$ is Lebesgue measurable; and for each compact subset $\mathcal{C}$ of $\mathcal{D}\times R$, there exists a Lebesgue integrable function $M_{\mathcal{C}}(\cdot)$ such that

$$\text{for all } (x,t) \in \mathcal{C}, \ \|f(x,t)\| \le M_{\mathcal{C}}(t). \tag{5.3}$$

Definition 5.2 (Strongly Caratheodory function) A function $f(\cdot): \mathcal{D}\times R \to R^s$ is strongly Caratheodory iff it satisfies (5.3) with $M_{\mathcal{C}}(\cdot)$ replaced by a constant $M_{\mathcal{C}}$.

The following assumptions are made on each of the subsystems S_i:

Assumption 5.1. Uncertain parameters $\sigma^i(\cdot): R \to \Sigma_i$, $v^i(\cdot): R \to \Lambda_i$, $w^i(\cdot): R \to \Omega_i$ are Lebesgue measurable, where $\Sigma_i \subset R^{p_i}$, $\Lambda_i \subset R^{q_i}$, $\Omega_i \subset R^{n_i}$ are prescribed compact subsets of the appropriate spaces.

Assumption 5.2. The functions $\Delta A(\cdot): \Sigma_i \to R^{n_i \times n_i}$, $e^i(\cdot): \Sigma_i \times \Lambda_i \to R^{n_i}$ and $g_{ij}(\cdot): R^{n_j} \times \Sigma_i \times R \to R^{n_i}$ are Caratheodory functions.

Assumption 5.3[5]. The matrix A_i is asymptotically stable.

Assumption 5.4. There exist matrix functions $E_i(\cdot): \Sigma_i \to R^{n_i \times n_i}$ and constant matrices F_i such that for all $\sigma^i \in \Sigma_i$

$$\Delta A_i(\sigma^i) = B_i E_i(\sigma^i)$$

$$D_i = B_i F_i \quad .$$

For $\varepsilon_i > 0$, consider a strongly Caratheodory function $p_{\varepsilon_i}(\cdot): R^{n_i} \to R^{m_i}$ such that

$$\begin{aligned} p_{\varepsilon_i}(y^i) &= -\frac{\mu_i(y^i)}{\|\mu_i(y^i)\|}\rho_i(y^i) \qquad \text{if } \|\mu_i(y^i)\| > \varepsilon_i \\ \|p_{\varepsilon_i}(y^i)\| &\le \rho_i(y^i) \qquad\qquad\quad \text{if } \|\mu_i(y^i)\| \le \varepsilon_i \end{aligned} \tag{5.4}$$

where

$$\mu_i(y^i) \triangleq B_i^T P_i y^i \rho_i(y^i)$$

and P_i is the solution of the Lyapunov equation

$$P_i A_i + A_i^T P_i + M_i = 0$$

for a given constant positive definite symmetric $n_i \times n_i$ matrix M_i.

The known function $\rho_i(\cdot): R^{n_i} \to R_+$ is chosen to satisfy

$$\rho_i(y^i) > \rho_{io}(y^i) \triangleq \max_{\sigma^i \in \Sigma_i} \|E_i(\sigma^i) y^i\| + \max_{\substack{\sigma^i \in \Sigma_i \\ w^i \in \Omega_i}} \|E_i(\sigma^i) w^i\| + \max_{\substack{\sigma^i \in \Sigma_i \\ v^i \in \Lambda_i}} \|F_i e^i(\sigma^i, v^i)\| .$$

A particular example of (5.4) is

---

[5]Alternatively, we may assume that (A_i, B_i) is stabilizable. In that event, the nonlinear feedback control must be supplemented by a linear one which results in a stable system matrix.

$$p_{\varepsilon_i}(y^i) = \begin{cases} -\dfrac{\mu_i(y^i)}{\|\mu_i(y^i)\|}\,\rho_i(y^i) & \text{if } \|\mu_i(y^i)\| > \varepsilon_i \\ \\ -\dfrac{\mu_i(y^i)}{\varepsilon_i}\,\rho_i(y^i) & \text{if } \|\mu_i(y^i)\| \leq \varepsilon_i\,. \end{cases} \tag{5.5}$$

We now give the definition of a feedback control that renders the uncertain system S of (5.2) globally practically stable.

Definition 5.3 (Globally Practically Stable) A feedback control $p_\varepsilon(\cdot) = (p_{\varepsilon 1}(\cdot)^T, p_{\varepsilon 2}(\cdot)^T, \ldots, p_{\varepsilon N}(\cdot)^T)^T$, $p_{\varepsilon i}(\cdot): R^{n_i} \to R^{m_i}$, renders the uncertain system S globally practically stable (g.p.s) iff there exists $\underline{d} > 0$ such that the following properties hold:

(i) Existence of Solutions. Given $(x^0, t_0) \in R^n \times R$, the closed loop system

$$\dot{x}(t) = [A + \Delta A(\sigma(t))]x(t) + Bp_\varepsilon(y) + De(\sigma(t), v(t)) + G(x(t), \sigma(t), t) \tag{5.6}$$

possesses a solution $x(\cdot): [t_0, t_1) \to R^n$, $x(t_0) = x^0$, $t_1 > t_0$.

(ii) Extension of Solutions. Every solution $x(\cdot): [t_0, t_1) \to R^n$ of (5.6) can be continued over $[t_0, \infty)$.

(iii) Uniform Boundedness. Given any $r \in (0,\infty)$, there exists a positive $d(r) < \infty$ such that for all solutions $x(\cdot): [t_0, \infty) \to R^n$, $x(t_0) = x^0$, of (5.6),

$$\|x^0\| \leq r \Rightarrow \|x(t)\| \leq d(r) \quad \forall t \in [t_0,\infty).$$

(iv) Uniform Ultimate Boundedness. Given any $\bar{d} > \underline{d}$ and any $r \in (0, \infty)$, there is a $T(\bar{d},r) \in [0, \infty)$ such that for every solution $x(\cdot)$: $[t_0, \infty) \to R^n$, $x(t_0) = x^0$, of (5.6),

$$\|x^0\| \leq r \Rightarrow \|x(t)\| \leq \bar{d} \quad \forall t \geq t_0 + T(\bar{d},r)\,.$$

(v) Uniform Stability. Given any $\bar{d} > \underline{d}$, there is a positive $\delta(\bar{d})$ such that for every solution $x(\cdot)$: $[t_0, \infty) \to R^n$, $x(t_0) = x^0$, of (5.6)

$$\|x^0\| \leq \delta(\bar{d}) \qquad \|x(t)\| \leq \bar{d} \quad \forall t \geq t_0\,.$$

To assure that $p_\varepsilon(\cdot)$ renders S globally practically stable, we introduce the following assumptions.

Assumption 5.5 $\rho_i(y^i)$ is cone-bounded; that is, there exist constants a_i, $b_i \in R_+$ such that, for all $y^i \in R^{n_i}$,

$$\|\rho_i(y^i)\| \leq a_i \|y^i\| + b_i \quad .$$

Assumption 5.6 There exist constants $c_{ij} \in R_+$ such that, for all

$$(x^j, t) \in R^{n_j} \times R, \; \sigma^i \in \Sigma_i \quad ,$$

$$\|g_{ij}(x^j, \sigma^i, t)\| \leq c_{ij} \|x^j\| \quad .$$

Note that, in view of (5.1), $c_{ij} = 0 \quad \forall i = j$.

Theorem 5.1 (see [16] for proof) Subject to Assumptions 5.1 - 5.6, the control

$$p_\varepsilon(\cdot) \triangleq (p_{\varepsilon_1}(\cdot)^T, \; p_{\varepsilon_2}(\cdot)^T, \; \ldots, \; p_{\varepsilon_N}(\cdot)^T)^T \quad ,$$

with $p_{\varepsilon_i}(\cdot)$ given in (5.5), renders the system S globally practically stable if the successive principal minors of the test matrix $T = [t_{ij}]_{N \times N}$ are all positive, where

$$t_{ij} = \begin{cases} \lambda_m(M_i) & i = j \\ -2\lambda_M(P_i)c_{ij} & i \neq j \end{cases}$$

where $\lambda_m(\Phi)$ and $\lambda_M(\Phi)$ denote the minimum and maximum eigenvalues of Φ respectively.

Lemma 5.1 (See [16] for proof) Suppose $c_{ij} = 0$ for all $j > i$ (or for all $i > j$ equivalently), $i,j \in I$, then all the successive principal minors of T are positive.

Appendix B

Consider an uncertain dynamical system S which consists of N interconnected subsystems S described by

$$S_i: \dot{x}^i(t) = [A_i + \Delta A_i(\sigma^i(t))]x^i(t) + Bu^i(t) + D_i e^i(\sigma^i(t), v^i(t))$$

$$+ \sum_{\substack{j=1 \\ j \neq i}}^{N} g_{ij}(x^j(t), \sigma^i(t), t) \; , \qquad x^i(t_o) = x^{io} \qquad (5.7)$$

with output

$$\hat{y}^i(t) = H_i x^i(t) + \delta^i(t) \tag{5.8}$$

for all $i \in I \triangleq \{i \mid i = 1, \ldots, N\}$. The notation used here is the same as that in Appendix A, except that $\hat{y}^i(t) \in R^{r_i}$ and $\delta^i(t) \in R^{r_i}$ replace $y^i(t)$ and $w^i(t)$, respectively. Thus in Assumption 5.1, we assume that $\delta^i(\cdot)$: $R \to \Delta_i$ is Lebesgue measurable where $\Delta_i \subset R^{r_i}$ is a known compact set. In Assumption 5.5, $\rho_i(y^i)$ and y^i are replaced by $\hat{\rho}_i(x_e^i)$ and x_e^i, respectively. In addition to Assumptions 5.1 - 5.6, we also need:

Assumption 5.7 (A_i, H_i) is observable.

Consider the observer equation

$$\dot{\zeta}^i(t) = \overline{D}_i\zeta^i(t) + T_iB_iu^i(t) + L_i\hat{y}^i(t) \tag{5.9}$$

$$\zeta^i(t_o) = \zeta_o^i$$

and the estimated state $x_e^i(t)$ defined as

$$x_e^i(t) = W_1^i\zeta^i(t) + W_2^i\hat{y}^i(t) \tag{5.10}$$

where $\zeta^i(t) \in R^{s_i}$ and $x_e^i(t) \in R^{n_i}$. The matrices and the dimensions that appear in (5.7) - (5.10) must satisfy the following observer constraints:

$$\begin{aligned} &(i) \quad n_i - r_i \leq s_i \leq n_i \\ &(ii) \quad \overline{D}_iT_i - T_iA_i + L_iH_i = 0 \\ &(iii) \quad W_1^iT_i + W_2^iH_i = I_i \\ &(iv) \quad \text{Re}\,\lambda_k(\overline{D}_i) < 0; \quad k = 1, \ldots, s_i \quad . \end{aligned} \tag{5.11}$$

Now, we introduce the estimated state feedback saturation control $\hat{p}_{\varepsilon_i}(x_e^i)$ as follows:

For any $\varepsilon_i > 0$, $\hat{p}_{\varepsilon_i}(\cdot)$: $R^{n_i} \to R^{m_i}$ is any continuous function defined by

$$\begin{aligned} &\hat{p}_{\varepsilon_i}(x_e^i) = -\frac{\hat{\mu}_i(x_e^i)}{\|\hat{\mu}_i(x_e^i)\|}\hat{\rho}_i(x_e^i) \qquad \text{if } \|\hat{\mu}_i(x_e^i)\| > \varepsilon_i \\ &\|p_{\varepsilon_i}(x_e^i)\| \leq \hat{\rho}_i(x_e^i) \qquad \text{if } \|\hat{\mu}(x_e^i)\| \leq \varepsilon_i \end{aligned} \tag{5.12}$$

where $\hat{\mu}_i(x_e^i) \triangleq B_i^T P_i x_e^i \hat{\rho}_i(x_e^i)$, and P_i is the solution of the Lyapunov equation.

The function $\hat{\rho}_i(\cdot): R^{n_i} \to R_+$ is taken to be

$$\hat{\rho}_i(x_e^i) = \max_{\substack{\sigma^i \in \Sigma_i \\ \delta^i \in \Delta_i \\ v^i \in \Lambda_i}} [\| E_i(\sigma^i)x_e^i \| + \| E_i(\sigma^i)W_2^i \delta^i \| + \| F_i e^i(\sigma^i, v^i) \|]$$

Theorem 5.2 (see [21] for proof) Subject to Assumptions 5.1 - 5.7 and the observer constraints (5.11), the control

$$\hat{P}_\varepsilon(\cdot) \triangleq (\hat{P}_{\varepsilon_1}(\cdot)^T, \hat{P}_{\varepsilon_2}(\cdot)^T, \ldots, \hat{P}_{\varepsilon_N}(\cdot)^T)^T \quad ,$$

with $\hat{P}_{\varepsilon i}(\cdot)$ given in (5.12), renders the system S globally, practically stable if

$$\frac{\lambda_m(M_i)\lambda_m(M_{\zeta i})}{4\lambda_M(P_i)\lambda_M(P_{\zeta i})} > \max_{\sigma^i \in \Sigma_i} \| T_i B_i E_i(\sigma^i) \| \max_{\sigma^i \in \Sigma_i} \| B_i E_i(\sigma^i) W_1^i \| \tag{5.13}$$

where $P_{\zeta i}$ and $M_{\zeta i}$ are positive definite matrices that appear in the Lyapunov equation

$$P_{\zeta i}\overline{D}_i + \overline{D}_i P_{\zeta i} + M_{\zeta i} = 0, \quad M_{\zeta i} > 0 \ . \tag{5.14}$$

References

[1] Jamshidi, M., "Large Scale Systems: Modeling and Control," North-Holland, N.Y., Amsterdam, Oxford, 1983.

[2] Singh, M.G., "Dynamical Hierarchical Control," North-Holland Publ., Amsterdam, 1977.

[3] Haimes, Y.Y., "Hierarchical Analyses of Water Resources Systems," McGraw-Hill Int'l., 1977.

[4] Singh, M.G. and Titli, A., "Systems: Decomposition, Optimisation and Control," Pergamon Press, 1978.

[5] Singh, M.G. and Hassan, M., "Closed Loop Hierarchical Control for River Pollution," Automatica, Vol. 12, 261-264, 1976.

[6] Mahmoud, M.S., Hassan, M.F., and Saleh, S.J., "Decentralized Structures for Stream Water Quality Control Problems," Optimal Control Applications and Methods, Vol. 6, 167-186, 1985.

[7] Singh, M.G., "River Pollution Control," Int'l J. Systems Sci., Vol. 6, No. 1, 9-21, 1975.

[8] Singh, M.G., "Hierarchical Methods in River Pollution Control," in: Halfon, E. (ed.), Theoretical Systems Ecology, 419-451, Academic Press, N.Y., 1979.

[9] Tamur, H., "A Discrete Dynamic Model with Distributed Transport Delays and Its Hierarchical Optimization for Preserving Stream Quality," IEEE Trans. Syst., Man, Cybern., Vol. SMC-4, No. 5, 424-431, 1974.

[10] Haimes, Y.Y. and Macko, D., "Hierarchical Structures in Water Resources Systems Management," IEEE Trans. Syst., Man, Cybern., Vol. SMC-3, 396-402, 1973.

[11] Olenik, S.C. and Haimes, Y.Y., "A Hierarchical Multi-objective Framework for Water Resources Planning," IEEE Trans. Syst., Man, Cybern., Vol. SMC-9, No. 9, 534-544, 1979.

[12] Lee, C.S. and Leitmann, G., "Uncertain Dynamical Systems: An Application to River Pollution Control," Second NSF Workshop on Renewable Resource Management, Honolulu, Hawaii, December 1985.

[13] Chen, Y.H. and Lee, C.S., "On the Control of An Uncertain Water Quality System," submitted for publication in Optimal Control Applications and Methods.

[14] Gutman, S. and Leitmann, G., "Stabilizing Feedback Control for Dynamical Systems with Bounded Uncertainty," Proceed. IEEE Conf. Decision and Control, Phoenix, Arizona, 1986.

[15] Corless, M. and Leitmann, G., "Continuous State Feedback Guaranteeing Uniform Ultimate Boundedness for Uncertain Dynamic Systems," IEEE Trans. Autom. Contr., Vol. AC-23, 1139, 1981.

[16] Leitmann, G., "On the Efficacy of Nonlinear Control in Uncertain Linear Systems," J. Dyn. Syst. Meas. Control, Vol. 102, 95, 1981.

[17] Chen, Y.H., "Deterministic Control of Large-Scale Uncertain Dynamical Systems," to appear in the J. Franklin Institute.

[18] Leitmann, G., Lee, C.S., and Chen, Y.H., "Hierarchical Control of Uncertain Systems: An Application to Water Quality Control," Optimization Days 1986, University of Montreal, Montreal, Canada, April 1986.

[19] Leitmann, G., Lee, C.S., and Chen, Y.H., "Decentralized Control for a Large Scale Uncertain River System," Proceed. of IFAC Workshop on Modelling, Decisions and Games for Social Phenomena, Beijing, China, 539-552, 1986.

[20] Hassan, M.F. and Younis, M.I., "Stream Quality Modelling: A Discussion," in: Lainiotis, D.G. and Tzannes, N.S. (eds), Applications of Information and Control Systems, D. Reidel Publ. Co., Dordrecht, Holland, 1980.

[21] Chen, Y.H., "Deterministic Control of Large-Scale Uncertain Systems Under State Detection," 1986 (to appear).

A NEW APPROACH FOR OPTIMIZING HYDROPOWER SYSTEM OPERATION WITH A QUADRATIC MODEL[1]

S.A. SOLIMAN and G.S. CHRISTENSEN
Electrical Engineering Department, University of Alberta, Edmonton, Alberta, Canada

1. Introduction

The problem of determining the optimal long-term operation of a multireservoir power system has been the subject of numerous publications over the past 40 years, and yet no completely satisfying solution has been obtained, since in every publication the problem has been simplified in order to be solved.

Aggregation of the multireservoir hydroplant into a single complex equivalent reservoir and solution by Stochastic Dynamic Programming (SDP) is one of the earlier approaches that has been used [1] and [2]. Obviously, such a representation of the reservoirs cannot take into account all local constraints on the contents of the reservoir, water flows, and hydroplant generation. This method can perform satisfactory for systems where reservoirs and inflow characteristics are sufficiently "similar" to justify aggregation into a single reservoir and hydroplant model [3].

Turgeon has proposed two methods for the solution of the problem. The first is really an extension of the aggregation method, and it breaks the problem into a two level problem. At the second level, the problem is to determine the monthly generation of the valley. This problem is solved by Dynamic Programming. The problem at the first level is to allocate that generation to the installation. This is done by finding functions that relate that water level of each reservoir to the total amount of potential energy stored in the valley, [4].

The second method [6] is the decomposition method by combining many reservoirs into one reservoir for the purpose of optimization and using the Dynamic Programming for solving n-1 problems of two state variables each.

The solution obtained by this method is a function of the water content of that reservoir and the total energy content of the downstream reservoirs. The main drawback is that the approach avoids answering the basic question as to how the individual reservoirs in the system are to be operating in a optimal fashion. Also the inflows to some reservoir may be periodic in phase with the annual demand cycle, while other reservoirs have an inflow cycle which lags by a certain time [6].

The objectives of this paper are to develop all the mathematical expressions to be used in the optimization, and to obtain, analyze, and compare results with the non-linear model developed in [16]. The optimization problem is described and formulated as the optimal control of a multivariable state-space model in which the state and

[1]This work was supported by the Natural Sciences and Engineering Research Council of Canada, Grant No. A4146. The authors would like to acknowledge data obtained from B.C. Hydro, Vancouver, B.C.

control vectors are constrained by sets of equality and inequality constraints to satisfy the multipurpose stream use requirements such as flood control and water supply. Lagrange and Kuhn-Tucker multipliers are used to adjoin these constraints to the objective function. The resulting cost functional is maximized by using the minimum norm formulation of functional analysis.

2. Problem Formulation

2.1 System Under Study

The system under study consists of m independent rivers with one or several reservoirs and power plants in series on each, and interconnection lines to the neighbouring system through which energy may be exchanged (Figure 1). Denote by

I^k_{ij} A random variable representing the natural inflow to the reservoir i on river j during a period k in Mm^3. It is assumed that no correlation exists between flows of independent rivers at different periods of time. These random variables are statistically independent. ($1Mm^3=10^6m^3$).

x^k_{ij} The storage of reservoir i on river j at the end of period k in Mm^3.

u^k_{ij} The discharge from reservoir i on river j during a period k in Mm^3.

s^k_{ij} The spill from reservoir i on river j during a period k in Mm^3.

c^k_j The value in dollars of one MWh produced anywhere on river j.

$\bar{x}_{ij}$ The maximum storage of reservoir i on river j in Mm^3.

$\underline{x}_{ij}$ The minimum storage of reservoir i on river j in Mm^3.

$\bar{u}^k_{ij}$ The maximum discharge through the turbines in Mm^3.

$\underline{u}^k_{ij}$ The minimum discharge through the turbines in Mm^3.

$G_{ij}(u^k_{ij},\frac{1}{2}(x^k_{ij}+x^{k-1}_{ij}))$ The generation of plant i on river j during a period k in MWh. It is a nonlinear function of the discharge u^k_{ij} and the average storage between two successive months.

$V_{ij}(x^K_{ij})$ Value in dollars of the water left in storage at the end of the planning horizon.

n_j Number of reservoirs on river j; i=1, ..., n_j, j=1, ..., ..., m

m The total number of rivers

k The superscript denoting the period; k=1, ..., ..., K.

2.2 The Objective Function

The long-term optimal operating problem aims to find the discharge u^k_{ij} i=1,...,..., n_j;j=1, ..., ..., m that maximizes the total expected benefits from the system (benefits from the generation and benefits from the amount of water left in storage at the end of the planning period),while satisfying certain constraints. In mathematical terms, the problem of the power system in Figure 1 is to find the discharge

u^k_{ij} that maximizes

$$J=E[\sum_{j=1}^{m}\sum_{i=1}^{n_j} V_{ij}(x^K_{ij})+\sum_{j=1}^{m}\sum_{i=1}^{n_j}\sum_{k=1}^{K} c^k_j G_{ij}(u^k_{ij},\tfrac{1}{2}(x^k_{ij}+x^{k-1}_{ij}))] \text{ in } \$ \tag{1}$$

Subject to satisfying the following constraints:

(1)The water conservation equation (continuity equation) for each reservoir may adequately be described by the following difference equation

$$x^k_{ij}=x^{k-1}_{ij}+I^k_{ij}+u^k_{(i-1)j}-u^k_{ij}+s^k_{(i-1)j}-s^k_{ij} \tag{2}$$

where

$$s^k_{ij}= \begin{cases} (x^{k-1}_{ij}+I^k_{ij}+u^k_{(i-1)j}+s^k_{(i-1)j}-x^k_{ij})-\bar{u}^k_{ij}; \text{If } (x^{k-1}_{ij}+I^k_{ij}+u^k_{(i-1)j}+s^k_{(i-1)j}- x^k_{ij}) > \bar{u}^k_{ij} \\ 0, \text{ otherwise.} \end{cases} \tag{3}$$

water is spilt when the reservoir is filled to capacity, and the inflow to the reservoir exceeds $\bar{u}^k_{ij}$

(2) To satisfy multipurpose stream use requirements, such as flood control, irrigation, fishing and other purposes if any, the following upper and lower limits on the variables should be satisfied.

(a) upper and lower bounds on the storage

$$\underline{x}_{ij}\leq x^k_{ij}\leq \bar{x}_{ij} \tag{4}$$

(b) upper and lower bounds on the discharge

$$\underline{u}^k_{ij}\leq u^k_{ij}\leq \bar{u}^k_{ij} \tag{5}$$

The first set of the inequality constraints simply states that the reservoir storage may not exceed a maximum level, nor be lower than a minimum level. For the maximum level this is determined by the elevation of the spillway crest or the top of the spillway gates. The minimum level may be fixed by the elevation of the lowest outlet in the dam or by conditions of operating efficiency for the turbines. The second set is determined by the discharge capacity of the power plant as well as its efficiency [13].

The initial storage x^0_{ij} and the expected value of the natural inflows into each stream during each month are assumed to be known.

2.3 Modelling of the System

The value in dollars of water left in storage at the end of the planning period is obtained as the following:

(a) multiply the amount of water left in storage by the conversion factor of at-a-site and downstream hydroplants; i.e., we convert this amount of water to the equivalent electrical energy MWh

(b) Since no one knows when this energy will be used in the future, we assume the cost of this energy is 1$/MWh, (the average value during a year). Following the above two steps we may choose the following for the function $V_{ij}(x^K_{ij})$.

$$V_{ij}(x^K_{ij}) = \sum_{u=i}^{n_j} x^K_{ij}(\alpha_{ij}+\beta_{uj}x^K_{uj}+\gamma_{uj}(x^K_{uj})^2) \text{ in } \$ \tag{6}$$

In equation (6), we assumed that the water conversion factor (MWh/Mm^3) at-a-site has a quadratic relation with the storage. (The storage-elevation curve is quadratic.) The constants α_{ij}, β_{ij} and γ_{ij} were obtained by least square curve fitting to typical plant data available.

The generation of a hydroelectric plant is a nonlinear function of the water discharge u^k_{ij} and the reservoir head, which itself is a function of the storage. To avoid underestimation of production for rising water levels and overestimation for falling water levels, an average of begin and end-of-time step (month) storage is used. We may choose the following for the function $G_{ij}(u^k_{ij},\frac{1}{2}(x^k_{ij}+x^{k-1}_{ij}))$

$$G_{ij}(u^k_{ij},\tfrac{1}{2}(x^k_{ij}+x^{k-1}_{ij}))=\alpha_{ij}u^k_{ij}+\tfrac{1}{2}\beta_{ij}u^k_{ij}(x^k_{ij}+x^{k-1}_{ij})+\tfrac{1}{4}\gamma_{ij}u^k_{ij}(x^k_{ij}+x^{k-1}_{ij})^2 \tag{7}$$

Substituting for x^k_{ij} from equation (2) into equation (7), one obtains

$$\begin{aligned}G_{ij}(u^k_{ij},\tfrac{1}{2}(x^k_{ij}+x^{k-1}_{ij}))=&\,b^k_{ij}u^k_{ij}+u^k_{ij}d^k_{ij}x^{k-1}_{ij}+u^k_{ij}f^k_{ij}(u^k_{(i-1)j}-u^k_{ij})+\gamma_{ij}u^k_{ij}\cdot\\ &\cdot(x^{k-1}_{ij})^2+\tfrac{1}{4}u^k_{ij}\gamma_{ij}((u^k_{ij})^2+(u^k_{(i-1)j})^2)+\gamma_{ij}u^k_{ij}x^{k-1}_{ij}(u^k_{(i-1)j}-u^k_{ij})\\ &-\tfrac{1}{2}\gamma_{ij}u^k_{(i-1)j}(u^k_{ij})^2\end{aligned} \tag{8}$$

where

$$q^k_{ij}=I^k_{ij}+s^k_{(i-1)j}-s^k_{ij} \tag{9}$$

$$b^k_{ij}=\alpha_{ij}+\tfrac{1}{2}\beta_{ij}q^k_{ij}+\tfrac{1}{4}\gamma_{ij}(q^k_{ij})^2 \tag{10}$$

$$d^k_{ij}=\beta_{ij}+\gamma_{ij}q^k_{ij} \tag{11}$$

$$f^k_{ij}=\tfrac{1}{2}\beta_{ij}+\tfrac{1}{2}\gamma_{ij}q^k_{ij} \tag{12}$$

It will be noticed that the generating function in equation (8) is a highly nonlinear function. If one defines the following pseudo-state variables such that [14]

$$y^k_{ij}=(x^k_{ij})^2;\ i=1,\ \ldots,\ n_j;\ j=1,\ \ldots,\ m,\ k=1,\ \ldots,\ K \tag{13}$$

$$z^k_{ij}=(u^k_{ij})^2;\ i=1,\ \ldots,\ n_j;\ j=1,\ \ldots,\ \ldots,\ m;\ k=1,\ \ldots,\ \ldots,\ K \tag{14}$$

$$r^{k-1}_{ij}=u^k_{ij}x^{k-1}_{ij},\ i=1,\ \ldots,\ \ldots,\ n_j;\ i=1,\ \ldots,\ \ldots,\ m;\ k=1,\ \ldots,\ \ldots,\ K \tag{15}$$

Then the function G_{ij} in equation (8) becomes

$$G_{ij}(u^k_{ij},\tfrac{1}{2}(x^k_{ij}+x^{k-1}_{ij}))=b^k_{ij}u^k_{ij}+u^k_{ij}d^k_{ij}x^{k-1}_{ij}+u^k_{ij}f^k_{ij}(u^k_{ij}-u^k_{ij})+\gamma_{ij}u^k_{ij}y^{k-1}_{ij}$$

$$+\frac{1}{4}u^{k}_{ij}\gamma_{ij}(z^{k}_{(i-1)j}+z^{k}_{ij})+\gamma_{ij}r^{k-1}_{ij}(u^{k}_{(i-1)j}-u^{k}_{ij})-\frac{1}{2}\gamma_{ij}z^{k}_{ij}u^{k}_{(i-1)j} \tag{16}$$

Now, the cost functional in equation (1) becomes

$$J=E[\sum_{j=1}^{m}\sum_{i=1}^{n_j}\sum_{u=i}^{n_j}x^{K}_{ij}(\alpha_{uj}+\beta_{uj}x^{K}_{ij}+\gamma_{uj}y^{K}_{uj})+\sum_{j=1}^{m}\sum_{i=1}^{n_j}\sum_{k=1}^{K}\{c^{k}_{j}b^{k}_{ij}u^{k}_{ij}+c^{k}_{j}u^{k}_{ij}d^{k}_{ij}x^{k-1}_{ij}$$

$$+c^{k}_{j}u^{k}_{ij}f^{k}_{ij}(u^{k}_{(i-1)j}-u^{k}_{ij})+c^{k}_{j}\gamma_{ij}u^{k}_{ij}y^{k-1}_{ij}+\frac{1}{4}c^{k}_{j}u^{k}_{ij}\gamma_{ij}(z^{k}_{(i-1)j}+z^{k}_{ij})+$$

$$+c^{k}_{j}\gamma_{ij}r^{k-1}_{ij}(u^{k}_{(i-1)j}-u^{k}_{ij})-\frac{1}{2}c^{k}_{j}\gamma_{ij}z^{k}_{ij}u^{k}_{(i-1)j}\}] \tag{17}$$

Subject to satisfying the following constraints.

$$y^{k}_{ij}=(x^{k}_{ij})^{2} \tag{18}$$

$$z^{k}_{ij}=(u^{k}_{ij})^{2} \tag{19}$$

$$r^{k-1}_{ij}=u^{k}_{ij}x^{k-1}_{ij} \tag{20}$$

$$x^{k}_{ij}=x^{k-1}_{ij}+q^{k}_{ij}+u^{k}_{(i-1)j}-u^{k}_{ij} \tag{21}$$

$$\underline{x}_{ij}\le x^{k}_{ij}\le\bar{x}_{ij} \tag{22}$$

$$\underline{u}^{k}_{ij}\le u^{k}_{ij}\le\bar{u}^{k}_{ij} \tag{23}$$

Now, the problem is that of maximizing (17) subject to satifying the constraints (18-23).

3. Mathematical Solution

3.1 A Minimum Norm Formulation

We can form an augmented cost functional $\tilde{J}$, by adjoining to the cost function in equation (17) the equality constraints (18-21) via Lagrange's multipliers, and the inequality constraints (22 and 23) via Kuhn-Tucker multipliers [13], on thus obtains

$$\tilde{J}=E[\sum_{j=1}^{m}\sum_{i=1}^{n_j}\sum_{u=i}^{n_j}x^{K}_{ij}(\alpha_{uj}+\beta_{uj}x^{K}_{uj}+\gamma_{uj}y^{k}_{uj})+\sum_{j=1}^{m}\sum_{i=1}^{n_j}\sum_{k=1}^{K}\{c^{k}_{j}b^{k}_{ij}u^{k}_{ij}+c^{k}_{j}u^{k}_{ij}d^{k}_{ij}x^{k-1}_{ij}+$$

$$c^{k}_{j}u^{k}_{ij}f^{k}_{ij}(u^{k}_{(i-1)j}-u^{k}_{ij})+c^{k}_{j}\gamma_{ij}u^{k}_{ij}y^{k-1}_{ij}+\frac{1}{4}c^{k}_{j}u^{k}_{ij}\gamma_{ij}(z^{k}_{(i-1)j}+z^{k}_{ij})+c^{k}_{j}\gamma_{ij}r^{k-1}_{ij}\;.$$

$$.(u^{k}_{(i-1)j}-u^{k}_{ij})-\frac{1}{2}c^{k}_{j}\gamma_{ij}z^{k}_{ij}u^{k}_{(i-1)j}+\mu^{k}_{ij}(-y^{k}_{ij}+(x^{k}_{ij})^{2})+\phi^{k}_{ij}+(u^{k}_{ij})^{2})$$

$$+\psi^{k}_{ij}(-r^{k-1}_{ij}+u^{k}_{ij}x^{k-1}_{ij})+\lambda^{k}_{ij}(-x^{k}_{ij}+x^{k-1}_{ij}+q^{k}_{ij}+u^{k}_{(i-1)j}-u^{k}_{ij})+e^{k}_{ij}(\underline{x}_{ij}-x^{k}_{ij})+$$

$$e^{1k}_{ij}(x^{k}_{ij}-\bar{x}_{ij})+g^{k}_{ij}(\underline{u}^{k}_{ij}-u^{k}_{ij})+g^{1k}_{ij}(u^{k}_{ij}-\bar{u}^{k}_{ij})\}] \tag{24}$$

where μ^{k}_{ij}, ϕ^{k}_{ij}, ψ^{k}_{ij} and λ^{k}_{ij} are Lagrange's multipliers, they are to be determined, so that the equality constraints (18-21) are satisfied and e^{k}_{ij}, e^{1k}_{ij}, g^{k}_{ij} and g^{1k}_{ij} are Kuhn-Tucker multiplier, these are equal to zero if the constraints are not violated, and greater than zero if the constraints are violated [13].

Define the following column vectors:

$$A_{ij} = \sum_{u=i}^{n_j} \alpha_{uj};\ i=1, \ldots, \ldots, n_j;\ j=1, \ldots, \ldots, m \tag{25}$$

$$A = \text{col.}\ (A_1, \ldots, \ldots, \ldots, A_m) \tag{26}$$

$$A_1 = \text{col.}\ (A_{11}, \ldots, \ldots, \ldots, A_{n_1 1}) \tag{27}$$

$$A_m = \text{col.}\ (A_{1m}, \ldots, \ldots, A_{n_m m}) \tag{28}$$

$$b(k) = \text{col.}\ (b_1(k), \ldots, \ldots, b_m(k)) \tag{29}$$

$$b_1(k) = \text{col.}\ (c_1^k b_{11}^k, \ldots, \ldots, c_1^k b_{n_1 1}^k) \tag{30}$$

$$b_m(k) = \text{col.}\ (c_m^k b_{1m}^k), \ldots, \ldots, c_m^k b_{n_m m}^k) \tag{31}$$

$$x(k) = \text{col.}\ (x_1(k), \ldots, \ldots, x_m(k)) \tag{32}$$

$$x_1(k) = \text{col.}\ (x_{11}^k, \ldots, \ldots, x_{n_1 1}^k) \tag{33}$$

$$x_m(k) = \text{col.}\ (x_{1m}^k, \ldots, \ldots, x_{n_m m}^k) \tag{34}$$

$y(k)$, $v(k)$, $z(k)$, $r(k)$, $\mu(k)$, $\phi(k)$, $\psi(k)$, $\lambda(k)$, $q(k)$, $s(k)$, $\nu(k)$ and $\sigma(k)$ are similarly defined, where

$$\nu_{ij}^k = e_{ij}^{1k} - e_{ij}^k \tag{35}$$

$$\sigma_{ij}^k = g_{ij}^{1k} - g_{ij}^k \tag{36}$$

Furthermore, define the following matricies

$$B = \text{diag.}\ (B_1, \ldots, \ldots, B_m) \tag{37}$$

B_1 is $n_1 \times n_1$ matrix whose elements are given by

$$\left.\begin{array}{l} \text{(i) } b_{ii} = \beta_{i1};\ i=1, \ldots, \ldots, n_1 \\ \text{(ii) } b_{(v+1)v} = b_{v(v+1)} = \frac{1}{2}\beta_{(v+1)1};\ v=1, \ldots, \ldots, n_1-1 \end{array}\right\} \tag{38}$$

B_m is $n_m \times n_m$ matrix whose elements are given by

$$\left.\begin{array}{l} \text{(i) } b_{ii} = \beta_{im};\ i=1, \ldots, n_m \\ \text{(ii) } b_{(v+1)v} = b_{v(v+1)} = \frac{1}{2}\beta_{(v+1)};\ v=1, \ldots, \ldots, n_m-1 \end{array}\right\} \tag{39}$$

$$C = \text{diag.}\ (C_1, \ldots, \ldots, C_m) \tag{40}$$

C_1 is upper triangular matrix whose elements are given by:

$$\left.\begin{array}{l} \text{(i) } c_{ii} = \gamma_{i1};\ i=1, \ldots, \ldots, n_1 \\ \text{(ii) } c_{v(v+1)} = \gamma_{(v+1)1};\ v=1, \ldots, \ldots, n_1-1 \end{array}\right\} \tag{41}$$

C_m is upper triangular matrix whose elements are given by:

$$\left.\begin{array}{ll}\text{(i)} & c_{ii} = \gamma_{im};\ i=1,\ \ldots,\ n_m \\ \text{(ii)} & c_{v(v+1)} = \gamma_{(v+1)m};\ v=1,\ \ldots,\ n_m-1\end{array}\right\} \quad (42)$$

$$d(k) = \text{diag.}\ (d_1(k),\ \ldots,\ \ldots,\ d_m(k)) \quad (43)$$

$$d_1(k) = \text{diag.}\ (c_1^k d_{11}^k,\ \ldots,\ \ldots,\ c_m^k d_{n_1 1}^k) \quad (44)$$

$$d_m(k) = \text{diag.}\ (c_m^k d_{1m}^k,\ \ldots,\ \ldots,\ c_m^k d_{n_m m}^k) \quad (45)$$

$$f(k) = \text{diag.}\ (f_1(k),\ \ldots,\ \ldots,\ f_m(k)) \quad (46)$$

$$f_1(k) = \text{diag.}\ (c_1^k f_{11}^k,\ \ldots,\ \ldots,\ c_1^k f_{n_1 1}^k) \quad (47)$$

$$f_m(k) = \text{diag.}\ (c_m^k f_{1m}^k,\ \ldots,\ \ldots,\ c_m^k f_{n_m m}^k) \quad (48)$$

$$C(k) = \text{diag.}\ (C_1(k),\ \ldots,\ \ldots,\ C_m(k)) \quad (49)$$

$$C_1(k) = \text{diag.}\ (c_1^k \gamma_{11},\ \ldots,\ \ldots,\ c_1^k \gamma_{n_1 1}) \quad (50)$$

$$C_m(k) = \text{diag.}\ (c_m^k \gamma_{1m},\ \ldots,\ \ldots,\ c_m^k \gamma_{n_m m}) \quad (51)$$

$$M = \text{diag.}\ (M_1,\ \ldots,\ \ldots,\ \ldots,\ M_m) \quad (52)$$

where any matrix M_1, ..., ..., M_m is a lower triangular matrix, whose elements are given by:

$$\left.\begin{array}{ll}\text{(i)} & m_{ii} = -1;\ i=1,\ \ldots,\ \ldots,\ n_j \\ \text{(ii)} & m_{(v+1)v}=1,\ v=1;\ \ldots,\ n-1\end{array}\right\} \quad (53)$$

$$N = \text{diag.}\ (N_1,\ \ldots,\ \ldots,\ \ldots,\ N_m) \quad (54)$$

where any matrix N_1, ..., ..., N_m is a lower triangular matrix whose elements are given by:

$$\left.\begin{array}{ll}\text{(i)} & n_{ii} = 1;\ i=1,\ \ldots,\ \ldots,\ n_j \\ \text{(ii)} & n_{(v+1)v} =1;\ v=1,\ \ldots,\ \ldots,\ n_j-1\end{array}\right\} \quad (55)$$

$$L = \text{diag.}\ (L_1,\ \ldots,\ \ldots,\ L_m) \quad (56)$$

where any matrix L_1, ..., ..., L_m is a lower triangular matrix whose elements are given by:

$$\text{(i)}\ \ell_{(v+1)v} = 1;\ v=1,\ \ldots,\ n_j-1 \quad (57)$$

(ii) the rest of the elements are equal to zero

Using all the above definitions, the augmented cost functional in equation (24) becomes

$$\tilde{J}=E[A^T x(K)+x^T(K)Bx(K)+X^T(K)Cy(K)+\sum_{k=1}^{K}\{b^T(k)u(k)+u^T(k)d(k)x(k-1)\ +$$

$$u^T(k)f(k)Mu(k)+u^T(k)C(k)y(k-1)+\frac{1}{4}u^T(k)C(k)Nz(k)+r^T(k-1)C(k)Mu(k)$$

$$-\frac{1}{2}z^T(k)C(k)Lu(k)+\mu^T(k)(-y(k)+x^T(k)\vec{H}x(k))+\phi^T(k)(-z(k)+u^T(k)\vec{H}u(k))+$$

$$\psi^T(k)(-r(k-1)+u^T(k)\vec{H}x(k-1))+\lambda^T(k)(-x(k)+x(k-1)+q(k)+Mu(k))+$$

$$\nu^T(k)(x(k-1)+q(k)+Mu(k))+\sigma^T(k)u(k)\}] \quad (58)$$

This expression is optimized by using the minimum norm formulation as in [16]. One then obtains the following equations:

$$E[\lambda(k)-\lambda(k-1)+\nu(k)+2\mu^T(k-1)\vec{H}x(k-1)+d(k)u(k)+\Psi^T(k)\vec{H}u(k)] = [0] \quad (59)$$

$$E[-\mu(k-1)+C(k)u(k)] = [0] \quad (60)$$

$$E[-\phi(k)+\frac{1}{4}N^TC(k)u(k)-\frac{1}{2}C(k)Lu(k)] = [0] \quad (61)$$

$$E[-\Psi(k)+C(k)Mu(k)] = [0] \quad (62)$$

$$E[b(k)+M^T\lambda(K)+M^T\nu(k)+\sigma(K)+d(k)x(k-1)+\Psi^T(k)\vec{H}x(k-1)+C(k)y(k-1)+$$

$$\frac{1}{4}C(k)N^Tz(k)-\frac{1}{4}L^TC(k)z(k)+M^TC(k)r(k-1)+2f(k)Mu(k)+2\phi^T(k)\vec{H}u(k)] = [0] \quad (63)$$

$$E[-x(k)+x(k-1)+q(k)+Mu(k)] = [0] \quad (64)$$

$$E[-y(k)+x^T(k)\vec{H}x(k)] = [0] \quad (65)$$

$$E[-z(k)+u^T(k)\vec{H}u(k)] = [0] \quad (66)$$

$$E[-r(k-1)+\mu^T(k)\vec{H}x(k-1)] = [0] \quad (67)$$

5. Practical Example

The algorithm of the last section has been used to solve the long-term optimal optimal operating problem for the system mentioned in [16]. The characteristics of the installations are given in Table 1. The optimization is done on a monthly time basis for a period of a year. If d^k denotes the number of days in a month k, then the maximum and minimum discharges through the turbine are given by

$$\bar{u}^k_{ij} = 0.0864d^k(\text{Maximum effective discharge in } m^3/\text{sec})$$

$$\underline{u}^k_{ij} = 0.0864d^k(\text{Minimum effective discharge in } m^3/\text{sec})$$

where the maximum and minimum effective discharges in m^3/sec are given in Table 1. The expected natural inflows to the sites for a year of high flow, which we call Year 1, and the cost of energy in \$/MWh in that year are given in Table 2. In Table 3, we give the optimal montly releases from each reservoir and the total benefits realized at the end of year 1 for the optimal global-feedback solution. In Table 4 we give the optimal storage scheduling for each reservoir during year 1. We have simulated the monthly operation of the system for widely different water conditions. Year 2 represents a year of low flow. We started both years with the reservoirs full.

The computing time to get the optimal solution for a period of a year for the system just described above was 3.5 sec. in CPU units, which is very small compared to other approaches.

It will be noticed from Table 3 using this model that the total benefits during Year 1 is \$179,837,406. However, by using another model in which the water conversion factor is a function of the storage of the previous month as mentioned [16], the total benefits during the same year for the same system is \$165,998,560. The percentage increase in the total benefits by using a more accurate model is 8.4 percent.

Table 1. The Characteristics of the installations

River	Site Name	Capacity of the Reservoir Mm^3	Minimum Storage Mm^3	Maximum effective discharge m^3/sec	Minimum effective discharge m^3/sec	Reservoir Constants α_{ij}	β_{ij}	γ_{ij}
1	R_{11}	24763	9949	1119	85	212.11	146.96×10^{-4}	$-30503142.65\times10^{-14}$
	R_{21}	5304	3734	1583	85	117.20	569.71×10^{-4}	$-368119890.48\times10^{-14}$
2	R_{12}	74255	33196	1877	283	232.46	359.45×10^{-4}	$-1603544.32\times10^{-14}$
	R_{22}	0	0	1930	283	100.74	0	0
3	R_{13}	45672	24467	1632	283	176.28	105.626×10^{-4}	$-10022665.72\times10^{-14}$
	R_{23}	9132	8886	1876	283	131.44	200.897×10^{-4}	$-34741725.6\times10^{-14}$

On the other hand, the total benefits at the end of Year 2 for the same system using this model were $152,846,864. But the total benefits from the same system using the model of [16] is $142,506,846. The percentage increase in the total benefits is 7.3 percent.

Table 2. The expected natural inflows to the sites in Year 1 and the cost of energy

Month k	I_{11}^k Mm^3	I_{21}^k Mm^3	I_{12}^k Mm^3	I_{22}^k Mm^3	I_{13}^k Mm^3	I_{23}^k Mm^3	c^k \$/MWh
1	948	326	2526	30	2799	318	1.4
2	482	189	1226	15	1632	192	1.4
3	350	148	1001	15	1380	221	1.4
4	300	113	849	8	1035	234	0.8
5	229	80	699	7	796	158	0.8
6	225	78	644	8	766	170	0.8
7	385	160	962	7	794	229	0.8
8	1388	910	4558	53	2017	374	0.8
9	4492	2143	17322	147	15509	1920	0.8
10	5028	2025	7660	76	7453	999	0.8
11	2685	963	5195	68	4953	711	1.1
12	1402	580	2349	29	3376	413	1.1

Table 3. The optimal releases from the turbines and the profits realized in Year 1

Month k	u_{11}^k Mm^3	u_{21}^k Mm^3	u_{12}^k Mm^3	u_{22}^k Mm^3	u_{13}^k Mm^3	u_{23}^k Mm^3	Profits $
1	1649	3451	3821	3851	3691	4010	9098372
2	1589	1453	3703	3718	3538	3730	8685940
3	1552	1578	3839	3855	3601	3871	8970313
4	1318	401	3690	3698	3294	3725	4554491
5	1177	1257	3304	3311	2931	3089	4214638
6	1322	1400	3695	3703	3236	3406	4658819
7	1293	1453	3567	3574	3092	3321	4491206
8	1387	2298	3699	3752	3185	3559	4886224
9	1318	3561	3583	3730	3150	4824	5598097
10	1637	3662	4540	4616	4371	5025	6627686
11	2266	3229	5027	5169	4371	5025	9571348
12	2706	3286	3416	3445	4029	4442	8345312
	Value of water left in storage at the end of the year						99,334,960
	Total benefits from the system						179,837,406

Table 4. Optimal reservoir storage during Year 1

Month k	x_{11}^k Mm^3	x_{21}^k Mm^3	x_{12}^k Mm^3	x_{13}^k Mm^3	x_{23}^k Mm^3
1	24062	3828	72959	44779	9132
2	22955	4152	70481	42873	9132
3	21752	4274	67642	40652	9083
4	20734	5304	64801	38393	8886
5	19786	5304	62196	36258	8886
6	18689	5304	59145	33788	8886
7	17780	5304	56539	31489	8886
8	17780	5304	57397	30320	8886
9	20953	5304	71135	42678	9132
10	24345	5304	74255	45672	9132
11	24763	5304	74255	45672	9132
12	23458	5304	73187	450189	9132

6. Conclusion

We present in this paper a new attractive approach for solving the long-term optimal operating problem for a multireservoir power system. The problem is formulated as a minimum norm problem using functional analysis. A set of discrete optimizing equations is obtained. These equations are solved forward and backward in time to get the optimal solution.

The model used for each reservoir is a nonlinear model of the discharge and the average storage to avoid underestimation in the production for rising water levels and overestimation for falling water levels. The total benefits using this model is larger than the total benefits obtained using the model in [16], for the same system.

The computing time to get the optimal solution for a period of a year was 3.5 sec. in CPU units, which is very small compared to what has been done so far using other approaches. (In [17] the computing time for NCVP system using dynamic programming was 10.3 min. in CPU units for a system of nine reservoirs for a period of one year optimization interval.)

7. References

[1] Arvanitidies, N.V., and Rosing, J., "Composite Representation of a Multi-reservoir Hydroelectric Power System", IEEE Transactions, Vol. PAS-89, No. 2, 1971.

[2] Arvanitidies, N.V., and Rosing, J., "Optimal Operation of Multireservoir Systems Using a Composite Representation", IEEE Transactions, Vol. PAS-89, No. 2, 1971.

[3] Duran, H. et.al., "Optimal Operation of Multireservoir Systems Using an Aggregation-Decomposition Approach", Paper submitted to IEEE PES., Winter Meeting, February, 1985.

[4] Turgeon, A., "A Decomposition/Projection Method for the Multireservoir Operating Problem", Paper presented at the Joint National TIMS/ORSA Meeting, Los Angeles, Calif., November, 1978.

[5] Turgeon, A., "Optimal Operation of Multireservoir Power Systems with Stochastic Inflows", Water Resources Research, Vol. 16, No. 6, 1981.

[6] Turgeon, A., "A Decomposition Method for the Long-Term Scheduling of Reservoirs in Series", Water Resources Research, Vol. 17, No. 6, 1981.

[7] Olcers, S., et.al., "Application of Linear and Dynamic Programming to the Optimization of the Production of Hydroelectric Power", Optimal Control Application and Methods, Vol. 6, 1985.

[8] Grygier, J.C. and Stedinger, J.R., "Algorithms for Optimizing Hydropower System Operation", Water Resources Research, Vol. 21, No. 1, 1985.

[9] Halliburton, T.S., and Sirisena, H.R., "Development of a Stochastic Optimization for Multireservoir Scheduling", IEEE Transaction on AC, Vol. AC-29, No. 1, 1984.

[10] Marino, M.A., and Loaiciga, H.A., "Quadratic Model for Reservoir Management: Application to the Central Valley Project", Water Resources Research, Vol. 21, No. 5, 1985.

[11] Christensen, G.S., and Soliman, S.A., "On the Application of Functional Analysis to the Optimization of the Production of Hydroelectric Power", Accepted for IEEE Trans. PAS, August 1986, IEEE paper No. 86-JPGC-660-5.

[12] Sage, A.P., "Optimal Systems Control", Prentice-Hall, Inc., Englewood Cliffs, New Jersey, 1968.

[13] El-Hawary, M.A. and Christensen, G.S., "Optimal Economic Operation of Electric Power System", Academic Press, New York, New York, 1979.

[14] Shamaly, A., Christensen, G.S., and El-Hawary, M.A., "A Transformation for Necessary Optimality Conditions for Systems with Polynomial Nonlinearities", IEEE Transactions on AC, Vol. AC-24, No. 6, 1979.

[15] Shamaly, A., Christensen, G.S., and El-Hawary, M.A., "Optimal Control of Large Turboalternator", Journal of Optimization Theory and Application,Vol.34,No.1, 1981.

[16] Christensen, G.S., and Soliman, S.A., "New Analytical Approach for Long-Term Optimal Operation of a Parallel Multireservoir Power System Based on Functional Analysis", Canadian Electrical Engineering Journal,Vol.11, No.3, 1986.

[17] Marino, M.A., and Loaiciga, H.A., "Dynamic Model for Multireservoir Operation", Water Resources Research, Vol. 21, No. 5, 1985.

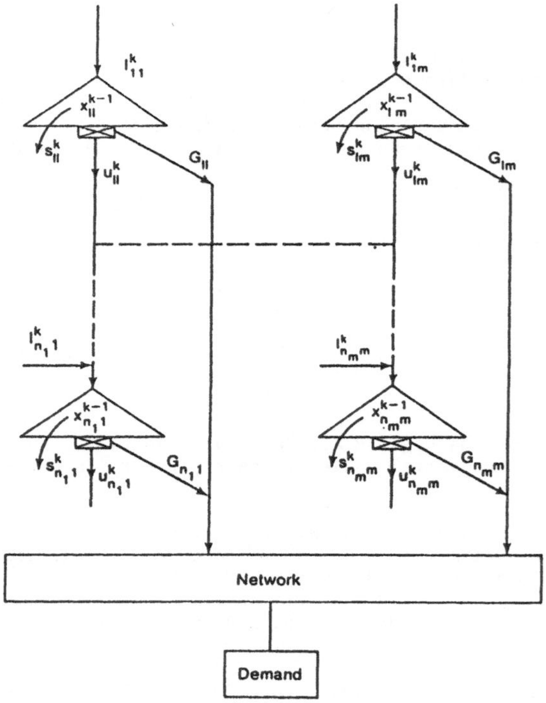

Figure 1.The Hydroelectric System

CONTROL OF FLEXIBLE STRUCTURES

SOME PROBLEMS ASSOCIATED WITH THE CONTROL OF DISTRIBUTED STRUCTURES

Leonard Meirovitch
Department of Engineering Science and Mechanics
Virginia Polytechnic Institute and State University
Blacksburg, VA 24061 USA

1. Introduction

Structures represent distributed-parameter systems, described by partial differential equations (Ref. 1). In some form or another, control of structures is carried out by modal control, whereby the structure is controlled by controlling its modes. Control of the entire infinity of modes requires in general a distributed actuator and a distributed sensor. If the control is such that the modes are coupled, then determination of the control gains is not possible. However, a solution is possible if the modes are controlled independently. Indeed, the independent modal-space control method is able to produce a globally optimal solution by preserving the independence of the modal equations (Ref. 2).

Modal control requires estimation of the modal states for feedback. This can present a problem, particularly for two- and three-dimensional structures. One approach that does not require modal state estimation is direct feedback control, whereby the control is carried out by collocated sensors and actuators. Direct feedback implies a gain matrix consisting of two diagonal submatrices, one for displacement and the other for velocity feedback. The fact that the off-diagonal gains are zero can be regarded as placing constraints on the controls. As a result determination of control gains by pole allocation or by optimal control experiences difficulties.

This paper examines some of the problems encountered in the control of distributed structures, concentrating on the problem of using direct feedback control in conjunction with pole placement. A perturbation technique permits the computation of control gains for multi-input control. The paper demonstrates that difficulties experienced in using direct feedback in conjunction with pole placement to control distributed structures are endemic to the approach and are not merely mathematical in nature. The difficulties can be attributed to the insistence on selecting the closed-loop poles in advance, as no problem exists if the control gains are selected first and the closed-loop eigenvalues are computed later.

2. Modal Equations

We are concerned with the problem of controlling a distributed structure whose behavior is governed by the partial differential equation (pde) (Ref. 1)

$$\mathcal{L}w(P,t) + m(P)\ddot{w}(P,t) = f(P,t), \quad P\varepsilon D \tag{1}$$

where $w(P,t)$ is the displacement of a typical point P inside domain D and at time t, $\mathcal{L}$ is a homogeneous, self-adjoint, positive definite differential operator, referred to as stiffness operator, $m(P)$ is the mass density and $f(P,t)$ is a distributed control force. The displacement $w(P,t)$ is subject to given boundary conditions to be satisfied at every point of the boundary S of D.

The open-loop eigenvalue problem has the form

$$\mathcal{L}\phi(P) = \omega^2 m(P)\phi(P), \quad P\varepsilon D \tag{2}$$

where $\phi(P)$ is subject to given boundary conditions. The solution of Eq. (2) consists of a denumerably infinite set of eigenvalues ω_r^2, where ω_r are the natural frequencies, and associated eigenfunctions ϕ_r $(r = 1, 2,\ldots)$. The eigenfunctions are orthogonal and can be normalized so that $(\phi_s, m\phi_r) = \delta_{rs}$, $(\phi_s, \mathcal{L}\phi_r) = \omega_r^2\delta_{rs}(r,s, = 1,2,\ldots)$, where (,) denotes an inner product. Using the expansion theorem (Ref. 1), the displacement $w(P,t)$ can be expressed as the linear combination

$$w(P,t) = \sum_{r=1}^{\infty} \phi_r(P)q_r(t) \tag{3}$$

where $q_r(t)$ $(r = 1,2,\ldots)$ are generalized coordinates ordinarily known as modal coordinates. Similarly, we can expand the distributed force $f(P,t)$ in the series

$$f(P,t) = \sum_{r=1}^{\infty} m(P)\phi_r(P)f_r(t), \quad f_r(t) = (\phi_r(P), f(P,t)), \quad r = 1,2,\ldots \tag{4a,b}$$

where $f_r(t)$ are known as modal forces. Then, inserting Eqs. (3) and (4) into Eq. (1), multiplying through by $\phi_s(P)$, integrating over the domain D and considering the orthonormality conditions, we obtain the modal equations

$$\ddot{q}_r(t) + \omega_r^2 q_r(t) = f_r(t), \quad r = 1,2,\ldots \tag{5}$$

We refer to control of a distributed structure by using Eqs. (5) to control the modes of the structure as modal control.

It will prove convenient to cast the modal equations in state form. To this and, we define the rth modal state vector $\underset{\sim}{x}_r(t) = [q_r(t)\ \dot{q}_r(t)]^T$. Then, adjoining the identities $\dot{q}_r(t) = \dot{q}_r(t)$ $(r = 1,2,\ldots)$, Eqs. (5) can be written in the state form

$$\dot{\underset{\sim}{x}}_r(t) = A_r\underset{\sim}{x}_r(t) + \underset{\sim}{B}_r f_r(t), \quad r = 1,2,\ldots \tag{6}$$

where

$$A_r = \begin{bmatrix} 0 & 1 \\ -\omega_r^2 & 0 \end{bmatrix}, \quad \underset{\sim}{B}_r = \begin{bmatrix} 0 \\ 1 \end{bmatrix}, \quad r = 1,2,\ldots \tag{7a,b}$$

are coefficient matrices. We note that complete modal controllability implies that each and every modal control $f_r(t)$ is nonzero. An infinity of modal controls $f_r(t)$ is tantamount to an actual distributed control function $f(P,t)$, as indicated by Eq. (4a).

Next, we assume that the modal states are related to the modal measurements $y_r(t)$ by

$$y_r(t) = \underset{\sim}{c}_r^T \underset{\sim}{x}_r(t), \quad r = 1,2,\ldots \tag{8}$$

where in the case of displacement measurements $\underset{\sim}{c}_r^T = [1\ 0]$ and in the case of velocity measurements $\underset{\sim}{c}_r^T = [0\ 1]$. It should be pointed out that the system is in general modal observable with either displacement measurements or velocity measurements. Notable exceptions are semidefinite systems, which admit rigid-body modes with zero eigenvalues. Indeed, semidefinite systems are not observable with velocity measurements alone. Note that an infinity of modal displacement or modal velocity observations implies distributed displacement measurement $w(P,t)$ or distributed velocity measurement $\dot{w}(P,t)$, respectively.

3. Feedback Control

Let us consider the distributed linear feedback control

$$f(P,t) = -\mathcal{G}(P)w(P,t) - \mathcal{H}(P)\dot{w}(P,t) \tag{9}$$

where $\mathcal{G}(P)$ and $\mathcal{H}(P)$ are control gain operators. Inserting Eq. (9) into Eq. (1), we obtain the closed-loop pde

$$\mathcal{L}^* w(P,t) + \mathcal{H}(P)\dot{w}(P,t) + m(P)\ddot{w}(P,t) = 0, \quad P \epsilon D \tag{10}$$

where $\mathcal{L}^* = \mathcal{L} + \mathcal{G}$ is a closed-loop stiffness operator. Retracing the steps leading from Eq. (1) to Eqs. (5), we obtain the closed-loop modal equations, which can be written in the compact form

$$\ddot{\underset{\sim}{q}}(t) + H\dot{\underset{\sim}{q}}(t) + (\Lambda + G)\underset{\sim}{q}(t) = \underset{\sim}{0} \tag{11}$$

where $\underset{\sim}{q}(t)$ is the infinite-dimensional modal configuration vector, Λ is the infinite-order diagonal matrix of eigenvalues and G and H are square control gain matrices of infinite order with entries given by

$$g_{sr} = (\phi_s, \mathcal{G}\phi_r),\ h_{sr} = (\phi_s, \mathcal{H}\phi_r), \quad r,s = 1,2,\ldots \tag{12a,b}$$

In the general case, the matrices G and H are not diagonal, so that the effect of feedback control is to couple the modal equations. Physically, the term $g_{sr}q_r(t)$ implies a generalized spring force and the term $h_{sr}\dot{q}_r(t)$ a generalized damping force. Hence, the fact that the matrices G and H are not diagonal implies that the feedback control provides nonproportional stiffness and damping (Ref. 1), respectively. We refer to the case in which G and H are not diagonal as <u>coupled modal control</u>. Note that in this case the matrices G and H may not be even symmetric.

Before the behavior of the closed-loop system can be established, it is necessary to determine the gain operators $\mathcal{G}$ and $\mathcal{H}$ or the gains matrices G and H. However, there are no algorithms capable of producing the operators $\mathcal{G}$ and $\mathcal{H}$ or the infinite-order matrices G and H. Hence, distributed feedback control realized through coupled modal control is not possible.

Next, we introduce the 2∞ -dimensional modal state vector $\underset{\sim}{x}(t) = [\underset{\sim}{q}^T(t) \,|\, \dot{\underset{\sim}{q}}^T(t)]^T$, so that Eq. (11) can be rewritten in the state form

$$\dot{\underset{\sim}{x}}(t) = A\underset{\sim}{x}(t) \tag{13}$$

where

$$A = \left[\begin{array}{c|c} 0 & I \\ \hline -(\Omega^2 + G) & -H \end{array}\right] \tag{14}$$

in which $\Omega = \text{diag}\,[\omega_r]$. The problem of determining the control gain matrices G and H remains. In this regard, one can consider pole allocation and optimal control. In the pole allocation method, the problem reduces to the solution of a set of nonlinear algebraic equations (Ref. 3), which is not feasible for infinite-dimensional systems. Similarly, for optimal control using a quadratic performance index, one is faced with the solution of a matrix Riccati equation of order 2∞, which is not possible.

4. Independent Modal-Space Control

There is one special case in which distributed feedback control is possible, namely the one in which the operators $\mathcal{G}$ and $\mathcal{H}$ satisfy the eigenvalue problems

$$\mathcal{G}\phi_r(P) = g_r m(P)\phi_r(P), \mathcal{H}\phi_r(P) = h_r m(P)\phi_r(P), \quad r = 1,2,... \tag{15a,b}$$

which imply that $\mathcal{G}$ and $\mathcal{H}$ are such that

$$(\phi_s, \mathcal{G}\phi_r) = g_r\delta_{rs}, \ (\phi_s, \mathcal{H}\phi_r) = h_r\delta_{rs}, \quad r,s = 1,2,... \tag{16a,b}$$

In this case the closed-loop modal equations reduce to the <u>independent</u> set

$$\ddot{q}_s(t) + h_s\dot{q}_s(t) + (\omega_s^2 + g_s)q_s(t) = 0, \quad s = 1,2,... \tag{17}$$

Because of the independence of the closed-loop modal equations, this type of control is called <u>independent modal-space control (IMSC)</u>. It is characterized by modal control forces of the form

$$f_s(t) = - g_s q_s(t) - h_s\dot{q}_s(t), \quad s = 1,2,... \tag{18}$$

In open-loop response problems, the coordinates $q_s(t)$ corresponding to independent equations of motion are called <u>natural</u>. Because IMSC guarantees the independence of the closed-loop equations, we refer to IMSC as <u>natural control</u>.

The fact that both the open-loop and closed-loop modal equations are independent has very important implications. Indeed, this implies that the open-loop eigenfunctions ϕ_s are closed-loop eigenfunctions as well. Hence, in natural control, the control effort is directed entirely to altering the eigenvalues, leaving the eigenfunctions unaltered. In this regard, it should be recalled that the stability of a linear system is determined by the system eigenvalues, with the eigenfunctions playing no role, so that in natural control no control effort is used unnecessarily.

The question remains as to how to determine the modal gains g_s and h_s ($s = 1,2,\ldots$). Two of the most widely used techniques are pole allocation and optimal control:

i. Pole allocation

In the pole allocation method, the closed-loop poles are selected in advance and the gains are determined so as to produce these poles. In the IMSC, the procedure is exceedingly simple. Denoting the closed-loop eigenvalue associated with the sth mode by $-\alpha_s + i\beta_s$, the solution of Eqs. (17) can be written as

$$q_s(t) = c_s \exp(-\alpha_s + i\beta_s)t, \quad s = 1,2,\ldots \tag{19}$$

Inserting Eqs. (19) into Eqs. (17) and separating the real and imaginary parts, we obtain the modal gains

$$g_s = \alpha_s^2 + \beta_s^2 - \omega_s^2, \; h_s = 2\alpha_s, \quad s = 1,2,\ldots \tag{20}$$

To guarantee asymptotic stability, however, it is only necessary to impart the open-loop eigenvalues some negative real part and it is not necessary to alter the frequencies. This can be achieved by letting $\beta_s = \omega_s$ ($s = 1,2,\ldots$). Hence, the frequency-preserving control gains are

$$g_s = \alpha_s^2, \; h_s = 2\alpha_s, \quad s = 1,2,\ldots \tag{21}$$

ii. Optimal control

In optimal control, the closed-loop poles are determined by minimizing a given performance index. Consistent with previous developments, we are interested in constant gains and, to this end, we consider the performance functional

$$J = \int_0^\infty [(\dot{w},m\dot{w}) + (w,\mathcal{L}w) + (f,rf)]dt \tag{22}$$

where the various quantities are as defined in Eq. (1), except for $r = r(P)$ which is a weighting function assumed to satisfy (Ref. 2)

$$(f,rf) = \sum_{r=1}^{\infty} R_r f_r^2 \tag{23}$$

where R_r are modal weights. Inserting Eqs. (3) and (23) into Eq. (22) and recalling the orthonormality conditions, we obtain

$$J = \sum_{r=1}^{\infty} J_r = \sum_{r=1}^{\infty} \int_0^\infty (\dot{q}_r^2 + \omega_r^2 q_r^2 + R_r f_r^2)dt \tag{24}$$

Where J_r are modal performance indices. Because in IMSC the modal control f_r is independent of any other modal control, it follows that

$$\min J = \min \sum_{r=1}^{\infty} J_r = \sum_{r=1}^{\infty} \min J_r \tag{25}$$

so that the minimization can be carried out independently for each mode.

The minimization of J_r leads to a 2×2 matrix Riccati equation that can be solved in closed form (Ref. 4), yielding the modal control gains

$$g_r = -\omega_r^2 + \omega_r(\omega_r^2 + R_r^{-1})^{1/2},\ h_r = [-2\omega_r^2 + R_r^{-1} + 2\omega_r(\omega_r^2 + R_r^{-1})^{1/2}]^{1/2},\quad r = 1,2,... \tag{26}$$

Because no constraint has been imposed on the control function $f = f(P,t)$, the solution defined by Eqs. (4), (18) and (26) is globally optimal, and is unique because the solution to the linear optimal control problem is unique (Ref. 5).

It should be pointed out that the solution presented above requires distributed sensors and actuators. Indeed, inserting Eqs. (18) into Eq. (4a), we obtain the distributed feedback control force

$$f(P,t) = -\sum_{r=1}^{\infty} m(P)\phi_r(P)[g_r q_r(t) + h_r \dot{q}_r(t)] \tag{27}$$

Equation (27) indicates that control implementation requires the entire infinity of modal displacements $q_r(t)$ and modal velocities $\dot{q}_r(t)$ $(r = 1,2,...)$ for feedback. This, in turn, implies a distributed sensor. Note that, inserting Eq. (3) into Eq. (9) and comparing the results with Eq. (27), we can verify Eqs. (15). At this point, we observe that the gain operators $\mathcal{G}$ and $\mathcal{H}$ are never determined explicitly, nor is it necessary to do so, as the determination of the modal gains g_r and h_r $(r = 1,2,...)$ is sufficient to produce the feedback control density function $f(P,t)$.

5. Control by Point Actuators

As pointed out in Sec. 4, globally optimal control of a distributed structure requires a distributed actuator. On the assumption that distributed actuation is not feasible, we seek control by means of a finite number p of discrete actuators acting at the points $P = P_i$ $(i = 1,2,...,p)$ of the structure. Discrete actuators can be treated as distributed by writing

$$f(P,t) = \sum_{i=1}^{p} F_i(t)\delta(P - P_i),\quad P \in D \tag{28}$$

where $F_i(t)$ are force amplitudes and $\delta(P - P_i)$ are spatial Dirac delta functions. Introducing Eq. (28) into Eq. (4b), we obtain the relation between the modal forces and the actuator forces in the form

$$f_r(t) = (\phi_r(P), f(P,t)) = \sum_{i=1}^{p} F_i(t) \int_D \phi_r(P)\delta(P - P_i)dD = \sum_{i=1}^{p} \phi_r(P_i)F_i(t),\quad r = 1,2,... \tag{29}$$

which can be written in the compact form

$$\underset{\sim}{f}(t) = \Phi \underset{\sim}{F}(t) \tag{30}$$

where $\underset{\sim}{f}(t)$ is the infinite-dimensional modal vector, Φ is the $\infty \times p$ modal participation matrix and $\underset{\sim}{F}(t)$ is the p-vector of actuator forces.

Considering the feedback control

$$\underset{\sim}{F}(t) = -G\underset{\sim}{q}(t) - H\dot{\underset{\sim}{q}}(t) \tag{31}$$

where this time G and H are $p \times \infty$ control gain matrices, the closed-loop state equations can once again be written in the form (13), but this time the coefficient matrix is

$$A = \left[\begin{array}{c|c} 0 & I \\ \hline -(\Omega^2 + \Phi G) & -\Phi H \end{array}\right] \tag{32}$$

The difficulties cited in Sec. 4 in conjunction with the determination of the gain matrices G and H remain. Some of these difficulties can be reduced by controlling a finite number of modes. This raises the question of control spillover into the uncontrolled modes (Ref. 6), particularly if the number of controlled modes is small.

6. Direct Feedback Control

One problem that can prove troublesome in modal control is the estimation of the modal states for feedback. To this end, one can consider a Luenberger observer (Ref. 7), but the question of observation spillover is potentially more serious than the problem of control spillover, as it can lead to instability (Ref. 6). Hence, a procedure not requiring modal state estimation appears desirable.

One approach not requiring modal state estimation is direct feedback control, whereby the sensors are collocated with the actuators and a given actuator force is a linear function of the sensor output at the same point. We consider p discrete actuators acting at the points $P = P_i$ $(i = 1,2,\ldots,p)$, where the force amplitudes are

$$F_i(t) = -g_i w(P_i,t) - h_i \dot{w}(P_i,t), \quad i = 1,2,\ldots,p \tag{33}$$

in which g_i and h_i $(i = 1,2,\ldots,p)$ are actual control gains. Clearly, the gains must be positive. As before, the discrete actuators can be regarded as distributed by writing

$$f(P,t) = -\sum_{i=1}^{p} [g_i w(P,t) + h_i \dot{w}(P,t)]\delta(P - P_i), \quad P \varepsilon D \tag{34}$$

To make the connection with Eq. (9), we can regard $\mathfrak{G}$ and $\mathcal{H}$ as operators having the expressions

$$\mathfrak{G}(P) = \sum_{i=1}^{p} g_i \delta(P - P_i), \; \mathcal{H}(P) = \sum_{i=1}^{p} h_i \delta(P - P_i), \quad P \varepsilon D \tag{35a,b}$$

so that, inserting Eqs. (35) into Eqs. (12), we obtain the entries of the control gain matrices G and H in the explicit form

$$g_{sr} = \sum_{i=1}^{p} g_i \phi_s(P_i)\phi_r(P_i), \; h_{sr} = \sum_{i=1}^{p} h_i \phi_s(P_i)\phi_r(P_i), \quad r,s, = 1,2,\ldots \tag{36a,b}$$

The state equations remain in the form (13) and the coefficient matrix A remains in the form (14).

Once again the problem is that of determining the control gains. The problem is different here because there is only a finite number of gains g_i and h_i ($i = 1,2,\ldots,p$) and the system is infinite-dimensional. There is no computational algorithm permitting the computation of the control gains in conjunction with either pole allocation or optimal control, so that one must consider modal truncation. Even for the truncated model, the situation remains questionable. The reason for this is that pole allocation and optimal control most likely will require gain matrices with entries independent of each other while direct feedback control implies that the entries of G and H are not independent, as can be seen from Eqs. (36). In fact, there is some question whether arbitrary pole placement is possible for direct feedback control. Moreover, because the entries of G and H are not independent, there is some question whether optimal control is possible in the presence of constraints on the control gains.

7. A Perturbation Approach to Pole Allocation for Direct Feedback

Application of the pole allocation method to multi-input control can cause serious difficulties. Moreover, the method is suitable for discrete systems only. In this section, we present an approach suitable for distributed systems, and in the process we reveal some limitations of the pole allocation method.

The eigenvalue problem corresponding to the closed-loop equation, Eq. (13), is

$$A\underset{\sim}{u} = \lambda \underset{\sim}{u} \tag{37}$$

where A is given by Eq. (14). We propose to determine the control gains by a perturbation approach. To this end, we assume that A can be expressed in the form

$$A = A_0 + A_1, \quad A_0 = \left[\begin{array}{c|c} 0 & I \\ \hline -\Omega^2 & 0 \end{array}\right], \quad A_1 = \left[\begin{array}{c|c} 0 & 0 \\ \hline -G & -H \end{array}\right] \tag{38a,b,c}$$

in which A_1 is "small" relative to A_0 in some sense, so that the open-loop matrix A_0 represents the unperturbed coefficient matrix and A_1 is the perturbation due to closing of the loop.

The zero-order eigenvalue problem, i.e., the unperturbed open-loop eigenvalue problem is characterized by the eigenvalues $\lambda_{0j} = i\omega_j$ $(j = 1,2,\ldots)$, where $i = \sqrt{-1}$, and by the right and left eigenvectors

$$\underset{\sim}{u}_{0j} = \begin{bmatrix} \underset{\sim}{e}_j \\ i\omega_j \underset{\sim}{e}_j \end{bmatrix}, \quad \underset{\sim}{v}_{0j} = \frac{1}{2}\begin{bmatrix} \underset{\sim}{e}_j \\ -(i/\omega_j)\underset{\sim}{e}_j \end{bmatrix}, \quad j = 1,2,\ldots \tag{39a,b}$$

in which $\underset{\sim}{e}_j$ is a standard unit vector. The two sets of eigenvectors satisfy the biorthonormality relations $\underset{\sim}{v}_{0k}^T \underset{\sim}{u}_{0j} = \delta_{jk}$, $\underset{\sim}{v}_{0k}^T A_0 \underset{\sim}{u}_{0j} = i\omega_j \delta_{jk}$ $(j,k = 1,2,\ldots)$. Equations (39) specify only one half of the right and left eigenvectors. The other half consists of the complex conjugates $\overline{\underset{\sim}{u}}_{0j}$ and $\overline{\underset{\sim}{v}}_{0j}$ corresponding to the eigenvalue $-i\omega_j$. Then, the first-order perturbation solution of Eq. (37) can be expressed as (Ref. 8)

$$\lambda_j = \lambda_{0j} + \lambda_{1j} = i\omega_j + \lambda_{1j}, \quad \underset{\sim}{u}_j = \underset{\sim}{u}_{0j} + \underset{\sim}{u}_{1j}, \quad j = 1,2,\ldots \tag{40a,b}$$

where

$$\lambda_{1j} = \underset{\sim}{v}_{0j}^T A_1 \underset{\sim}{u}_{0j}, \quad \underset{\sim}{u}_{1j} = \sum_{k=1}^{\infty} \frac{\underset{\sim}{v}_{0k}^T A_1 \underset{\sim}{u}_{0j}}{\lambda_{0j} - \lambda_{0k}} \underset{\sim}{u}_{0k}, \quad j = 1,2,\ldots; \; k \neq j \tag{41a,b}$$

Inserting Eqs. (39) into Eq. (41a), we obtain

$$\lambda_{1j} = \frac{i}{2\omega_j} \underset{\sim}{e}_j^T (G + i\omega_j H)\underset{\sim}{e}_j^T, \quad j = 1,2,\ldots \tag{42}$$

so that

$$\text{Re}\,\lambda_{1j} = -\frac{1}{2}\underset{\sim}{e}_j^T H \underset{\sim}{e}_j = -\frac{1}{2}h_{jj} = -\frac{1}{2}\sum_{i=1}^{p} h_i \phi_j^2(P_i), \quad j = 1,2,\ldots \tag{43}$$

and similar expressions exist for $I_m\,\lambda_{1j}$. Introducing the notation $R_e\,\lambda_{1j} = -\alpha_j$, $\phi_j^2(P_i) = b_{ji}$, Eqs. (43) can be rewritten as

$$\sum_{i=1}^{p} b_{ji} h_i = 2\alpha_j \quad j = 1,2,\ldots \tag{44}$$

Equations (44) represent an infinite set of algebraic equations. We note that all b_{ji} $(i = 1,2,\ldots,p;\; j = 1,2,\ldots)$ are positive. If the gains h_i $(i = 1,2,\ldots,p)$ are selected in advance, and if we recall that they must be positive, we conclude that all α_j $(j = 1,2,\ldots)$ are positive, which guarantees that, to the first approximation, direct feedback leads to asymptotic stability.

In pole placement, however, the closed-loop poles rather than the gains are selected in advance. If all the poles are to be placed, which implies that all α_j $(j = 1,2,\ldots)$ must be selected in advance, then Eqs. (44) represent an infinite set of equations and p unknowns, namely h_i $(i = 1,2,\ldots,p)$. Clearly, no solution is possible, so that we consider placing only a finite number of poles. Physically, this presents no problem as higher modes are seldom excited. In general, the object is to place a larger number of poles than the number p of actuators. Hence, let us assume that we wish to place the first n poles, $n > p$, and write Eqs. (44) in the matrix form

$$B\underset{\sim}{h} = 2\underset{\sim}{\alpha} \tag{45}$$

where B is an n×p matrix, $\underset{\sim}{h}$ is the p-dimensional gain vector and $\underset{\sim}{\alpha}$ is the n-dimensional vector of preselected pole shifts along the real axis. A least-squares solution of Eq. (45) yields

$$\underset{\sim}{h} = 2B^{\dagger}\underset{\sim}{\alpha},\ B^{\dagger} = (B^TB)^{-1}B^T \tag{46a,b}$$

where $B^{\dagger}$ is the pseudo-inverse of B. Then, the shifts of the remaining poles along the real axis can be obtained from Eqs. (44) corresponding to j = n + 1, n + 2,.... For stability, $\alpha_j(j = n + 1, n + 2,...)$ must all be nonnegative.

The fact that all α_j $(j = n + 1, n + 2,...)$ must be nonnegative implies that all the gains h_i $(i = 1,2,...,p)$ must be positive. Indeed, if some h_i are negative, then the left sides of Eqs. (44) corresponding to j > n represent indefinite forms, so that some α_j, j > n, can be negative, which implies destabilization of some of the higer modes. Yet, the solution (46a) cannot guarantee that all the components of $\underset{\sim}{h}$ are positive for any choice of $\underset{\sim}{\alpha}$. It follows that <u>in direct feedback control the poles cannot be placed arbitrarily</u>. This fact can be explained easily if we recognize that direct feedback is a special type of control in which a given actuator force depends only on the state at the same location, as expressed by Eqs. (33). As a result, the gain matrix contains no cross-products. The zero entries in the gain matrix can be regarded as constraints on the control, limiting the freedom to choose the poles. Hence, <u>direct feedback control and pole allocation are incompatible</u>.

It must be stressed that the difficulties encountered above do not exist when the control gains are selected first and the closed-loop poles are computed subsequently, so that <u>the problem lies not with direct feedback control but with pole allocation used in conjunction with direct feedback</u> to control a reduced number of modes.

The preceding analysis was based on linear approximation. In reality, the poles are likely to differ from the ones based on the first-order approximaton, but the question is whether the difference is significant. It can be demonstrated that a second-order perturbation solution does not lead to different conclusions than the ones based on the first-order perturbation solution.

8. <u>Numerical Example</u>

Let us consider the problem of controlling a cantilever beam by means of three equally-spaced actuators, $x_i = iL/3$ $(i = 1,2,3)$. The eigenfunctions are given by (Ref. 1)

$$\phi_r(x) = A_r[\cos\beta_r x - \cosh\beta_r x + \frac{\sin\beta_r L - \sinh\beta_r L}{\cos\beta_r L + \cosh\beta_r L}(\sin\beta_r x - \sinh\beta_r x)],$$
$$r = 1,2,... \tag{47}$$

where $\beta_r L$ are the roots of the characteristic equation $\cos \beta_r L \cosh \beta_r L = -1$. Normalizing the eigenfunctions so that $\int_0^L m\phi_r^2\, dx = 1$, we obtain $A_1 = 0.99803\ m^{-1/2}$, $A_2 = 0.99803\ m^{-1/2}$, $A_3 = 0.99802\ m^{-1/2}$, $A_4 = 1.0230\ m^{-1/2}$, $A_5 = 1.0177\ m^{-1/2}$, $A_6 = 1.0143\ m^{-1/2}$, ... Moreover, the roots of the characteristic equation are $\beta_1 L = 1.87510$, $\beta_2 L = 4.69409$, $\beta_3 L = 7.85476$, $\beta_4 L = 10.99550$, $\beta_5 L = 14.13720$, $\beta_6 L = 17.27879$, ..., and note that as the mode number increases the roots approach odd multiples of $\pi/2$.

Letting $r = 3$ and using Eq. (46a), we obtain the control gains

$$
\begin{aligned}
h_1 &= (-1.2274\ \alpha_1 + 0.6000\ \alpha_2 + 0.6276\ \alpha_3)m \\
h_2 &= (0.9036\ \alpha_1 - 2.5720\ \alpha_2 + 1.6686\ \alpha_3)m \\
h_3 &= (0.2654\ \alpha_1 + 0.7530\ \alpha_2 - 0.5164\ \alpha_3)m
\end{aligned}
\tag{48}
$$

It is clear that, because the gains must be positive, <u>the poles cannot be placed arbitrarily</u>. We recall that α_1, α_2 and α_3 must also be positive. To develop a feel for the restrictions on the pole placement, let us imagine a three-dimensional space defined by α_1, α_2 and α_3. The pole shifts must be such that $\alpha_1 > 0$, $\alpha_2 > 0$ and $\alpha_3 > 0$, which restricts the placement to the positive one eighth of the three-dimensional space. Then, we consider a typical equation from the set (48) and write it in the form

$$
h = a\alpha_1 + b\alpha_2 + c\alpha_3 \tag{49}
$$

For $h = 0$, Eq. (49) represents a plane through the origin of the three-dimensional space $\alpha_1, \alpha_2, \alpha_3$. Hence, the inequality $h > 0$ implies that the acceptable points lie in one half of the space. Denoting by S_0 the space defined by $\alpha_1 > 0$, $\alpha_2 > 0$ and $\alpha_3 > 0$ and by S_1 the space corresponding to $h > 0$, we conclude that the closed-loop poles must be such that α_1, α_2 and α_3 lie in the intersection of S_0 and S_1. In our case, there are three inequalities, $h_i > 0$ ($i = 1,2,3$), to be satisfied. Denoting the associated spaces by S_i ($i = 1,2,3$), we conclude that α_1, α_2 and α_3 must lie in the intersection of the spaces S_0, S_1, S_2 and S_3. This intersection defines a cone with the vertex at the origin of the space $\alpha_1, \alpha_2, \alpha_3$ (Fig. 1). Whereas this region may provide many choices, it is obvious that <u>a choice of α_1, α_2 and α_3 cannot be made arbitrarily</u>. In fact, it can be verified by inspecting Eqs. (48) that it is very easy to choose values of α_1, α_2 and α_3 such that h_1, h_2, or h_3 becomes negative. The reason for this is that the cone has a narrow base. For values of α_1, α_2 and α_3 corresponding to points lying outside the cone, the first three modes are asymptotically stable, but some of the higher modes are likely to be destablized.

As an illustration of the case in which arbitrarily chosen poles destabilize the higher modes, let us consider the shifts in the first three poles

$$
\alpha_1 = 3\alpha,\ \alpha_2 = 2\alpha,\ \alpha_3 = \alpha \tag{50}
$$

Inserting Eqs. (50) into Eqs. (48), we obtain the control gains

$$h_1 = -1.8546\ \alpha m,\ h_2 = -0.7646\ \alpha m,\ h_3 = 1.7858\ \alpha m \tag{51}$$

To determine the shift in the poles 4, 5 and 6, we insert Eqs. (50) and (51) into Eqs. (44), we obtain

$$\alpha_4 = 2.9257\ \alpha,\ \alpha_5 = -1.6563\ \alpha,\ \alpha_6 = -6.3490\ \alpha \tag{52}$$

so that modes 5 and 6 are destablized by the choice (50).

One suitable choice, i.e., one lying inside the cone, is that in which the shifts in the first three poles are

$$\alpha_1 = \alpha,\ \alpha_2 = 2\alpha,\ \alpha_3 = 3\alpha \tag{53}$$

In this case, the control gains become

$$h_1 = 1.8551\ \alpha m,\ h_2 = 0.7651\ \alpha m,\ h_3 = 0.2223\ \alpha m \tag{54}$$

Because $h_i > 0$ ($i = 1,2,3$), it follows from Eqs. (44) that all the expressions on the left side represent positive definite quadratic forms, so that all the closed-loop poles are shifted to the left of the imaginary axis. Inserting Eqs. (53) and (54) into Eqs. (44), we obtain

$$\alpha_4 = 0.3190\ \alpha \quad , \quad \alpha_5 = 1.4547\ \alpha \quad , \quad \alpha_6 = 2.6212\ \alpha \tag{55}$$

indicating that now the modes 4, 5 and 6 are damped adequately in comparison to the first three modes.

It will prove of interest to examine the accuracy of the pole-placement technique based on the perturbation scheme. To this end, we propose to solve the closed-loop eigenvalue problem for the successful choice, i.e., for the case in which the gains are given by Eqs. (54). Because the solution of the eigenvalue problem is strictly a numerical problem, we must assign values to the system parameters. For convenience, we choose $\alpha = 1$, $m = 1$, $EI = 1$, $L = 1$, where EI is the bending stiffness. Using Eqs. (36b), in conjunction with the gains given by Eqs. (54), we obtain

$$H = \begin{bmatrix} 2.0000 & 0.5407 & 0.6956 & 0.1049 & 0.6833 & -2.4547 \\ & 4.0000 & 1.4311 & 1.5707 & -1.9965 & -3.4627 \\ & & 6.0000 & 0.2487 & -3.2446 & -2.7830 \\ & & & 0.6379 & -0.6891 & -1.1943 \\ \text{symm.} & & & & 2.9093 & 1.2780 \\ & & & & & 5.2424 \end{bmatrix} \tag{56}$$

On the other hand, because we are only using velocity feedback, $G = 0$. Moreover, the matrix of natural frequencies is

$$\Omega = \text{diag}[3.516\quad 22.034\quad 61.697\quad 120.901\quad 199.860\quad 298.557] \tag{57}$$

The eigensolution was obtained by truncating A to a 4×4, a 5×5 and a 6×6 matrix. The corresponding closed-loop eigenvalues are displayed in Table I. Comparing the values in Eqs. (53) and (54) with the corresponding ones in Table

I, we conclude that the results obtained by the perturbation approach are accurate to the fourth significant figure. It is also easy to verify that truncation of the matrix A does not affect the eigenvalues materially. Hence, the perturbation approach to the computation of the control gains for pole allocation in conjunction with direct feedback control gives sufficiently accurate results, at least in this particular example.

9. Conclusions

Control of distributed structures requires distributed actuators and sensors. Practical considerations dictate that control implementation be carried out by means of discrete actuators and sensors. Moreover, it is impossible to control or estimate the entire infinity of modes, so that control must be limited to a finite number of modes. Problems of modal control and estimation remain when the natural frequencies are closely spaced, as is often the case with two- and three-dimensional structures.

One approach not requiring modal state estimaton is direct feedback control, in which an actuator at a given point of a structure generates a force input depending on the sensor output at the same point. For linear control, the gain matrix consists of two diagonal submatrices. The question remains as to how to produce the control gains. Two widely used techniques are pole allocation and optimal control. The diagonal nature of the gain matrix characterizing direct feedback control is likely to cause difficulties.

In the pole allocation method, the closed-loop poles are selected first and the gains matching these poles are computed subsequently. There are two factors that may limit the freedom to choose closed-loop poles in direct feedback. In the first place, the gain matrix has a special nature, characterized by the off-diagonal entries being equal to zero, which can be intepreted as placing constraints on the gains. In the second place, the control gains must be such that the uncontrolled modes are not destabilized. We recall that for a distributed structure there are always uncontrolled modes.

This paper develops a perturbation approach to the computation of control gains corresponding to given closed-loop poles, whereby in the first approximation the problem reduces to the solution of linear algebraic equations for the control gains. The approach reveals an inherent difficulty in the use of pole placement in conjunction with direct feedback control. In particular, whereas in computing gains for a discrete system in which all the modes are controlled the problem can be regarded as solved provided controllability is satisfied, here the gains are constrained by the requirement that the higher modes not be destabilized. This can be guaranteed by requiring that all the gains be positive. Hence, physical considerations dictate that the only admissible solutions of the algebraic equations for the control gains are those in which all the components of the solution vector are positive. Because this cannot be guaranteed for any preselected closed-loop poles, it follows that the closed-loop poles cannot be chosen arbitrarily. If we envision a space defined by the real part of the closed-loop poles, then the admissible controls lie in a certain cone-shaped subregion of constraint of that space.

The question can be raised as to whether it is possible to draw such sweeping conclusions from a first-order perturbation analysis. The answer must be affirmative. Indeed, for small real parts of the closed-loop poles, the first-order perturbation yields accurate results. As the real parts increase in magnitude, the constraints on the control gains remain, so that the nature of the problem does not change. The likely outcome of a higher-order perturbation is to affect the boundaries of the cone of constraint, in the sense that the boundaries become curved surfaces tangent to the hyperplanes of constraint at the origin, but cannot negate the existence of such subdomains of constraint. It should be pointed out that, in the absence of displacement feedback, a second-order perturbation does not affect the real parts of the eigenvalues.

The ideas presented in this paper are demonstrated via a numerical example in which an attempt is made to control a cantilever beam by means of three point actuators while placing three poles. Placing the poles so that the real parts lie outisde the cone of constraint yields instability, thus showing that poles cannot be placed arbitrarily. On the other hand, placing the poles so that the real parts lie inside the cone yields stability. Then, using the computed gains to generate the matrix of coefficients A, the closed-loop eigenvalue problem corresponding to the stable case is solved "exactly," i.e., without the use of a perturbation analysis. The first six computed eigenvalues agree to the fourth significant figure with those achieved by the perturbation approach to pole placement, so that the perturbation approach yields sufficiently accurate results, at least for the example at hand.

10. References

1. Meirovitch, L. (1980), "Computational Methods in Structural Dynamics", Sijthoff & Noordhoff, The Netherlands.

2. Meirovitch, L. and Silverberg, L. M., "Globally Optimal Control of Self-Adjoint Distributed Systems," Optimal Control Applications and Methods, Vol. 4, 1983, pp. 365-386.

3. Porter, B. and Crossley, T. R. (1972), "Modal Control-Theory and Applications", Taylor and Francis, London.

4. Meirovitch, L. and Baruh, H., "Control of Self-Adjoint Distributed-Parameter Systems, Journal of Guidance, Control, and Dynamics, Vol. 5, No. 1, 1982, pp. 60-66.

5. Curtain, R. F. and Pritchard, A. J., "The Infinite-Dimensional Riccati Equation," Journal of Mathematical Analysis and Applications, Vol. 47, 1974, pp. 43-57.

6. Balas, M. J., "Active Control of Flexible Systems," Journal of Optimization Theory and Applications, Vol. 24, No. 3, 1978, pp. 415-436.

7. Brogan, W. L. (1974), Modern Control Theory, QPI Publishers, New York.

8. Wilkinson, J. H. (1965), The Algebraic Eigenvalue Problem, Oxford University Press, London.

TABLE I - Closed-Loop Eigenvalues from Truncated A

	A is 4×4		A is 5×5		A is 6×6	
r	Im γ_r	Re λ_r	Im λ_r	Re λ_r	Im λ_r	Re λ_r
1	-1.00068	±3.37175	-1.00071	±3.37173	-1.00075	±3.37175
2	-2.00108	±21.94475	-2.00113	±21.94574	-2.00141	±21.94696
3	-2.99873	±61.59982	-2.99991	±61.60840	-3.00046	±61.61201
4	-0.31851	±120.88660	-0.31857	±120.88730	-0.31889	±120.88810
5			-1.45320	±199.80580	-1.45360	±199.80940
6					-2.61968	±298.48120

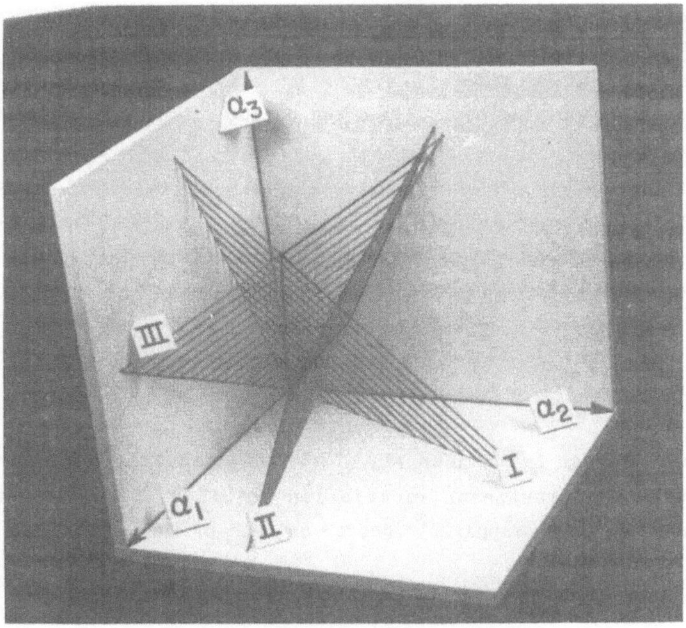

OPTIMAL CONTROL OF A DISTRIBUTED SYSTEM

J. A. Burns
Department of Mathematics

E. M. Cliff
Department of Aerospace and Ocean Engineering
Virginia Polytechnic Institute and State University
Blacksburg, Virginia 24061 USA

1. Introduction

We study a control problem for a distributed system consisting of a rigid hub with a cantilevered beam attached to it. The hub moves about a fixed axis and the beam carries a tip-mass at its free end. Newtonian mechanics are used to arrive at a linearized model consisting of three ordinary differential equations describing the motions at the rigid-hub and tip-mass and a second-order partial differential equation describing the elastic motions at the beam (Euler-Bernoulli).

The formal model suggests a formulation as a abstract differential equation in a Hilbert space ($R^4 \times L_2 \times L_2$) with the inner-product induced by the energy-norm. The Lumer-Phillips theorem is used to show that the abstract Cauchy-problem is well-posed.

A control problem is formulated by introducing an operator which reads-out certain "states", including the strain or velocity at selected points on the beam. The control problem is to take the system from some initial state and minimize a quadratic functional of the control and the output on the semi-infinite interval.

In the next section we sketch the development of the mathematical model. Following this a control problem is formulated. A subsequent section discusses numerical approximations; this requires introduction of a weak-formulation of the system dynamics. A final section presents numerical results.

2. Model Formulation

The system of interest is shown in Fig. 1. Formal models can be derived by any of several methods. A Newtonian formulation requires expressions for the inertial accelerations and for the applied "forces" on each of the components (hub, beam and tip-mass). For the system of interest this leads to:

$$w_{tt}(t,x) + x\dot{\omega}(t) = -\frac{EI}{\rho} w_{xxxx}(t,x) \quad \text{(beam deflection)} \tag{2.1}$$

$$I_A\dot{\omega}(t) = -EI\, w_{xx}(t,0) + u(t) \quad \text{(hub rotation)} \tag{2.2}$$

$$I_c[\dot{\omega}(t) + \dot{\xi}(t)] = EI\, w_{xx}(t,L) \quad \text{(tip-mass rotation)} \tag{2.3}$$

$$M_c[L\dot{\omega}(t) + \dot{\eta}(t)] = EI\, w_{xxx}(t,L) \quad \text{(tip-mass transverse motion)} \tag{2.4}$$

In these equations EI is the beam rigidity; ρ is the mass per unit length; I_A is the hub moment-of-inertia; I_c is the tip-mass moment of inertia; and, M_c is the tip-mass. The term $w(t,x)$ is the transverse deflection of the beam; $\omega(t)$ is the hub angular velocity; $u(t)$ is the applied control torque; $\xi(t)$ is the angular velocity of the tip-mass with respect to the hub-fixed x-axis; and, $\eta(t)$ is the transverse displacement of the tip-mass. A more detailed analysis of these equations can be found in [2].

In addition to the differential equations (2.1) - (2.4) there are important boundary conditions. Since the beam remains joined to the hub one has $w(t,0) = w_t(t,0) = 0$. The cantilever nature of the connection also requires that $w_x(t,0) = w_{xt}(t,0) = 0$, while integrity at the upper joint requires $w_t(t,L) = \eta(t)$ and $w_{xt}(t,L) = \xi(t)$. Finally, note that the angular orientation of the hub is related to its angular velocity by $\dot{\theta}(t) = \omega(t)$.

The above collection of differential equations and boundary conditions provide a <u>formal</u> description of the system dynamics. To complete this analysis one is obligated to demonstrate that the model is well-posed (existence, uniqueness and 'continuous' dependence on initial data).

While this might be done directly, a favored approach is to make use of the mathematical theory of semi-groups [4, 8, 10]. Briefly, this requires that the differential equations be written in the form

$$\dot{z}(t) = Az(t) + Bu(t) \tag{2.5}$$

where A is a linear operator on an appropriate space (Z) of initial data (i.e. state-space). For our system the state can be formulated as:

$$z(t) = \begin{bmatrix} \theta(t) \\ \omega(t) \\ \eta(t) + L\omega(t) \\ \xi(t) + \omega(t) \\ w_{xx}(t,x) \\ w_t(t,x) + x\omega(t) \end{bmatrix} \tag{2.6}$$

in the state-space $Z = R^4 \times L_2 \times L_2$. The operator A is given by

$$A = \begin{bmatrix} 0 & 1 & 0 & 0 & 0 & 0 \\ 0 & 0 & 0 & 0 & \frac{EI}{I_A}\delta_0 & 0 \\ 0 & 0 & 0 & 0 & \frac{EI}{M_c}\delta_L \cdot D & 0 \\ 0 & 0 & 0 & 0 & -\frac{EI}{I_c}\delta_L & 0 \\ 0 & 0 & 0 & 0 & 0 & D^2 \\ 0 & 0 & 0 & 0 & -\frac{EI}{\rho}D^2 & 0 \end{bmatrix} \tag{2.7}$$

where δ_ρ denotes evaluation [i.e. $\delta_\rho(\phi) = \phi(\rho)$], and D denotes differentiation [i.e. $D\phi = \phi'$]. The boundary conditions are incorporated into the domain of the operator. In particular,

$$D(A) = \{(z_1,\ldots,z_6)\varepsilon R^4 \times L_2 \times L_2 \,|\, z_5, z_6 \;\varepsilon\; H^2$$

$$z_6(0) = 0,\; z_6(L) = z_3,\; z_6'(0) = z_2, \quad z_6'(L) = z_4\} \tag{2.8}$$

where H^2 is the usual space of real-valued functions with the function and its derivative in L_2 (see [1]). As a distance measure on the space Z we introduce the inner-product

$$\langle z,y\rangle = z_1 \cdot y_1 + I_A z_2 y_2 + M_c z_3 y_3$$

$$+ I_c z_4 y_4 + \int_0^L EI z_5(x) y_5(x)dx + \int_0^L \rho z_6(x) y_6(x)dx$$

It can be verified that $||z|| = \sqrt{\langle z,z\rangle}$ is a norm and that Z is a Hilbert space. The inner product $\langle z,z\rangle$ is essentially the mechanical energy in the system at state z. The control operator is given by

$$B = [0,\, 1/I_A,\, 0,\, 0,\, 0,\, 0]. \tag{2.9}$$

Formally, one can write the solution to (2.5) in terms of a variation of constants formula

$$z(t) = e^{At}z(0) + \int_0^t e^{A(t-s)}Bu(s)ds. \tag{2.10}$$

Since B is bounded, the solution (2.10) will have the desired properties (existence, uniqueness, continuous dependence) if, and only if, $\{e^{At}\}$ is a C_o-semigroup. Hence, we must show that the operator A generates such a semigroup. This is the central matter in abstract formulations such as (2.5). In the report [2] we make use of a special form of the Lummer-Phillips theorem [8] to prove the following:

Theorem: The operator A defined above generates a C_o-semigroup on the state-space Z.

3. A Control Problem

Now that it has been shown that the dynamics for our system are well-posed it is meaningful to consider a control problem. In this section we formulate such a problem. In terms of the formal model we suppose that one can detect the angular position and velocity of the hub [i.e $\theta(t)$ and $\omega(t)$] as well as the lateral and angular velocities of the tip-mass [i.e. $\eta(t)$ and $\xi(t)$, respectively]. In addition, we suppose that the beam has been instrumented so that one can measure the lateral velocity at selected points along the beam and the strain at certain points. For concreteness we suppose that beam velocity is measured at x=0.5 L and x=0.7 L and that the strain is measured at x=0 and x=.3 L. These eight quantities are to be controlled.

In terms of our abstract model one has $y(t) = C z(t)$, where C is an operator from Z to R^8. Specifically

$$C z = \mathrm{col}[z_1, z_2, z_3, z_4, z_5(0), z_5(.3\ L), z_6(.5\ L), z_6(.7\ L)] \tag{3.1}$$

In our topology C is unbounded because it involves (in the last four entries) evaluation of L_2 "functions". To avoid this difficulty we presume that in lieu of evaluation we use, for example,

$$y_8 = \int_0^L \delta_\varepsilon(s;\ .7\ L) z_6(s) ds$$

where $\delta_\varepsilon(\cdot;p)$ is an approximation to the Dirac-delta at p.

Our control problem is to find a feedback control $u(t) = [F(t)]z(t)$ to minimize

$$J[u,z_o] = \int_0^T \{||y(t)||^2\ dt + \beta[u(t)]^2\}dt \tag{3.2}$$

for arbitrary initial data z_o. Here $\beta>0$ is given, as is $T>0$. In fact, we shall consider only the infinite horizon ($T = \infty$) and will restrict attention to the steady-state gain operator F (i.e. independent of t).

The problem we have defined is a natural extension of the finite-dimensional linear, quadratic regulator problem [6] to the infinite-dimensional setting. The theoretical results for the problem parallel those for the finite-dimensional case. It is known (see [3]) that the optimal control u(t) is given in feedback form and that for the infinite horizon case this feedback tends to a time-invariant gain. For our problem this feedback form is given by:

$$u(t) = k_1\theta(t) + k_2\omega(t) + k_3[\eta(t) + L\cdot\omega(t)] + k_4[\xi(t) + \omega(t)] + \int_0^L k_5(x)w_{xx}(t,x)dx + \int_0^L k_6(x)[w_t(t,x) + x\omega(t)]dx \tag{3.3}$$

The goal is to compute the optimal gains k_1, k_2 k_3, k_4 and the 'functional gains' $k_5(x)$, $k_6(x)$.

4. Approximation Ideas

The approximation theory which we employ is rooted in the state-space formulation of the dynamics. In particular, we shall be led to construct a sequence of subspaces $\{Z^N\}$ and a sequence of operators $\{A^N\}$ with $A^N: Z^N \to Z^N$. These sequences are constructed so that they converge (in an appropriate sense) to Z and to A respectively. While the approach is based on the 'abstract' formulation, it is worth noting that our state-space is infinite-dimensional because of the distributed nature of the beam deflection, w(t,x). This is apparent in the fifth and sixth state coordinates. In principle, one can produce an approximation scheme by employing some favorite 'shape' functions to represent these coordinates (say):

$$[z_5(t)](x) = \sum \alpha_i(t) \quad \phi_i(x)$$

and

$$[z_6(t)](x) = \sum \beta L(t) \quad \psi_i(x)$$

It is helpful to keep in mind that $[z_5(t)](\cdot)$ and $[z_6(t)](\cdot)$ are (each) related to the beam deflection (i.e. $z_5 \sim u_{xx}$ and $z_6 \sim u_t$). Thus, if one thinks in terms of approximating the deflection

$$w(t,x) = \sum \gamma_i(t)\ h_i(x) \tag{4.1}$$

then 'compatibility' suggests using the shape functions $h_i(\cdot)$ to represent $u_t(\sim z_6)$ while using $h_i''(\cdot)$ to represent $u_{xx}(\sim z_5)$. Additionally, in describing a basis for

z^N one must impose the essential boundary conditions relating various components of $z(t)$. These conditions imply certain constraints on the admissible shape functions, $h_i(x)$.

While one can proceed along general lines it will simplify notation and shorten discussion if we make a specific choice for the shape functions. Accordingly, consider a uniform grid with N 'panels' on the interval [0,L]. To represent deflections we employ cubic B-splines, which form an (N+3)-parameter family of C^1 (piecewise C^2) functions (see [9], pp. 79-81). The boundary conditions require that both $w_t(t,0)$ and $w_x(t,0)$ should vanish. This reduces the admissible functions to an (N+1)-parameter family which we denote by $h_1^N(x), h_2^N(x),\ldots,h_{N+1}^N(x)$. With this family in mind we now proceed to construct a basis for Z^N.

As noted above, the beam velocity $[w_t]$ should be represented in terms of the shape functions, h_i^N. The remaining boundary conditions (see 2.8) lead one to choose the "beam-velocity" basis vectors:

$$e_i^N = \begin{bmatrix} 0 \\ 0 \\ h_i^N(L) \\ h_L^{N\prime}(L) \\ 0 \\ h_i^N(\cdot) \end{bmatrix}, \tag{4.2}$$

The 'beam-stress' basis vectors are:

$$e_{N+1+i}^N = \text{col}[0,\ 0,\ 0,\ 0,\ h_i^{N\prime\prime}(\cdot),\ 0]. \tag{4.3}$$

The hub velocity [i.e. $\omega(t)$] suggests the use of

$$e_o^N = \text{col}[0,\ 1,\ L,\ 1,\ 0,\ x], \tag{4.4}$$

while the hub-position [i.e. $\theta(t)$] adds

$$e_{-1}^N = \text{col}[1,\ 0,\ 0,\ 0,\ 0,\ 0]. \tag{4.5}$$

The subspace Z^N is generated as the span of the set $\{e_i^N\}$ and has dimension $(2\cdot N+4)$. In what follows we shall simplify (and abuse) the notation by suppressing the explicit appearance of N in describing shape functions, basis elements, etc. Thus, we write h_i in lieu of h_i^N, and e_i in lieu of e_i^N. The superscript N will be used in several places where we wish to emphasize the dependence on this mesh parameter.

One could proceed to apply these approximation ideas directly to the abstract system (2.5). Note, however, that our A operator [given by (2.7)] requires that the z_5 coordinate [i.e. w_{xx}] be twice differentiable. The basis we have constructed allows only that z_5 be continuous and piecewise (once) differentiable. Apparently, to proceed further one should go back and re-define our 'shape' functions [$h_i(x)$] to include more smoothness [e.g. $h_i(x)$ could be taken as quintic splines]. A second, more subtle, way to proceed involves 'weak' formulation of the problem. To understand this reformulation one notes that (2.5) holds if and only if

$$\langle \dot{z}(t),v\rangle = \langle Az(t),v\rangle + \langle Bu(t),v\rangle \tag{4.6}$$

holds for all $v \varepsilon \hat{\mathcal{D}}$, where $\hat{\mathcal{D}}$ is any dense subset of Z. Consider the first term on the right side of (4.6) for $v,z \varepsilon \mathcal{D}(A)$

$$\langle Az,v\rangle = v_1 z_2 + EI\{v_2 z_5(0) + v_3 z_5'(L) - v_4 z_5(L)\}$$

$$+ \int_0^L EIz_6''(x)v_5(x)dx - \int_0^L EIz_5''(x)v_6(x)dy,$$

The last integral can be integrated by parts to yield

$$\langle Az,v\rangle = v_1 z_2 + EI\ \{(v_2 - v_6'(0)]z_5(0) + [v_3 - v_6(L)]z_5'(L) + v_6(0)z_5'(0)$$

$$+ [v_6'(L) - v_4]z_5(L)\} + \int_0^L EI\{z_6''(x)v_5(x) - z_5(x)v_6''\}dx$$

$$= v_1 z_2 + \int_0^L EI\{z_6''(x)v_5(x) - z_5(x)v_6''(x)\}dx.$$

However, the right hand side of this equation is valid for any $v,z\varepsilon \hat{\mathcal{D}}$ where

$$\hat{\mathcal{D}} = \{w\varepsilon Z/w_6 \varepsilon H^2,\ w_5 \varepsilon H^1,\ w_6(0) = 0,\ \ w_6'(0) = w_2,\ w_6(L) = w_3,\ w_6'(L) = w_4\}$$

Therefore, we define the bilinear form $\sigma:\hat{\mathcal{D}} \times \hat{\mathcal{D}} \to R$ by

$$\sigma(v,z) = v_1 z_2 + \int_0^L EI\{z_6''(x)v_5(x) - z_5(x)v_6''(x)\}dx$$

and note that σ is continuous on $\hat{\mathcal{D}} \times \hat{\mathcal{D}}$. Also, $Z^N \times Z^N \subseteq \hat{\mathcal{D}} \times \hat{\mathcal{D}}$ so that σ restricted to $Z^N \times Z^N$ is bounded. This machinery allows us to define an operator A^N: $Z^N \to Z^N$ by the rule:

$$\langle v^N,\ A^N z^N\rangle = \sigma(v^N,z^N)$$

for all v^N, $z^N \varepsilon Z^N$. This is, in essence, the way one restricts the weak form (4.6) to Z^N.

The weak-form (4.6) when restricted to the subspace Z^N leads to the Galerkin approximation

$$\langle \sum \dot{x}_j(t)e_j,\ e_i\rangle = \sigma(\sum x_j(t)e_j,\ e_i) + \langle\ u(t),\ e_i\rangle \qquad (4.7)$$

$$i,j \in \{-1,\ 0,..2N+2\}$$

This can be written

$$Q^N\ \dot{x}^N(t) = F^N x^N(t) + H^N u(t) \qquad (4.8)$$

where $x^N = \text{col}[x_{-1},\ x_o,\ldots,x_{2N+2}]$. Q^N is an $(2N+4) \times (2N+4)$ symmetric (in fact, positive definite) matrix with 'entries'

$$Q^N(i,j) = \langle e_j,\ e_i\rangle;$$

F^N is a $(2N+4) \times (2N+4)$ matrix with entries

$$F^N(i,j) = \sigma(e_j,\ e_i)$$

and; finally, H^N is a $(2N+4)$ column matrix with entries

$$H^N_i = \langle B\ ,\ e_i\rangle.$$

With the choice of basis elements described above one may show that

$$Q^N = \left[\begin{array}{c|c|ccc|c} 1 & 0 & & 0 & & 0 \\ \hline 0 & f_o & f_1 & \cdot\ \cdot & f_{N+1} & 0 \\ \hline & f_1 & & & & \\ 0 & \vdots & & G_1 & & 0 \\ & f_{N+1} & & & & \\ \hline 0 & 0 & & 0 & & G_2 \end{array}\right]$$

where

$$f_o = \rho L^3/3 + I_A + M_c L^2 + I_c,$$

$$f_i = M_c\ h_i(L)\ L + I_c\ h_i'(L) + \rho \int_o^L x\ h_i(x)dx,$$

$$G_1(i,j) = M_c\ h_i(L)\ h_j(L) + I_c\ h_i'(L)\ h_j'(L) + \rho \int_o^L h_i(x)\ h_j(x)dx,$$

and, finally,

$$G_2(i,j) = E\cdot I \int_o^L h_i''(x)h_j''(x)dx.$$

It will be convenient for later discussion to 'normalize' these quantities as follows:

$\hat{f}_i = f_i/\rho L^3$,

$\hat{G}_1 = G_1/\rho L^3$,

$\hat{G}_2 = G_2/(EI/L)$

The F^N and H^N matrices may be shown to be given by

$$F^N = \left[\begin{array}{cc|c|c} 0 & 1 & 0 & 0 \\ 0 & 0 & & \\ \hline \multicolumn{2}{c|}{0} & 0 & -G_2 \\ \hline \multicolumn{2}{c|}{0} & G_2 & 0 \end{array}\right]$$

and

$$H^N = \mathrm{col}[0, 1, 0, 0]$$

From the structure of the Q^N, F^N, and H^N matrices it is clear that the system (4.8) is equivalent to the system

$$\hat{Q}^N \dot{x}^N(t) = \hat{F}^N x^N(t) + \hat{H}^N u(t) \tag{4.9}$$

where

$$\hat{Q}^N = \left[\begin{array}{c|c|c|c} 1 & 0 & 0 & 0 \\ \hline 0 & \hat{f}_0 & \hat{f}_1 \;\cdot\;\cdot\; \hat{f}_{N+1} & 0 \\ \hline 0 & \begin{array}{c}\hat{f}_1 \\ \hat{f}_{N+1}\end{array} & \hat{G}_1 & 0 \\ \hline 0 & 0 & 0 & I \end{array}\right],$$

$$\hat{F}^N = \left[\begin{array}{cc|c|c} 0 & 1 & 0 & 0 \\ 0 & 0 & & \\ \hline \multicolumn{2}{c|}{0} & 0 & -\alpha\hat{G}_2^N \\ \hline \multicolumn{2}{c|}{0} & I & 0 \end{array}\right],$$

and $\hat{H}^N = H^N$ is unchanged. The quantity α in $\hat{F}^N$ is a 'frequency' parameter given by

$$\alpha = \frac{EI}{\rho L^4}$$

The system dynamics can be expressed in normal form as

$$\dot{x}^N(t) = A^N x^N(t) + B^N u(t) \tag{4.10}$$

with A^N and B^N from

$$\hat{Q}^N \cdot A^N = \hat{F}^N$$

and

$$\hat{Q}^N \cdot B^N = \hat{H}^N$$

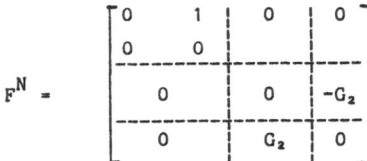

The special block structure of $\hat{Q}^N$ can be used in solving for A^N and $\mathcal{B}^N$.

The equation (4.10) is a coordinate representation of an approximating dynamical system for the original system (2.5). Thus, given an initial condition z_o for (2.5) and some control $u(\cdot)$ one first projects the initial data onto Z^N. This amounts to solving the normal equations

$$Q^N x^N_o = r^N$$

where r^N is a (2N+4) vector with entries $r^N_i = \langle z_o, e_i\rangle$.

At this point one can construct the following approximating control problem:

$$\min_u J^N[u, z^N_o] = \int_o^\infty \{||y^N(t)||^2 + \beta[u(t)]\}dt \qquad (4.11)$$

where y^N is the output of the approximate system (4.10) defined by

$$y^N(t) = Cz^N(t)$$

and

$$z^N(t) = \sum x^N_i(t)\, e^N_i. \qquad (4.12)$$

Of course, one may substitute the representation for $z^N(t)$ into the output equation and so construct a matrix C^N such that

$$y^N(t) = C^N x^N(t)$$

The system (4.10) with cost function (4.11) defines the approximate optimal control problem. This is a standard, finite-dimensional problem and the optimal control is again given by state-feedback

$$u^{N*}(t) = \sum_{i=-1}^{2N+2} g^N_i\, x^N_i(t) \qquad (4.13)$$

The gains g^N_i may be computed in several ways (see [6], pp 248-253). The feedback law given by (4.13) can also be interpreted in terms of an operator on Z^N. That is,

$$u^{N*}(t) = \mathcal{F}^N[z^N(t)] \qquad (4.14)$$

Making use of the coordinate representation (4.12) one finds from (4.13) and (4.14)

$$\mathcal{F}^N[e^N_i] = g^N_i \qquad (4.15)$$

A linear functional on Z^N, such as the operator $\mathcal{F}^N$, can be naturally associated with an element of Z^N by the duality map

$$\langle k^N, z^N\rangle = \mathcal{F}^N(z^N)$$

The coefficient (β^N_i) in the coordinate representation for $k^N \varepsilon Z^N$ in terms of our basis $\{e^N_i\}$ are found from

$$\sum Q^N(i,j)\, \beta^N_j = g^N_i \,. \tag{4.16}$$

With the β^N_j in hand one can construct k^N and, in the sense described above, k^N is the feedback operator for the N-th approximating problem. With one modification our scheme fits the framework of the theory in [3] and allows us to conclude that $k^N \to k$, the feedback operator for the distributed parameter system [see equation (3.3)]. The required change is that the system (2.5) must be modified to include some inherent damping. It is shown in [3] that in the absence of such damping the gain sequence $\{k^N\}$ will diverge. Indeed, without such damping the control problem given by (2.5), (3.1), (3.2) need not have a solution.

5. Numerical Results

In this section we present some numerical results to illustrate the procedures discussed above. As a first step we must select values for the system parameters. These are shown in Table 1. The values are appropriate for an aluminum beam that is 30 ft long and has a rectangular cross-section with height 1 ft and thickness .25 in. These are representative values for a solar panel (see Ref. [5]). The required damping is introduced by modifying the A^N matrix (see 4.10) so that the $\dot{x}^N_i(t)$ includes a term $-.1 \cdot x^N_i(t)$ for $i=1,2,..N+1$. This amounts to modifying the p.d.e. (2.1) to include a term $w_t(t,x)$.

The system matrices were assembled and the eigenvalues at A^N were computed for several values of N. The system has an eigenvalue at zero (multiplicity two) and complex eigenvalues at $\lambda = \sigma \pm i\omega$. With our value of the damping parameter it follows that the σ-values are all -.05. The first nine frequencies are given in Table 2 for several values of N.

The optimal control problem described in Section 3 was approximated, as discussed in Section 4. The eight outputs, given by (3.1), were each weighted before the norm was evaluated as in (3.2). The weights were (.1, .1, .1, .1, .01, .01, .1, .1). The control weight (β) was taken as unity. The approximating problem was solved by Potter's Method (see [6], pp 252) for several values of N. Results for the (four) gain values are given in Table 3. Graphs of the functional gain $k^N_s(\cdot)$ are shown in

Fig. 2 for N=8, 16 and 32. The N=64 graph is not distinguishable from the N=32 case. Similar results for the functional gain $k_6^N(\cdot)$ are shown in Fig. 3.

These results were computed on an IBM 3090/200 with vector facility. The code assembled the various matrices, computed the open-loop eigenvalues and then computed the feedback gains (g_i^N). With N=32 the resulting ode system is of 68th order. The calculations required approximately four cpu seconds. For N=64 the corresponding cpu time was approximately 24 seconds.

Acknowledgement: This work was supported, in part, under grant AFOSR 85-0287, Capt. J. Thomas, Program Director.

Table 1

Parameter	Value
EI	1085 slug ft²
L	30 ft
ρ	.109 slug/ft
I_A	981 slug-ft²
M_c	.327 slug
I_c	.3 slug-ft²

Table 2

Open-Loop Frequencies

Mode	N=8	N=16	N=32
1	.487	.489	.487
2	2.171	2.171	2.171
3	6.111	6.108	6.108
4	12.09	12.06	12.05
5	20.09	19.93	19.92
6	30.15	29.56	29.54
7	42.47	40.74	40.67
8	57.35	53.42	53.27
9	71.08	68.00	67.64

Table 3

Gain Parameters

N	k_1^N	k_2^N	k_3^N	k_4^N
8	.1000	13.68	-293.1	-37.72
16	.1000	13.70	-275.1	-59.60
32	.1000	13.70	-269.8	-61.35
64	.1000	13.70	-268.9	-60.95

References

1. Adams, R.A., Sobolev Spaces, Academic Press, New York, 1975.
2. Burns, J.A. et al., "State-Space Models and Advanced Control Concepts for Large Space Structures," USAFRPL Report, Optimization Inc., Blacksburg, VA, April 1985.
3. Gibson, J.S. and Adamian, A., "Approximation Theory for LQG Optimal Control of Flexible Structures," SIAM J. Control and Optimization, to appear.
4. Hille, E. and Phillips, R.S., Functional Analysis and Semi-groups, AMS Colloq. Pub., Vol. 31, 1957.
5. Juang, J.N. and Horta, L.G., "Effects of Atmosphere on Slewing Control of a Flexible Beam," AIAA Paper 86 100-CP, San Antonio, TX, May 1986.
6. Kwackernaak, H. and Sivan, R., Linear Optimal Control Systems, Wiley-Interscience, New York, 1972.
7. Oden, J.T. and Carey, G.F., Finite Elements, Vol. IV, Prentice Hall, Englewood Cliffs, NJ, 1983.
8. Pazy, A., Semigroups of Linear Operators and Applications to Partial Differential Equations, Springer-Verlag, New York, 1983.
9. Prenter, P., Splines and Variational Methods, John Wiley, New York, 1975.
10. Walker, J.A., Dynamical Systems and Evolution Equations, Plenum Press, New York, 1980.

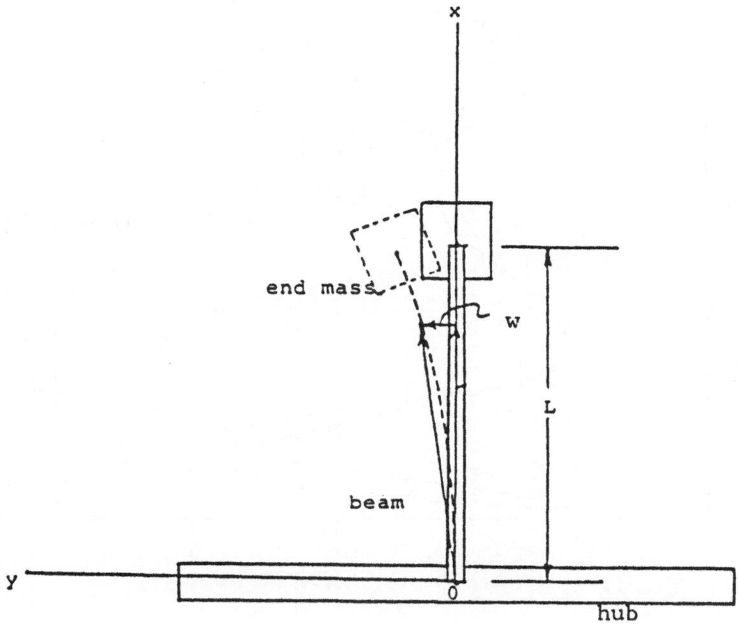

Figure 1 Flexible Structure

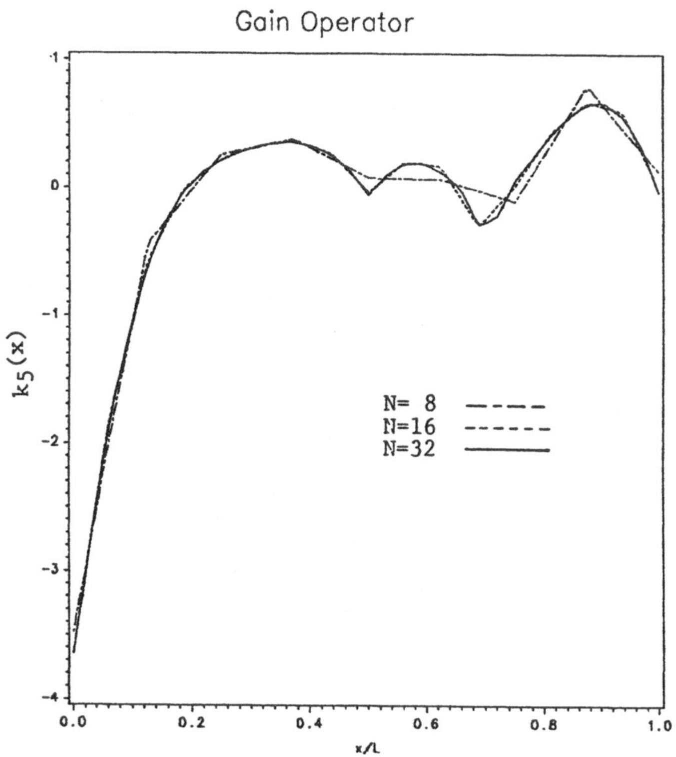

Figure 2 Feedback operator for beam strain

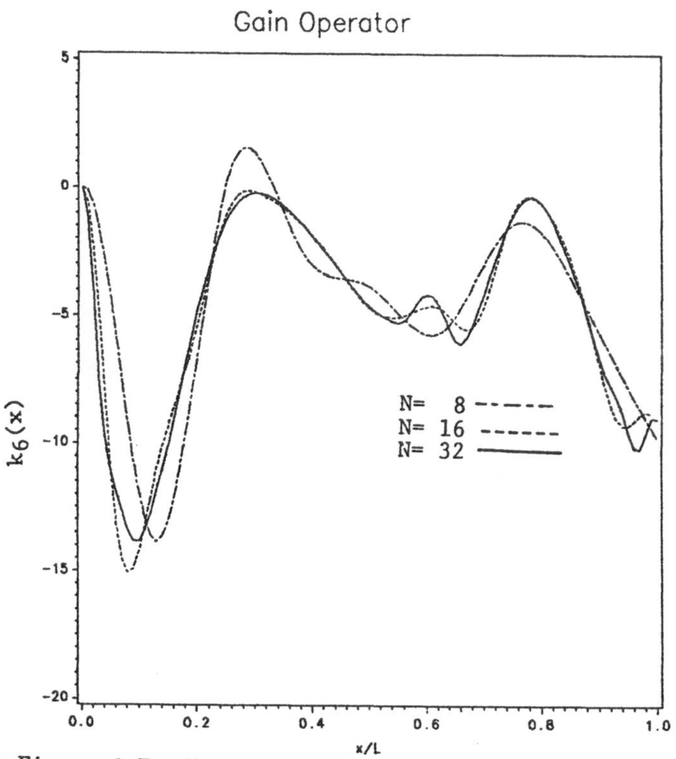

Figure 3 Feedback operator for beam velocity

OPTIMALE GESTALTUNG VON ELASTISCHEN BALKEN *

Leszek Mikulski

Stipendiat der Alexander von Humboldt-Stiftung
Mathematisches Institut der TU München
D-8000 München

1. EINLEITUNG

Das Thema dieser Arbeit ist die Bestimmung der optimalen Form von statisch bestimmten oder unbestimmten elastischen Balken mit rechteckigem oder I - förmigem Querschnitt. Unter verschiedenen Nebenbedingungen sollen dabei die Balken so gestaltet werden, daß entweder die Durchbiegung oder das Eigengewicht minimiert wird.

Im Falle des Balkens mit rechteckigem Querschnitt wählen wir die dimensionslose Höhe (oder die Breite) als Steuervariable. Im Falle eines I - Profils verwenden wir dazu die Breite.

Die Nebenbedingungen beschränken dabei den Querschnitt und die Normalspannungen. Darüber hinaus betrachten wir auch die Balken mit dünnwandigem Querschnitt.

Mit Hilfe der Theorie optimaler Steuerungsprobleme lassen sich diese Optimierungsaufgaben auf Randwertprobleme mit Schaltbedingungen für gewöhnliche Differentialgleichungen zurückführen.

Das Programm BNDSCO, das von Oberle entwickelt wurde, dient zur numerischen Lösung dieser Aufgaben. Es stellt einen Ausbau bereits vorhandener Randwertproblemlöser auf der Basis der Mehrzielmethode von Bulirsch [1] dar.

2. PROBLEMSTELLUNG

In einem rechtwinkligen Koordinatensystem betrachten wir die Durchbiegung eines Balkens unter seinem Eigengewicht und einer äußeren Belastung.
Die Statik und die Kinematik des Balkens werden durch die dimensionalen kanonischen Differentialgleichungen beschrieben.

---

* Diese Arbeit entstand mit Unterstützung der Alexander von Humboldt-Stiftung.

Dabei bezeichnen:

y_1, y_2	- die Komponente des Verschiebungsvektors
y_3	- das Biegemoment
y_4	- die Querkraft
y_5, y_6	- Drehwinkel
y_7	- das Bimoment
y_8	- das das Gesamtschnittdrillmoment
S	- Steuerung
q	- äußere Belastung
P	- Beulkraft
$a_1, \ldots, a_{16}$	- Konstanten

1. Fall: Balken mit rechteckigem Querschnitt

$$y_1' = y_2$$

$$y_2' = -\frac{12\, y_3\, a_1}{S^r} \qquad r = \begin{cases} 3 \\ 1 \end{cases}$$

$$y_3' = y_y\, a_2$$

$$y_4' = -q - a_3 \cdot S \tag{2.1}$$

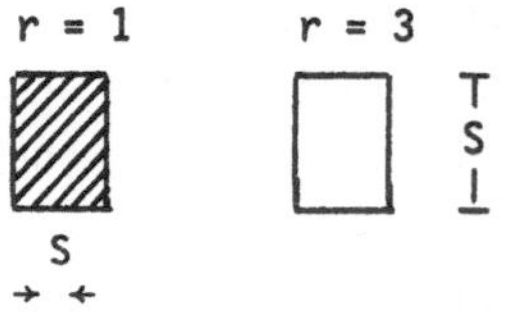

$$' = \left(\frac{d}{dz}\right) \qquad 0 \le Z \le 1$$

2. Fall: Balken mit I - Profil:

$$y_1' = y_2$$

$$y_2' = -\frac{12\, y_3}{a_4 + a_5 \cdot S}$$

$$y_3' = a_3 y_4$$

$$y_4' = -q - a_6 - a_7 \cdot S \tag{2.2}$$

3. Fall: Balken mit dünnwandigem I - Profil:

$$
\begin{aligned}
y_5' &= y_6 \\
y_6' &= -\frac{24\, y_7\, a_8}{S^3} \\
y_7' &= y_8 a_9 - y_6(a_{10} + a_{11} \cdot S) + y_6 \cdot P\, \frac{a_{12} + a_{13} \cdot S + a_{14} \cdot S^3}{a_{15} \cdot S + a_{16}} \\
y_8' &= 0\,.
\end{aligned} \tag{2.3}
$$

Wir nehmen als Steuervariable S die demensionslose Höhe oder Breite des rechteckigen Querschnitts bzw. die Breite im Fall eines I - Profils an.
Für die Steuerung führen wir folgende Beschränkungen ein:

$$U_1 \leq S \leq U_2 \tag{2.4}$$

Außerdem gelten folgende Beschränkungen für die Normalspannungen:

$$\sigma = \frac{|M|}{|W|} \leq \sigma_{ad} \qquad \text{(rechteckiger Querschnitt)} \tag{2.5}$$

$$\sigma = \frac{M}{W} + \frac{B}{W\,\Omega} \leq \sigma_{ad} \qquad \text{(I - Profil)} \tag{2.6}$$

Drückt man die Normalspannungen ebenfalls durch dimensionslose Größen aus, erhält man die folgenden Ungleichungsnebenbedingungen für die Steuervariable S.

$$
\left.\begin{aligned}
g_1 &= U_1 - S \leq 0 \\
g_2 &= S - U_2 \leq 0
\end{aligned}\right\} \text{ in allen drei Fällen} \tag{2.7}
$$

zusätzlich:

$$g_3 = \frac{|y_3|\, b_1}{S^r} - 1 \leq 0 \quad \text{im Fall des Balkens mit rechteckigem Querschnitt.}$$

$$g_4 = \frac{|y_3|}{b_2 + S(b_3 + b_4)} - 1 \leq 0 \quad \text{Im Fall des I - Profils}$$

$$g_5 = \frac{|y_7|}{S^3} \cdot b_1 - 1 \leq 0 \quad \text{Im Fall des dünnwandigen I - Profils.}$$

Wir betrachten die beiden folgenden Zielfunktionen:

(Z1) Die Verschiebung eines vorgegebenen Punktes soll minimiert werden, d.h.

$$\min_S J(S) \qquad J(S) = y_1(1) \tag{2.8}$$

(Z2) Das Volumen soll minimiert werden, d.h.

$$J(S) = \int_0^1 (W_1 + W_2 \cdot S)\, dz \, . \tag{2.9}$$

Damit hat man folgendes Problem der optimalen Steuerung:

1. Die Zustandsgleichungen
   (2.1) oder (2.2) oder (2.3)
   mit entsprechenden Randbedingungen.

2. Die Restriktionen (2.7), die den Bereich der zulässigen Steuerungen bestimmen.

3. Die Zielfunktion (2.8) oder (2.9) .

Gesucht werden die optimale Verteilung der Breite (Höhe) S(z) und der Zustand $\bar{y} = (y_1, y_2, y_3, y_4)$, so daß die Zustandsgleichungen mit den Randbedingungen sowie Restriktionen erfüllt sind und die Zielfunktion minimiert wird.

3. NOTWENDIGE BEDINGUNGEN

Zur Lösung der Aufgabe benutzen wir die Theorie der optimalen Steuerungen. Dies soll hier nur für den Fall des rechteckigen Balkens demonstriert werden.

Zunächst bezeichne

$$\begin{aligned} &H(y,\lambda,S) := \lambda^T f(y,S) \\ &H = \lambda_1 y_2 - \lambda_2 \frac{12 y_3\, a_1}{S^r} + \lambda_3\, y_4\, a_2 - \lambda_4\, (q + a_3 \cdot S) \end{aligned} \tag{3.1}$$

die Hamilton-Funktion für das Problem ohne die Nebenbedingungen (2.7).

Für die optimale Trajektorie $\bar{y}(z)$ und die zugehörige Steuerung $S(z)$ existieren Lagrange-Parameter $\lambda(z) \in R^n$. Diese sind Lösung der adjungierten Differentialgleichungen:

$$\begin{aligned} &\lambda'(z) = - H_y^T \big(y(z), \lambda(z), S(z)\big) \qquad && \lambda_3' = \lambda_2 \frac{12 a_1}{S^r} \\ &\lambda_1' = 0 && \lambda_4' = - \lambda_3\, a_2 \\ &\lambda_2' = - \lambda_1 && \end{aligned} \tag{3.2}$$

Für $S(z)$ gilt das Minimumprinzip

$$H\left(y(z),\ \lambda(z),\ S(z)\right) = \min_{v\in R^m} H\left(y(z),\ \lambda(z),\ v\right) \quad \text{für alle} \quad Z\in(0,1)\ .$$

Insbesondere hat man damit die beiden notwendigen Bedingungen von Euler-Lagrange bzw. Legendre-Clebsch

$$H_S = 0 \tag{3.4}$$

und H_{SS} ist positiv semidefinite Matrix

$$H_S = \frac{12r\,\lambda_2\,y_3\,a_5}{S^{r+1}} - \lambda_4\,a_3 = 0 \quad \text{im Fall des Balkens mit rechteckigem Querschnitt,}$$

$$H_S = \frac{12\,\lambda_2\,y_3\,a_5}{(a_4+a_5\cdot S)^2} - \lambda_4\,a_7 = 0 \quad \text{im Fall des I-Profils .}$$

Für den Fall des dünnwandigen I-Profils mit der Zielfunktion (Z2) und den Randbedingungen für den beidseitig eingespannten bzw. beidseitig gelenkig gelagerten Balken läßt sich zeigen, daß

$$\lambda_4 = y_5$$

$$\lambda_3 = -\frac{y_6}{a_9}$$

$$\lambda_2 = \frac{y_7}{a_9}$$

$$\lambda_1 = -y_8$$

gilt. Damit vereinfacht sich die notwendige Bedingung von Euler-Lagrange zu:

$$H_S = -W_2 + \frac{72a_8\,y_7^2}{a_9 S^4} + \frac{y_6^2}{a_9}\,a_{11} - \frac{y_6^2}{a_9}\cdot P\,\frac{2a_{14}a_{15}\cdot S^3 + 3a_{14}a_{16}S^2 + a_{13}a_{16} - a_{12}a_{15}}{(a_{15}\cdot S + a_{16})^2} = 0$$

wobei $g_i(y,S) < 0 \quad i = 1,5$.

Anfangs- und Endwerte für λ ergeben sich aus den Transversalitätsbedingungen.

Die Beschränkung (2.7) ist von der Steuerung abhängig. Nur derartige Beschränkungen werden in dieser Arbeit behandelt.

Die Beschränkung

$$g_i\left(y(z),S(z)\right) \le 0 \qquad g_{i,s} \neq 0 \tag{3.5}$$

läßt sich als Spezialfall einer Zustandsbeschränkung der Ordnung 0 interpretieren.

Die Zustandsbeschränkung (3.5) kann nur intervallweise aktiv werden.
Die Lagrange-Parameter λ sind stetig in den Verbindungspunkten z_i. Einzige Schaltbedingung im Punkte z_i ist die Stetigkeit der Hamiltonfunktion

$$H(z_i^+) = H(z_i^-) . \qquad (3.6)$$

Insbesondere ist die Bedingung (3.6) in diesem Fall äquivalent mit

$$S(z_i^+) = S(z_i^-) \qquad (3.7)$$

d.h.

$$U_1 = \sqrt[r+1]{\frac{12r\lambda_2 y_3 a_1}{\lambda_4 a_3}} \qquad g_1(y,\lambda,U)|_{z_i} = 0$$

$$U_2 = \sqrt[r+1]{\frac{12r\lambda_2 y_3 a_1}{\lambda_4 a_3}} \qquad g_2(y,\lambda,U)|_{z_i} = 0$$

$$|y_3|\, b_1 = S^r \qquad g_3(y,\lambda,U)|_{z_i} = 0$$

$$|y_3| = b_2 + S(b_3 + b_4) \qquad g_4(y,\lambda,U)|_{z_i} = 0$$

wobei $U = S^{frei}$ aus (3.4).

Bei den betrachteten Problemen waren die Nebenbedingungen $g_3 \le 0$, $g_4 \le 0$ und $g_5 \le 0$ stets erfüllt, wenn $g_1 \le 0$ und $g_2 \le 0$ berücksichtigt wurden. Aus diesem Grund braucht man die Hamiltonfunktion längs der beschränkten Extremalenbögen der zustandsabhängigen Nebenbedingungen $g_i \le 0$, $i = 3,4,5$ nicht zu modifizieren. Somit ändert sich auch nicht das System der adjungierten Differentialgleichungen längs dieser beschränkten Extremalenbögen.

Die optimale Steuerung hat also die Gestalt

U_1	$g_1(y,\lambda,s) = 0$	
U_2	$g_2(y,\lambda,s) = 0$	
(3.4)	$g_i(y,\lambda,s) < 0 \quad i = 3,4,5$	
$(\lvert y_3\rvert b_1)^{1/r}$	$g_3(y, \ ,s) = 0$	(3.8)
$\dfrac{\lvert y_3\rvert - b_2}{b_3 + b_4}$	$g_4(y, \ ,s) = 0$ im Falle des I - Profils	
$(\lvert y_7\rvert b_1)^{1/3}$	$g_5(y, \ ,s) = 0$ im Falle des dünnwandigen I - Profils	

4. BEISPIEL

Als Anwendungsbeispiel der hier präsentierten Methoden betrachten wir einen Stahlbetonbalken mit folgenden Randbedingungen:

a) einseitig eingespannter Balken:

(4.1)

$y_1(0)=0 \qquad y_3(1)=0$

$y_2(0)=0 \qquad y_4(1)=0$

b) beidseitig gelenkig gelagerter Balken:

(4.2)

$y_1(0)=0 \qquad y_2(1)=0$

$y_3(0)=0 \qquad y_4(1)=0$

c) beidseitig eingespannter Balken:

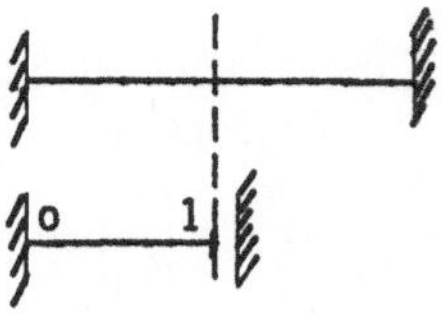

(4.3)

$y_1(0)=0 \qquad y_2(1)=0$

$y_3(0)=0 \qquad y_4(1)=0$

Anfangs- und Endwerte für λ bekommen wir aus Transversalitätsbedingungen [4] z.B. für die Zielfunktion (Z1)

Ad a

$$\lambda_3(0) = 0 \qquad \lambda_1(1) = 1$$
$$\lambda_4(0) = 0 \qquad \lambda_2(1) = 0 \qquad (4.4)$$

Ad b

$$\lambda_2(0) = 0 \qquad \lambda_1(1) = 1$$
$$\lambda_4(0) = 0 \qquad \lambda_3(1) = 0 \qquad (4.5)$$

Ad c

$$\lambda_3(0) = 0 \qquad \lambda_1(1) = 1$$
$$\lambda_4(0) = 0 \qquad \lambda_3(1) = 0 \qquad (4.6)$$

Die Aufgabe wird also auf die Lösung des Differentialgleichungssystems (2.1), (3.2) mit Randbedingungen (4.1), (4.4); (4.2), (4.5); (4.3), (4.6) und Schaltbedingungen (3.7) reduziert. Zur numerischen Lösung dieser Randwertaufgabe wurde die Mehrzielmethode [3] verwendet, die bisher mit großem Erfolg bei Problemen der optimalen Steuerung eingesetzt worden ist. Die Methode wurde in der Routine BNDSCO [2] implementiert, diese erlaubt die numerische Lösung von Randwertaufgaben mit Schaltbedingungen. Die Schaltstruktur, d.h. die Folge der verschiedenen beschränkten bzw. unbeschränkten Extremalbögen muß vorher bekannt sein. Die gewählten Beispiele sind von praktischer Bedeutung für die Baumechanik. Die Ergebnisse sind in Fig.1 bis Fig.8 dargestellt. Dabei wird auch die Abhängigkeit der Schaltstrukturen von den Beschränkungen gezeigt.

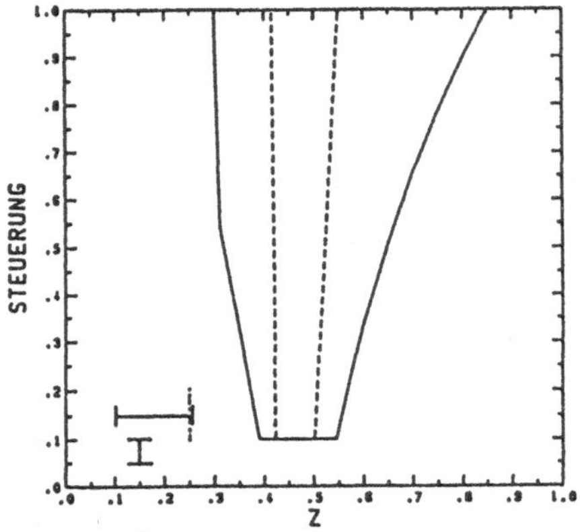

FIG. 1

——OHNE BELASTUNG
---MIT BELASTUNG

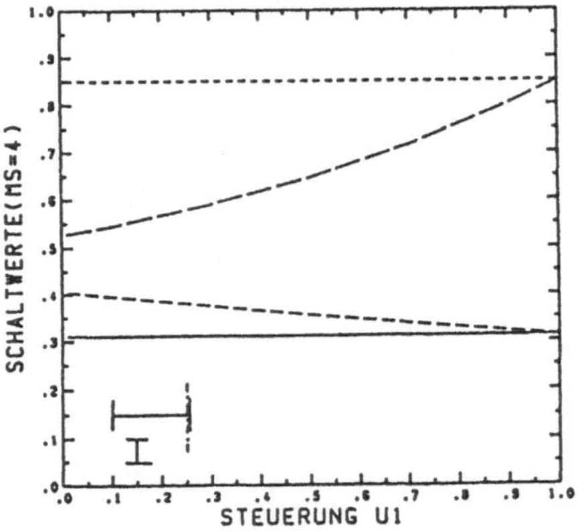

FIG. 2

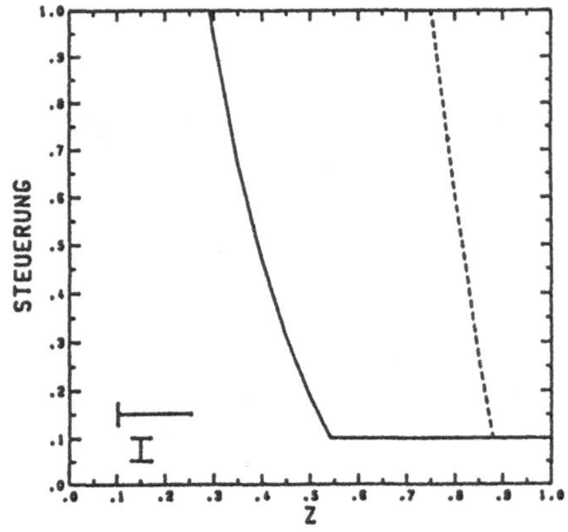

FIG. 3

——OHNE BELASTUNG
---MIT BELASTUNG

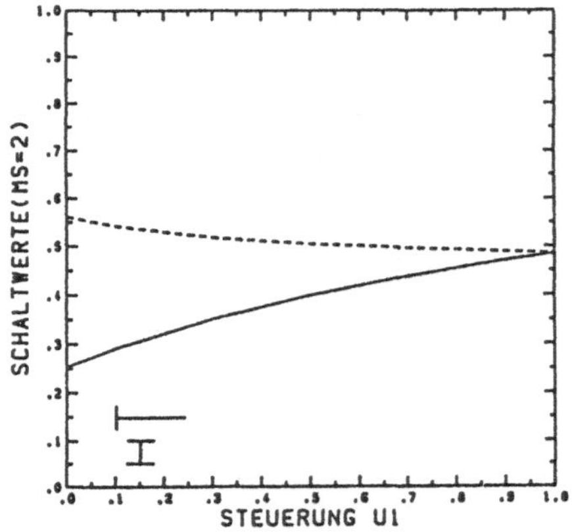

FIG. 4

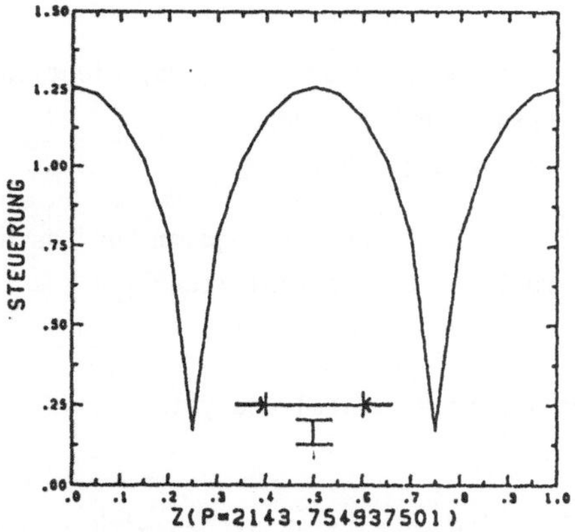

FIG. 5

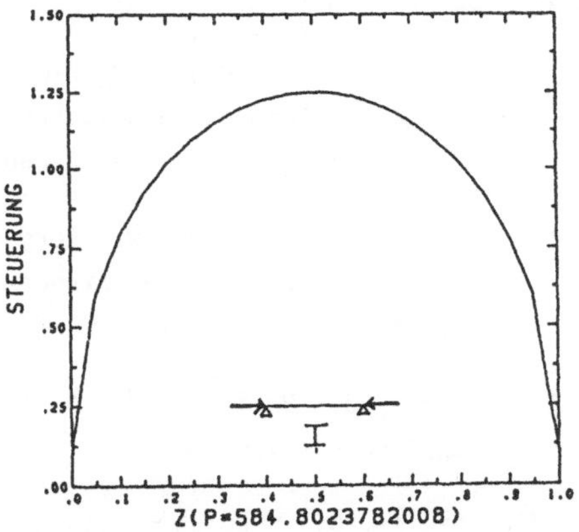

FIG. 6

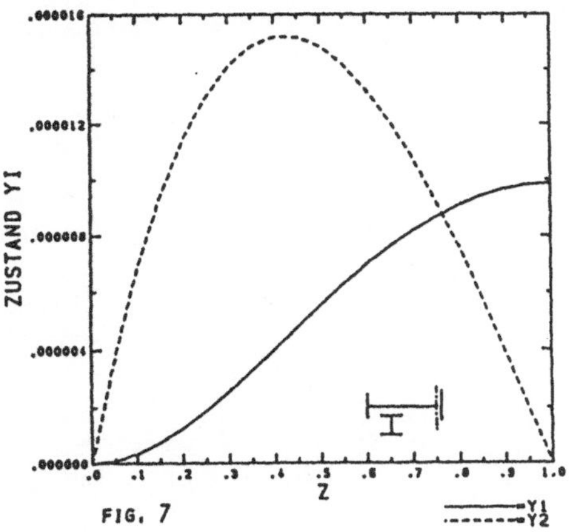

FIG. 7

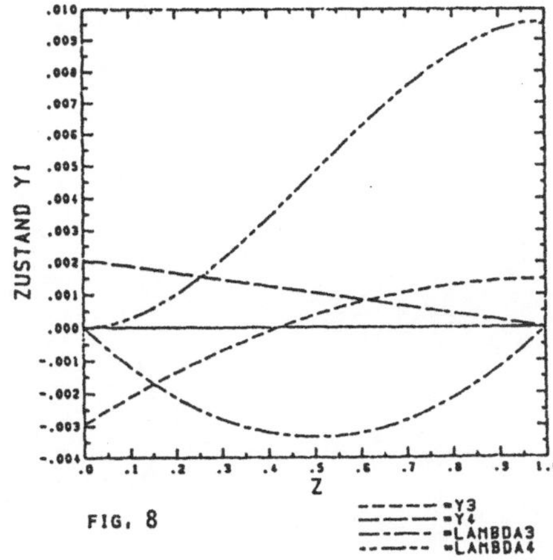

FIG. 8

5. SCHLUSSWORT

Die in der Arbeit durchgeführte Analyse beweist die Effektivität dieser Methode. Die formulierten Optimierungsaufgaben wurden effektiv gelöst. Dabei wurde gezeigt, daß die geometrischen Beschränkungen aktiv sind. Die numerischen Rechnungen wurden an der CDC Cyber des Leibniz-Rechenzentrums der Bayerischen Akademie der Wissenschaften durchgeführt. Die Arbeit bildet einen Teil eines Studiums der Anwendungsmöglichkeiten des Maximumprinzips zur Optimierung von Konstruktionselementen.

LITERATURVERZEICHNIS

[1] Bulirsch, R., "Die Mehrzielmethode zur numerischen Lösung von nichtlinearen Randwertaufgaben und Aufgaben der optimalen Steuerung". Carl-Cranz-Lehrgang über Flugbahnoptimierung 1971

[2] Oberle, H.J.; Grimm, W.; Berger, E.; "BNDSCO - Rechenprogramm zur Lösung beschränkter optimaler Steuerungsprobleme, Benutzeranleitung", TUM-M8509, August 1985

[3] Stoer, J.; Bulirsch, R.; "Introduction to Numerical Analysis", Springer-Verlag 1980

[4] Szefer, G.; Mikulski, L.; "Optimal design of elastic arches with I - cross-section", Engineering Transactions 32, 4 467 - 480 IPPT-PAN Warszawa.

APPENDIX

$$a_1 = \frac{\sigma_o \cdot W_w \cdot l_o}{E\ B_{max}\ h_o^3}$$

$$a_2 = \frac{F_w \cdot l_o}{W_w}$$

$$a_3 = \frac{\gamma \cdot h_o \cdot B_{max} \cdot l_o}{\sigma_o \cdot F_w}$$

$$q = \frac{q_o \cdot l_o}{\sigma_o \cdot F_w}$$

$$a_4 = \frac{E \cdot t_o \cdot h_o^3}{\sigma_o \cdot W_w \cdot l_o}$$

$$a_5 = \frac{E\left[2B_{max}\ t_o^3 + 6B_{max} \cdot t_o\,(t_o + h)^2\right]}{\sigma_o \cdot W_w \cdot l_o}$$

$$a_6 = \frac{\gamma \cdot t_o \cdot h_o \cdot l_o}{\sigma_o \cdot F_w}$$

$$a_7 = \frac{2\gamma\ t_o \cdot B_{max} \cdot l_o}{\sigma_o \cdot F_w}$$

$$Z = \frac{x}{l_o}$$

$$y_1 = \frac{w}{l_o}$$

$$y_3 = \frac{M}{\sigma_o\ W_w}$$

$$y_4 = \frac{Q}{\sigma_o\ F_w}$$

$$y_7 = \frac{B}{W\Omega \cdot \sigma_o}$$

$$y_8 = \frac{M_z}{\sigma_o \cdot WS}$$

$$WS = \frac{2J_s}{h_o}$$

$$W\Omega = \frac{J_{ww}}{w_{max}}$$

$$a_8 = \frac{\sigma_o\ W\Omega \cdot l_o^2}{E \cdot t_o \cdot h_o^2 \cdot B^3{}_{max}}$$

$$a_9 = \frac{WS \cdot l_o}{W\Omega}$$

$$a_{10} = \frac{G \cdot t_o^3 \cdot h_o \cdot l_o}{3 \cdot W\Omega \cdot \sigma_o}$$

$$a_{11} = \frac{2G \cdot t^3 \cdot B_{max} \cdot l_o}{3 \cdot W\Omega \cdot \sigma_o}$$

$$a_{12} = \frac{t_o \cdot h_o^3}{2 \cdot A_o \cdot l_o}$$

$$a_{13} = \frac{t_o \cdot h_o^2 \cdot B_{max}}{2 \cdot A_o \cdot l_o}$$

$$a_{14} = \frac{t_o \cdot B_{max}}{6 \cdot A_o \cdot l_o}$$

$$a_{15} = \frac{2 \cdot t_o \cdot B_{max}}{A_o}$$

$$a_{16} = \frac{h_o \cdot t_o}{A_o}$$

$$p = \frac{S_o \cdot l_o^2}{\sigma_o \; W\Omega}$$

$$a_{17} = \frac{m \cdot l_o}{\sigma_o \; WS}$$

$$b_1 = \frac{\sigma_o}{K_g}$$

$$b_2 = \frac{K_g \cdot t_o \cdot h^3}{12\left(\frac{h_o}{2} + t_o\right) W_w \sigma_o}$$

$$b_3 = \frac{K_g \cdot B_{max} \cdot t_o^3}{6\left(\frac{h_o}{2} + t_o\right) W_w \sigma_o}$$

$$b_4 = \frac{K_g \cdot B_{max} \cdot t_o (h_o + t_o)^2}{2\left(\frac{h_o}{2} + t_o\right) W_w \cdot \sigma_o}$$

$$A_o = h_o \cdot t_o + 2B_{max} \cdot t_o$$

$$J_{ww} = \frac{t_o \cdot B^3_{max} \, h_o^2}{24}$$

$$W_{max} = 0.25 \cdot B_{max} \cdot h_o$$

$$J_s = \frac{1}{3} \cdot \left(t_o^3 \cdot h_o + 2t_o^3 \, B_{max}\right)$$

$$WW = \frac{B_{max} \cdot h_o^2}{6}$$

$$F_w = A_o$$

BEZEICHNUNGEN

- B - Biomoment
- B_{max} - Breite des Profils
- E - Elastizitätsmodul
- G - Schubmodul
- h_o - Höhe des Profils
- K_g - zulässige Normalspannung
- l_o - Länge des Balkens
- M - Biegemoment
- M_z - Gesamtschnittdrillmoment
- m - stetige Torsionslast
- t_o - Dicke des Profils
- S_o - Normalkraft
- Q - Querkraft
- q_o - Normalkomponente der Belastung
- W - Normalkomponente des Verschiebungsvektors
- X - unabhängige Variable
- y_5 - Drehwinkel
- γ - spezielles Gewicht
- σ_o - Normalspannung

DATA

$B_{max} = 0.40\text{m}$, $E = 2{,}1 \cdot 10^{11} \frac{N}{m^2}$, $h_o = 1\text{m}$, $l_o = 5\text{m}$, $q_o = 60\,000 \frac{N}{m^2}$,

$t_o = 0.01\text{m}$, $\gamma = 70\,000 \frac{N}{m^3}$, $\sigma_o = 1{,}7 \cdot 10^8 \frac{N}{m^2}$, $G = 8{,}0 \cdot 10^{10} \frac{N}{m^2}$.

Lecture Notes in Control and Information Sciences

Edited by M. Thoma

Lecture Notes in Control and Information Sciences

Edited by M. Thoma and A. Wyner

Vol. 62: Analysis and Optimization
of Systems
Proceedings of the Sixth International
Conference on Analysis and Optimization
of Systems
Nice, June 19–22, 1984
Edited by A. Bensoussan, J. L. Lions
XIX, 591 pages. 1984.

Vol. 63: Analysis and Optimization
of Systems
Proceedings of the Sixth International
Conference on Analysis and Optimization
of Systems
Nice, June 19–22, 1984
Edited by A. Bensoussan, J. L. Lions
XIX, 700 pages. 1984.

Vol. 64: Arunabha Bagchi
Stackelberg Differential Games
in Economic Models
VIII, 203 pages, 1984

Vol. 65: Yaakov Yavin
Numerical Studies
in Nonlinear Filtering
VIII, 273 pages, 1985.

Vol. 66: Systems and Optimization
Proceedings of the Twente Workshop
Enschede, The Netherlands, April 16–18, 1984
Edited by A. Bagchi, H. Th. Jongen
X, 206 pages, 1985.

Vol. 67: Real Time Control of Large Scale Systems
Proceedings of the First European Workshop
University of Patras, Greece, Juli 9–12, 1984
Edited by G. Schmidt, M. Singh, A. Titli,
S. Tzafestas
XI, 650 pages, 1985.

Vol. 68: T. Kaczorek
Two-Dimensional Linear Systems
IX, 397 pages, 1985.

Vol. 69: Stochastic Differential Systems –
Filtering and Control
Proceedings of the IFIP-WG 7/1 Working Conference
Marseille-Luminy, France, March 12-17, 1984
Edited by M. Metivier, E. Pardoux
X, 310 pages, 1985.

Vol. 70: Uncertainty and Control
Proceedings of a DFVLR International Colloquium
Bonn, Germany, March, 1985
Edited by J. Ackermann
IV, 236 pages, 1985.

Vol. 71: N. Baba
New Topics in Learning Automata
Theory and Applications
VII, 231 pages, 1985.

Vol. 72: A. Isidori
Nonlinear Control Systems:
An Introduction
VI, 297 pages, 1985.

Vol. 73: J. Zarzycki
Nonlinear Prediction
Ladder-Filters for Higher-Order
Stochastic Sequences
V, 132 pages, 1985.

Vol. 74: K. Ichikawa
Control System Design based on
Exact Model Matching Techniques
VII, 129 pages, 1985.

Vol. 75: Distributed Parameter
Systems
Proceedings of the 2nd International
Conference, Vorau, Austria 1984
Edited by F. Kappel, K. Kunisch,
W. Schappacher
VIII, 460 pages, 1985.

Vol. 76: Stochastic Programming
Edited by F. Archetti, G. Di Pillo,
M. Lucertini
V, 285 pages, 1986.

Vol. 77: Detection of
Abrupt Changes in Signals
and Dynamical Systems
Edited by M. Basseville,
A. Benveniste
X, 373 pages, 1986.

Vol. 78: Stochastic
Differential Systems
Proceedings of the 3rd Bad Honnef
Conference, June 3–7, 1985
Edited by N. Christopeit, K. Helmes,
M. Kohlmann
V, 372 pages, 1986.

Vol. 79: Signal
Processing for Control
Edited by K. Godfrey, P. Jones
XVIII, 413 pages, 1986.

Vol. 80: Artificial Intelligence
and Man-Machine Systems
Edited by H. Winter
IV, 211 pages, 1986.

Lecture Notes in Control and Information Sciences

Edited by M. Thoma and A. Wyner

Vol. 81: Stochastic Optimization
Proceedings of the International Conference, Kiew, 1984
Edited by I. Arkin, A. Shiraev, R. Wets
X, 754 pages, 1986.

Vol. 82: Analysis and Algorithms of Optimization Problems
Edited by K. Malanowski, K. Mizukami
VIII, 240 pages, 1986.

Vol. 83: Analysis and Optimization of Systems
Proceedings of the Seventh International Conference of Analysis and Optimization of Systems
Antiba, June 26-27, 1986
Edited by A. Bensoussan, J. L. Lions
XVI, 901 pages, 1986.

Vol. 84: System Modelling and Optimization
Proceedings of the 12th IFIP Conference
Budapest, Hungary, September 2–6, 1985
Edited by A. Prékopa, J. Szelezsán, B. Strazicky
XII, 1046 pages, 1986.

Vol. 85: Stochastic Processes in Underwater Acoustics
Edited by Charles R. Baker
V, 205 pages, 1986.

Vol. 86: Time Series and Linear Systems
Edited by Sergio Bittanti
XVII, 243 pages, 1986.

Vol. 87: Recent Advances in System Modelling and Optimization
Proceedings of the IFIP-WG 7/1 Working Conference
Santiago, Chile, August 27-31, 1984
Edited by L. Contesse, R. Correa, A. Weintraub
IV, 199 pages, 1987.

Vol. 88: Bruce A. Francis
A Course in H_∞ Control Theory
XI, 156 pages, 1987.

Vol. 89: G. K. H. Pang/A. G. J. McFarlane
An Expert System Approach to Computer-Aided Design of Multivariable Systems
XII, 223 pages, 1987.

Vol. 90: Singular Perturbations and Asymptotic Analysis in Control Systems
Edited by P. Kokotovic, A. Bensoussan, G. Blankenship
VI, 419 pages, 1987.

Vol. 91 Stochastic Modelling and Filtering
Proceedings of the IFIP-WG 7/1 Working Conference
Rome, Italy, Decembre 10-14, 1984
Edited by A. Germani
IV, 209 pages, 1987.

Vol. 92: L. T. Grujić, A. A. Martynyuk, M. Ribbens-Pavella
Large-Scale Systems Stability Under Structural and Singular Perturbations
XV, 366 pages, 1987.

Vol. 93: K. Malanowski
Stability of Solutions to Convex Problems of Optimization
IX, 137 pages, 1987.

Vol. 94: H. Krishna
Computational Complexity of Bilinear Forms
Algebraic Coding Theory and Applications to Digital Communication Systems
XVIII, 166 pages, 1987.

Vol. 95: Optimal Control
Proceedings of the Conference on Optimal Control and Variational Calculus
Oberwolfach, West-Germany, June 15-21, 1986
Edited by R. Bulirsch, A. Miele, J. Stoer and K. H. Well
XII, 321 pages, 1987.